T0289830

CRITICAL CONNECTIONS

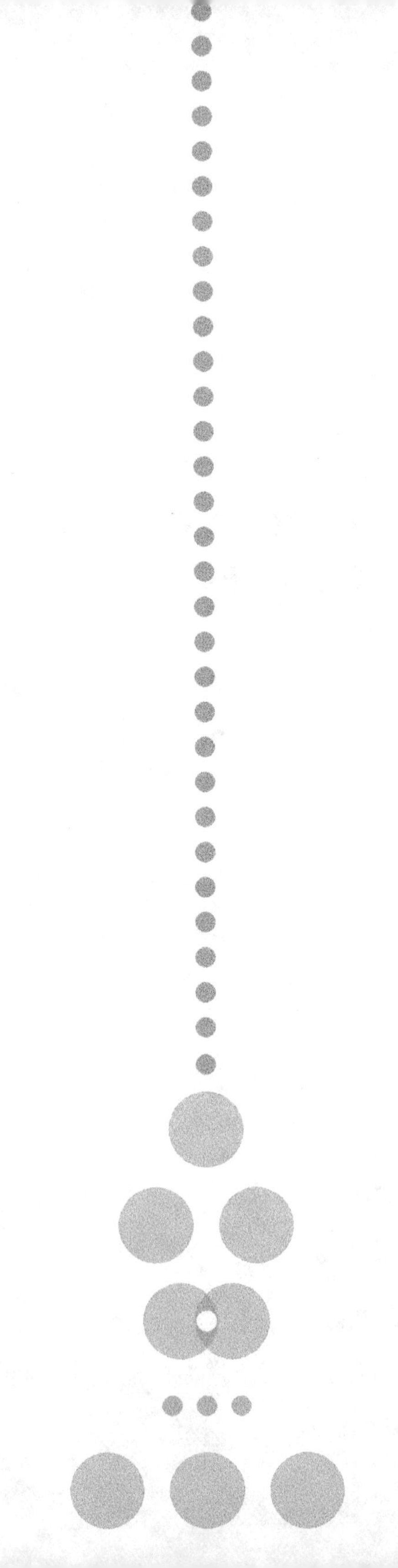

CRITICAL CONNECTIONS

The University of Tennessee and Oak Ridge from the Dawn of the Atomic Age to the Present

Lee Riedinger, Al Ekkebus,
Ray Smith, and William Bugg

With a Foreword by
Lamar Alexander and Homer Fisher

THE UNIVERSITY OF TENNESSEE PRESS
Knoxville

FIRST EDITION.

Library of Congress Cataloging-in-Publication Data

NAMES: Riedinger, Lee, author. | Ekkebus, Al, author. | Smith, David Ray (newspaper columnist), author. | Bugg, William, author. | Alexander, Lamar, 1940– writer of foreword. | Fisher, Homer, writer of foreword.
TITLE: Critical connections : the University of Tennessee and Oak Ridge from the dawn of the atomic age to the present / Lee Riedinger, Al Ekkebus, Ray Smith, and William Bugg ; with a foreword by Lamar Alexander, with a foreword by Homer Fisher.
DESCRIPTION: First edition. | Knoxville : The University of Tennessee Press, [2024] | Includes bibliographical references and index. |
SUMMARY: "This is a history of the long association of the University of Tennessee with Oak Ridge National Laboratory, dating back to the Manhattan Project. While large-scale partnerships between scientific laboratories and academic institutions are now common, in the aftermath of World War II it was not clear what role this huge research and development initiative would play in postwar America; but pioneering professors and administrators were determined that one option—dismantling the whole thing—would not happen. Thus began a now eight-decade long association that has flowered into one of the world's largest collaborations between a federal agency and a research university"—Provided by publisher.
IDENTIFIERS: LCCN 2024004409 (print) | LCCN 2024004410 (ebook) | ISBN 9781621906544 (cloth) | ISBN 9781621906551 (pdf) | ISBN 9781621906568 (kindle edition)
SUBJECTS: LCSH: Oak Ridge National Laboratory—History. | University of Tennessee, Knoxville—History—20th century. | University of Tennessee, Knoxville—History—21st century. | Y-12 National Security Complex (U.S.)—History. | Research institutes—Tennessee—Oak Ridge—History. | Nuclear weapons industry—Tennessee—Oak Ridge—History. | BISAC: Technology & Engineering / History | Education / History
CLASSIFICATION: LCC QC789.2.U62 O2575 2024 (print) | LCC QC789.2.U62 (ebook) | DDC 621.042072/076873—dc23/eng/20240226
LC record available at https://lccn.loc.gov/2024004409
LC ebook record available at https://lccn.loc.gov/2024004410

* * *

We acknowledge and appreciate the efforts of Dr. Lee Riedinger and his team of writers and editors, who, over the last several years, have painstakingly compiled and synthesized the joint interactions between the University of Tennessee, Oak Ridge National Laboratory, the City of Oak Ridge, and other institutions, which together have created and sustained an important and enduring partnership. This publication serves as a tribute to the numerous staff, faculty, administrators, and scientists who participated in the initial planning and establishment of ORNL and to the many staff, faculty, scientists, and joint faculty who continue to build an even stronger partnership between the university, ORNL, and the City of Oak Ridge for the benefit of Tennessee and the nation.

* * *

JIMMY G. CHEEK

Chancellor Emeritus—UT Knoxville (2009–2017)

WAYNE T. DAVIS

Chancellor Emeritus—UT Knoxville (2018–2019)

Dean Emeritus, Tickle College of Engineering—
UT Knoxville (2008–2018)

DONDE A. PLOWMAN

Chancellor—UT Knoxville (2019–present)

CONTENTS

ILLUSTRATIONS

FOREWORD

Lamar Alexander

WHEN JAPANESE AIRCRAFT attacked Pearl Harbor on December 7, 1941, twenty-two-year-old Sam Beall had no idea that he would soon be swept into the Manhattan Project, a massive effort to build a bomb to win a world war—an effort that also created the foundation for today's remarkable partnership between the nation's largest science and engineering laboratory and a major research university.

Sam came to Knoxville from Plains, Georgia, in 1938 to study chemical engineering at the University of Tennessee (UT). In the spring of 1943, he went to work for DuPont at the University of Chicago, where he helped design the graphite reactor that would create the fuel for a plutonium bomb. Then, in September, DuPont sent Sam back to East Tennessee to build the reactor.

Meanwhile, the U.S. government acquired nearly sixty thousand acres and evicted three thousand people, transforming several small farming communities into Oak Ridge, a secret city of seventy-five thousand residents with three hastily constructed industrial sites. Workers toiled overtime in what was then called Clinton Engineer Works to produce the material necessary to build a nuclear weapon before Germany did.

The reactor that Sam helped build at Oak Ridge's Clinton Laboratories produced enough plutonium to prove that a nuclear bomb was possible. Therefore, DuPont next sent Sam to Richland in Washington state, where huge reactors made plutonium that was then sent to Los Alamos, New Mexico, to make the "Fat Man" bomb, which the U.S. dropped over Nagasaki to end World War II.

After the war, Sam came back to Clinton Laboratories, now known as Oak Ridge National Laboratory (ORNL). The mission of the secret city was accomplished, and rumors spoke of the complex's closure. However, scientists at UT and Oak Ridge argued for continuation of the laboratory, and these scientists organized a coalition of other universities to make the case. Soon, scientists who worked at Oak Ridge began earning doctorates at UT and Sam continued to work on advanced reactors, which led to the development of commercial reactors that produce electricity.

Sam Beall celebrated his hundredth birthday in 2019. He has lived longer than the city of Oak Ridge, and he has seen UT grow from an institution that offered no doctoral programs to a major research university. The reason to read this carefully researched book by Lee Riedinger and his co-authors is to learn the rest of this fascinating story.

• • •

A succession of Tennessee governors, university officials, and community leaders have, for the past nearly eighty years, found it irresistible to try to deepen connections between UT and Oak Ridge. By the time I became governor in 1979, the first joint programs were already in place. And, as did those who came before and after me, I looked for every opportunity to strengthen those connections.

In 1983, I found such an opportunity. The United States Senate was about to cancel the breeder reactor at Oak Ridge in the home state of then-Senate Majority Leader Howard H. Baker, Jr. Sensing some sympathy in Washington, D.C. for Senator Baker's predicament, I asked for a meeting with U.S. Secretary of Energy Don Hodel in the Majority Leader's Capitol office. Baker said to the secretary, "The governor here says that we ought to put two dozen distinguished scientists at the Oak Ridge National Laboratory with joint appointments at the University of Tennessee. He says the state will pay for half of it, and he wants the federal government to pay for the other half."

The Secretary of Energy thought for a moment and replied, "I guess I have to do it, don't I?"

More opportunities followed. During the 1980s, the state created Chairs of Excellence and Centers of Excellence at Tennessee universities, which supported a new Science Alliance between UT and the laboratory. The state built Pellissippi Parkway, an interstate-quality, four-lane highway between the Knoxville airport and the laboratory, a topic of discussion that had gone on for years.

As president of UT in the late 1980s, I would tout the concentration of brainpower in the Knoxville-Oak Ridge area and compare the Oak Ridge Corridor to Massachusetts' Route 128, North Carolina's Research Triangle, and even California's Silicon Valley. Some people thought of this as a little far-fetched. But, thirty-some years later, the idea of an Oak Ridge Corridor does not seem far-fetched at all. The Knoxville airport has a new sign that says: "Welcome to Knoxville: Gateway to the Smokies and The Oak Ridge Corridor." It is about time we celebrate.

Route 128, the Research Triangle, and Silicon Valley are all popular brand names that symbolize concentrations of brainpower in those regions. There are about three thousand scientists and engineers who work at ORNL, UT, and the Tennessee Valley Authority (TVA)—a comparable concentration of brainpower to those other centers of science and engineering.

In 2018, Tennessee's new Governor Bill Lee, an engineer himself, told a Nashville conference of entrepreneurs that "in this day and time, you need a magnet to attract capital and jobs." The next day, after visiting Oak Ridge, the governor said, "Tennessee already has that magnet right here."

Oak Ridge already is a brand name that symbolizes a concentration of talent rivaling the Research Triangle. Additionally, beyond the talent, Oak Ridge is known for excellence in technology, housing one of the world's fastest supercomputers and the most advanced research facilities in new materials and 3D printing for manufacturing. Each year, federal taxpayers spend $2 billion at the laboratory to advance that excellence.

In the past, the idea of an Oak Ridge Corridor caused some in Knoxville and other area communities to say, "We don't live in Oak Ridge," so the concept went nowhere. Now I have come to think of the Oak Ridge Corridor not as something that requires renaming a town or highway, but as a concept—a regional brand that can include Maryville and Morristown, Kingston and Sevierville, and any other East Tennessee community that wants to claim it.

UT, which, with the Battelle Memorial Institute, now manages ORNL, is taking the next step to make the Oak Ridge Corridor a reality. Under the leadership of President Randy Boyd and Chancellor Donde Plowman, the university created the UT-Oak Ridge Innovation Institute. This will gather under one umbrella the joint programs between the nation's largest science and energy laboratory and a major research university. The institute will exist as a pipeline for a new supply of American-trained scientists and engineers, which our country sorely needs in this competitive world. Already, the

UT-Oak Ridge partnership has 250 joint faculty, five joint institutes, and 250 PhD students in jointly administered energy and data programs.

One of the joys of my public life was to see and encourage the UT-Oak Ridge alliance from the vantage point of governor, UT president, and chairman of the Senate Energy and Water Development Appropriations Subcommittee, which annually sends about $4 billion in federal dollars to the Oak Ridge complex.

With such a strong foundation and such strong current leadership, I bet that during the next eighty years, the brand Oak Ridge Corridor and the UT-Oak Ridge partnership will be clearly recognized as one of the most important science and engineering alliances in the world.

MARYVILLE, TENNESSEE
September 1, 2019

FOREWORD

Homer Fisher

IT WAS BIGGER than a University of Tennessee (UT) football victory over Alabama or Florida or even a national championship. Few thought it possible, yet it happened. In 1999, UT-Battelle LLC was awarded the contract to manage Oak Ridge National Laboratory (ORNL) and has done so with great success in the twenty-one years since.

Many years earlier, a capable young UT physics professor named Lee Riedinger dreamed of securing the ORNL management contract. This interesting and enjoyable book by Riedinger and his co-authors details how that dream came about and more about the long history of two great partnering institutions, a state university, and a national laboratory.

Years before I came to UT in 1977, I was well acquainted with ORNL. Friends in Florida State University's physics department talked about their engagement with the laboratory and the wonderful facilities and scientists there. When I arrived at UT to assume the post of vice chancellor for business and finance of the Knoxville campus, I anticipated finding multiple ties between the university and the laboratory. Yes, there existed some engagement but far less than I anticipated. But Lee, who over the years served in administrative roles at both institutions, visioned increasing the university's engagement with ORNL in ways that would benefit both entities.

In 1999, during the last two years of my ten years of service as senior vice president of UT's statewide system of campuses, I had the opportunity to lead the university's effort—with Lee as my foremost colleague—to team with Bill Madia of Battelle and form UT-Battelle LLC and to compete for and win the contract to manage ORNL. Our successful bid reflected just how far UT had evolved as an institution since the days of the Manhattan Project, and this book chronicles that evolution.

In assembling his writing team, Lee picked three distinguished and knowledgeable men to serve as co-authors, and they add richness to the story because of the major roles they played in the community, laboratory, and university. I have known and respected all of them for many years. William Bugg, who attended Oak Ridge High School during the Manhattan Project, built a strong physics department at UT and fully supported joint initiatives between his faculty and graduate students and ORNL scientists. His fine work substantially strengthened UT's contributions to and interactions with the laboratory's science programs. Oak Ridge historian D. Ray Smith, through his newspaper columns and public appearances, arguably contributed more to knowledge and awareness of the federal programs in the former Secret City than any other individual. Al Ekkebus served for many years as a key member of the ORNL director's staff and headed the laboratory's library. Al's extensive knowledge of the laboratory's history, leadership, and culture is evident throughout this book.

In this well-written volume, Lee and his co-authors, who collectively represent 180 years of professional engagement at one or both of the partnering institutions, detail with humor and inside knowledge the major UT-ORNL joint initiatives that launched and the people who enabled them.

Lee asking Senator Lamar Alexander to write the primary foreword for this book was most fitting. As governor, Lamar recognized the potential for the UT and ORNL relationship to become special, and Lee describes well the actions Lamar took as governor and UT president to provide the funding and leadership to substantially strengthen the partnership.

KNOXVILLE, TENNESSEE
February 2021

ACKNOWLEDGMENTS

THE DEVELOPMENT OF the partnership between the University of Tennessee and Oak Ridge has been important for the university, the federal facilities in Oak Ridge, and the city of Oak Ridge. The partnership developed over eighty years and started during the Manhattan Project in the 1940s. UT was not a research-oriented university in 1940 and had no doctoral programs. The university needed the formation and development of the laboratories in Oak Ridge to advance, especially in areas related to science and engineering. This development occurred incrementally, decade by decade, in part due to the vision and leadership of many people. One outcome of this step-by-step evolution was the formation of UT-Battelle LLC and its role in managing Oak Ridge National Laboratory since 2000, an outcome considered unachievable only twenty years earlier.

Oak Ridge largely shaped my career in nuclear physics. I did dissertation research there in the late 1960s and then utilized ORNL accelerator facilities over forty-nine years on the UT physics faculty. I was involved in this partnership development for decades in part by serving in administrative leadership posts both at UT as vice chancellor for research (three times, all interim) and at ORNL as deputy director for science and technology. In the 1980s, I led the Science Alliance, the first official UT/ORNL partnership program, and, in the 2010s, the Bredesen Center and two new doctoral programs organized between the university and the national laboratory. Twice, UT considered jumping into the competition for an Oak Ridge facility management contract, and I was deeply involved in both committees and thought processes. This did not proceed in 1983 as the university did not submit a bid. However, sixteen years later, joint UT-ORNL programs matured and the university was ready to compete. We chose Battelle as a partner, and the UT-Battelle management of ORNL proved extremely successful over the last twenty-three years.

Documenting and writing about this eighty-year partnership has been a goal for some years. New leaders at UT or in Oak Ridge come and go, and it is important for them to know about the past and understand how the partnership programs developed. I was deeply involved in, and sometimes driving, these developments over the last forty years, and writing this half of the book was aided considerably by personal notes, files, and anecdotes. Documenting in detail the first forty years was more difficult and was possible especially through the contributions of my co-authors. Al Ekkebus worked at ORNL for thirty-three years, has a background in library science, and conducted much research and writing that pertains especially to the early chapters. Ray Smith is a retired Y-12 employee and now the official Oak Ridge historian. His stories of the early days of Oak Ridge are remarkable, and we have included some in this book. Bill Bugg grew up in Oak Ridge, got his PhD at UT, and then served on the physics faculty for forty years. He contributed wonderful personal stories and much about UT before 1940. And, he hired me onto the physics faculty in 1971.

Writing many research papers over a career turned out easier than writing this book! This project was possible due to the contributions of my three co-authors and also two editors who read every sentence and corrected many. Mary Zuhr, a retired Oak Ridge school librarian and a voracious reader, was our first-pass editor, fixed many grammatical errors, and told me when the included science seemed too heavy. David Brill, an accomplished author and former UT employee, stressed style and flow, plus the inclusion of stories and happenings, instead of writing as a research physicist. He has vast experience in writing books and he encouraged us often to keep going and not get discouraged. Scot Danforth, head of UT Press, guided us throughout the process and actually read in detail the draft first submitted, offering many comments for improvement.

Help for this project came in many other forms. Monica Ihli of the UT library faculty provided access to the university archives, crucial for research on the early days of the partnership. David Keim, ORNL director of communications, helped us with access to laboratory archives. Bonnie Nestor, recently retired from ORNL, carefully read and corrected an earlier version of the book and also wrote some pages on Department of Energy GOCO (government owned contractor operated) contracts. Our historical research was greatly aided by people at ORNL (Anna Galyon, James Kidder, Genevieve Martin, and Joy Anderson) and at Oak Ridge Public Library (Teresa Fortney, Virginia Spence, William Gwin, and Michael Stallo).

Many people contributed to the development of the UT-Oak Ridge partnership, as this book recounts. Some stand out as especially important. Homer Fisher, as UT vice chancellor and then vice president, always supported the partnership and played lead roles a number of times, for example, as the UT "capture manager" of the ORNL contract that started in 2000; Bill Madia was the Battelle capture manager and then laboratory director. Joe Johnson was the UT president for a decade and always understood and supported the partnership. Joe DiPietro was a later UT president who was important for the development of the Bredesen Center. Two recent UT Knoxville chancellors stand out: Loren Crabtree and Jimmy Cheek. The former was key to keep the partnership developing as we endured turnover of three UT presidents during the decade of the 2000s. Cheek was absolutely crucial for the creation and early growth of the Bredesen Center.

In Tennessee, we benefited from the direct investments of six consecutive governors in the partnership. Two especially stand out: Lamar Alexander and Phil Bredesen. The former (1979–1987) helped fund the Science Alliance, the first joint institute building, and the first crop of high-level UT/ORNL joint hires. The latter (2003–2011) funded the second crop of hires of top joint faculty and then a new UT/ORNL interdisciplinary center and associated joint doctoral programs. We now call this the Bredesen Center. Wanda Davis and Mike Simpson helped me make this center attractive to graduate students coming from far and wide.

The writing of this book was aided by financial support from Wayne Davis as interim chancellor (2018–2019) and its production was aided by contributions from him, Jimmy Cheek (chancellor from 2009 to 2017), and current chancellor Donde Plowman.

On a personal level, two prominent nuclear physicists (Joe Hamilton, Vanderbilt, and Noah Johnson, ORNL) guided my dissertation research and were close colleagues and friends through my long career. Three colleagues (Russell Robinson, ORNL; Carrol Bingham, UT; and Hamilton) helped me start and operate the first UT/ORNL joint institute focused on nuclear physics. I worked closely with Paul Huray on partnership development until he left the university in the late 1980s. Senator Howard Baker was a joy to get to know and work with, as I served as his science advisor for a year in the U.S. Senate and then on the board of the Baker Center at UT. Jim Roberto and

Wayne Davis were always valued colleagues for partnership development, the former in leadership roles at ORNL and the latter at UT. My wife Tina was at my side for forty-eight years during career evolution until she died in 2015. My good fortune returned two years later when Mary Zuhr came into my life and also contributed to this book as an editor.

As someone said recently, "it takes a village." This book certainly did.

LEE RIEDINGER
November 2023

INTRODUCTION

THIS STORY BEGINS in 1940 when the University of Tennessee (UT) had no doctoral programs and sustained little faculty research, and the city of Oak Ridge did not yet exist. The story extends for eighty years to 2020, by which time the university had built robust graduate degree programs and boasted vibrant faculty research, Oak Ridge National Laboratory (ORNL) had attained the likely status of the best of Department of Energy's (DOE) science and energy laboratories, and Y-12 had become an excellent national security facility. Perhaps the most consequential aspect of this evolution was the close partnership built between the Oak Ridge facilities and the university. This partnership began as a broad cooperation between UT and three federal facilities in the new town of Oak Ridge. As the decades progressed and the missions of these three facilities evolved, the UT partnership also evolved to become increasingly focused on ORNL, with a renewed effort to partner with Y-12. The result was that this partnership is unique and responsible in part for the development of UT and ORNL as premier institutions.

World War II led to the Manhattan Project and the formation of three unique and first-of-a-kind facilities to produce materials for an atomic bomb in the newly formed wartime city of Oak Ridge. The Clinton Engineer Works sprang out of farmland in 1942 and 1943, with the X-10, Y-12, and K-25 sites separated by ridges for safety reasons, largely because the potentially hazardous work was conducted there for the first time. The purpose of the Y-12 site was to mass separate natural uranium (U) and enrich it in the isotope of mass 235; physicists in Europe had discovered that isotope ^{235}U fissions with a large release of energy. The K-25 site was built to develop gaseous diffusion as a means to enrich uranium in ^{235}U.

In 1941, no one knew which process would work more advantageously in efficiency or speed to generate the kilogram quantities of ^{235}U needed for a theoretically possible atomic bomb. The purpose of the X-10 site was to build the world's first industrial-scale nuclear reactor for a controlled and sustainable chain reaction of uranium to produce plutonium, viewed then as a fertile fissionable nucleus that might also fuel an atomic bomb. No one knew whether these developmental projects would work, but the U.S. government believed that Nazi Germany had already started a project to build an atomic weapon. By 1941, Germany and the Axis Power's battlefield gains far outpaced those of the Allies, and most strategists realized that if German scientists created an atomic bomb first, the Nazis and their allies would be all but invincible.

The story of the Manhattan Project has already been told in many ways for various purposes and audiences. The purpose of this book is to show how this wartime crash program led to the development of excellent government-funded facilities after World War II, the tremendous upgrade in capabilities and reputation of nearby UT, and the evolution of a unique partnership between large federal facilities and a large state university. No attempt is made in this book to address labor, environment, or health topics, as other publications have covered these extensively.

Whether this partnership, initiated during the war with some faculty joining the effort, would continue or even blossom was not at all clear after the surrender of Japan in August of 1945. Indeed, the initial assumption at the end of WWII was that the Oak Ridge facilities would close—the facilities' mission accomplished. If this had occurred, the university might well have languished for many years and even decades as an institution focused almost exclusively on educating undergraduates in an academic environment nearly bereft of any meaningful scientific research.

But through a combination of insightful leadership and awareness of complementary strengths and resources, the UT-Oak Ridge partnership evolved, ultimately took root as an extensive university-national laboratory linkage, and prospered over its eighty-year history. This book tells that story of the development of the university and the Oak Ridge facilities into robust and world-leading institutions in many fields, in large part as a result of the evolving nature of their partnership. Each institution leveraged the others to build expertise and status in various areas of science and engineering, and these achievements came in large part through the leadership of strong individuals and top officials who were present at each institution during succeeding phases of this evolution. Their stories are told on these pages to show how each person came prepared to lead the development of this unique partnership.

The following chapters present a decade-by-decade examination of the UT-Oak Ridge partnership. Throughout this temporal discussion, four important and recurrent themes emerge, and all undergird the many crucial steps along the partnership's pathway of development. One such theme is *people driving change*, which reflects the key role of highly motivated scientists and administrative leaders in creating opportunities and pushing boundaries. A second theme derives from individuals at the partnering institutions *responding to opportunities* that often have appeared on the distant horizon. A third embraces *political leadership's critical role*—including Tennessee governors' vital investments of time and resources—in helping this partnership progress. And a fourth theme—*recruiting talented people to create and join new programs*—is reflected in the Distinguished Scientist and Governor's Chair programs, both of which have sought to bring the best and brightest to high-level positions that jointly serve UT and ORNL, and new doctoral programs to attract top-notch graduate students.

Throughout this book, important critical connections between people and institutions are discussed; connections that greatly impacted UT and Oak Ridge institutions and flowered into new possibilities, new directions, and new partnership programs.[1] The many critical connections have played major roles in shaping the involved institutions and bringing the partnership to a high level of achievement eighty years after the story began.

The chapters on the 1940s describe the first consequential leaders to significantly impact the partnership and its subsequent development. Two faculty from the UT Department of Physics, William Pollard and Kenneth Hertel, significantly contributed to the voices calling for the continuation of the Manhattan Project Graphite Reactor and supporting laboratory facilities in Oak Ridge, in part for the benefit of southeastern universities. These universities generally lagged behind top universities in other parts of the country in research and graduate education. Pollard and Hertel's lobbying effort worked, and the U.S. government decided to keep the Oak Ridge laboratory open and to allow the formation of an official university presence at or near these facilities in Oak Ridge.

UT departments—chiefly physics, chemistry, and chemical engineering—responded by building expertise with larger faculties and starting doctoral programs in part to provide continuing education for many Manhattan

Project scientists whose graduate studies at universities around the country were interrupted by the crash program to build the atomic bomb. These UT departments suddenly welcomed a cadre of excellent part-time graduate students from Oak Ridge, many with the intention on earning doctorate degrees in the years after the end of World War II. In the 1950s, this influx of graduate students significantly impacted the development of various science and engineering departments at UT.

The decade of the 1950s also saw substantial growth at ORNL of strong programs in biology, chemistry, and physics, all supporting ORNL's mission as a nuclear energy lab. The first reactor training program offered in 1946 contributed to the development of nuclear reactors for the Navy and later for the commercial generation of electrical power. The corresponding departments built up in parallel at the university, with UT offering graduate courses in Oak Ridge and with a few dozen faculty working as consultants at the three facilities in Oak Ridge (ORNL, Y-12, and K-25). The nuclear engineering department was created at UT in part due to nearby people and programs at ORNL, and it grew to be one of the top departments in the country.

The decade of the 1960s saw the beginning of the first official joint programs between UT and ORNL, in part because of the building of larger and stronger faculties in corresponding departments at the university. A Ford Foundation grant brought the first cadre of Oak Ridge researchers to part-time faculty roles at UT. In 1967 the UT-ORNL Graduate School of Biomedical Sciences opened, with seed contributions provided by Governor Frank Clement matched by funds from Union Carbide, the contractor that managed the three Oak Ridge facilities for thirty-seven years.

The National Institutes of Health (NIH) initially funded this school, which operated for three decades and attracted a top-flight group of doctoral students to take courses in the ORNL Biology Division (located at Y-12), conduct research there, and live in rented housing in Oak Ridge. The small overlap with biology departments at UT, the ending of the NIH funding, and the changing nature of large-scale biology research led to this fine example of a university-laboratory partnership being reshaped into the UT-ORNL graduate program of Genome Science and Technology, established in 1998.

The 1970s saw a diversification in ORNL's scientific scope, as the laboratory-initiated research divisions devoted to energy and environmental science, thereby providing new areas of overlap with the university. The development of the first user-originated scientific instrument at a national laboratory occurred at ORNL early in this decade, when twelve southeastern universities (led by UT and Vanderbilt) combined resources to build and operate a mass separator on-line at the cyclotron at ORNL, facilitating a new wave of experiments in nuclear physics.

A significant leap forward in the partnership occurred in the 1980s. Union Carbide Corporation decided to end its thirty-seven-year tenure as the manager of the three Oak Ridge facilities, and a competition process to find a new managing contractor began in 1982. By then, DOE's landscape changed, with an increasing emphasis on including universities in managing contractor consortia at all national laboratories (beyond the three DOE laboratories operated by the University of California).

In fact, eleven universities announced their intent to compete in some way for the Oak Ridge management contract. Chancellor Jack Reese at UT, Knoxville saw the opportunity and formed a large committee to study this possibility from various angles. In the end, UT opted not to submit a management bid, alone or with a corporation. Ultimately, that proved a wise decision as, in many ways, the university was not ready for this big step. Joint programs were not yet robust enough and the UT administration was not fully aware of growth that could come from an enhanced Oak Ridge partnership.

Even though UT did not submit a bid, value came from this intense exploratory process and discussions with many companies about potential joint bids occurred. One positive result was a Distinguished Scientist program whereby top science and engineering faculty were hired into fifty-fifty joint appointments between UT and ORNL. Neither institution could, on its own, have attracted these science stars to its staff. The combined assets of the two institutions allowed the addition of more than a dozen top scientists, who, in turn, served as recruiting magnets and leaders of new research programs.

Martin Marietta, which submitted the winning bid in 1983, began managing ORNL, Y-12, and K-25 in 1984. Fourteen years later, in 1998, DOE announced that the ORNL management contract would be up for competition in 1999. By then, UT was ready, and the research staff at ORNL largely supported a university having a hand in the laboratory's management. At the time, some government officials wanted UT to sit on the sidelines in this competition and make a marriage with whichever bidding consortium was chosen. However, leadership at UT and in the state insisted on mounting a bid.

Around Thanksgiving of 1998, UT formed a partnership with the Battelle Memorial Institute based in Columbus, Ohio, forming UT-Battelle LLC. Amid much preparation and intrigue, UT-Battelle won the competition and became the ORNL management and operating contractor in April of 2000. This partnership has won high marks in scientific output, new facilities, and DOE annual reviews in the years since.

As with most long-term collaborations, the UT-ORNL partnership suffered several setbacks along the way. Between 2001 and 2009, three consecutive UT presidents were fired or left prematurely, and the debate over who controlled the UT-ORNL partnership was a factor in two of those departures. But even in those times of strain, the work and joint programs of many faculty and laboratory researchers kept the ball rolling and the partnership advancing, with leadership and support from UT chancellors at that time.

Overall, UT's presidents and chancellors saw the benefit of the partnership with Oak Ridge and promoted it. Meanwhile, the directors of ORNL and the Y-12 National Security Complex worked to develop the UT partnership while also promoting joint programs with a number of other universities. Further, seven Tennessee governors invested resources to support specific aspects of the UT-ORNL partnership and joint efforts with Y-12. This marks a remarkable combination of state and federal resources devoted to achieving joint goals, and Tennessee may be the only state that can claim this level of leadership.

The step-by-step development of this partnership has achieved many successes that increased in number over the partnership's eighty-year history—successes in people educated and trained, top-level faculty/laboratory researchers hired, joint institutes and centers began, and research and technology generated.

This book discusses not only the leaders who spurred and sponsored this successful evolution but also the people who inspired and nurtured in this joint university and federal facility milieu and who prospered in their careers in Tennessee and elsewhere. The output of science, engineering, and technology from this partnership has been profound. One recent example is the joint program in additive manufacturing, including the Institute for Advanced Composites Manufacturing Innovation, which received a $70 million federal grant in 2015. Perhaps the most obvious success in the partnership's development is the contract awarded to UT-Battelle LLC to manage ORNL. The year 2020 marks two decades of this widely acclaimed management role.

While Tennessee—one of five states with no income tax—does not particularly excel at funding education, the state exists as a place where governors recognize the potential to build university programs by leveraging what was once a secret set of facilities in Oak Ridge. These DOE operations benefited greatly from this partnership through joint faculty, state support in acquiring complex and costly instruments, and the presence of many graduate students working on research.

Further, UT played an expanding positive role in enriching the city of Oak Ridge. The early role of the university in providing graduate courses in Oak Ridge benefited all involved, and the later formation of UT-Battelle LLC resulted in substantial investments in the city.

The initial chapters of the book present a history of the early development of this partnership, while subsequent chapters also serve in part as a personal memoir of the authors, who contributed to or at times even led the step-by-step progression over much of the past sixty years. All four have long experience in things relating to Oak Ridge and UT. Lee Riedinger joined the UT faculty in 1971 and served in a variety of UT and ORNL roles for nearly fifty years. William Bugg attended high school in Oak Ridge just after the Manhattan Project and earned his graduate degree in physics at UT in the 1950s, before serving on the faculty for four decades. Al Ekkebus worked at ORNL in various roles for thirty-three years. Ray Smith was a longtime Y-12 employee and serves as the unofficial Oak Ridge historian. In some instances, this book presents their personal stories about a partnership that enhanced the university, the federal facilities in Oak Ridge, and the city of Oak Ridge.

UT has grown enormously over these eight decades, especially in science and engineering fields, in part due to the presence and proximity of a premier national laboratory and a nuclear weapons production plant located in nearby Oak Ridge. This partnership, with its many facets, demonstrates the positive results of dedicated and creative thinkers who showed determined purpose and undertook initiatives for the good of the nation, state, and local community.

An overriding goal of this book is to demonstrate clearly that UT and ORNL operate more effectively together than they might apart. This book also documents the pride individuals and high-performance work teams have taken and still take in their work at the university and in Oak Ridge. The nation and the world benefit from the substantial scientific achievements evolving from such partnerships, and, thus, this book provides a successful model that other scientific institutions and universities can learn from and emulate.

1

THE EARLY 1940S

A World War and an Urgent Secret Project Bring Major Change to East Tennessee

BEFORE OAK RIDGE came into existence, UT had a long history, tracing its roots to the establishment of Blount College in Knoxville in September 1794.[1] Blount College was re-chartered as East Tennessee College by the state legislature in 1807 and then as East Tennessee University in 1840. Around 1828 the campus was moved from downtown Knoxville to a hill just outside the city limits.[2] The university closed during the Civil War, and the small campus "on the hill" was occupied by the armies of both the North and South at different times during the conflict.

In 1869, the state legislature designated the university as the recipient of the funds provided by the Morrill Act of 1862. This federal act awarded land grants to states for the establishment of colleges and universities that would teach agriculture, the mechanical arts, and military science. President Abraham Lincoln signed the legislation in 1862, but the act did not include states that seceded from the union during the Civil War, including Tennessee. In 1867, Congress passed a special law that made these states eligible for the land-grant funds. The Tennessee legislature accepted the land grant, and East Tennessee University (renamed the University of Tennessee in 1879) ultimately received $396,000 in state bonds (from the sale of federal land granted to the state) bearing 6 percent interest. This set the university on a course to create wide-ranging agricultural programs and the path that resulted in UT's state-wide system of today. The second Morrill Act of 1890 established historically black colleges and universities as land-grant institutions, and Tennessee State University officially became Tennessee's second land-grant university in 1958.

During this period, there was great conflict over the curriculum between those wishing to preserve the classical educational tradition and those recommending emphasis on the mechanical and agricultural arts. Because UT President Thomas Humes stood resolutely on the side of the classical tradition, placing him at odds with the trustees, the trustees requested his resignation, and, on August 24, 1883, he acceded to their request. His departure signaled the chance for the university to significantly revise its educational offerings.

In 1887, after a four-year hiatus during which the university was left leaderless, UT trustees appointed as president Charles Dabney, a 32-year-old PhD chemist with a degree from the University of Göttingen in Germany. By the end of his seventeen-year tenure, Dabney was backed by the legislature and therefore able to bring about changes that would allow UT to attain a new status as a modern university. He immediately dismissed nearly all the faculty, hiring "outsiders" from Massachusetts, Maine, New York, Virginia, and North Carolina as their replacements. He reorganized the undergraduate curriculum into a single College of Agriculture, Mechanical Arts, and Sciences, dropping the terms "Liberal" and "Classical" Arts.

By 1890, the academic core of the university consisted of ten schools, including the School of Mechanical Engineering and Physics. Under construction was Science Hall, a well-equipped building located on the hill, with ample laboratory space for physics and chemistry. The building remained in use until the early 1960s. In 1898, mechanical engineering was elevated to a stand-alone school, and physics joined an important emerging discipline to become the School of Physics and Electrical Engineering.

Dabney actively recruited new students and involved state legislators in the selection process, asking them to suggest the best students for scholarships, without regard to the students' residence in their districts. He admitted women and appointed a Dean of Women in 1893. He was a nationally recognized expert in agricultural policy, serving as Assistant Secretary of Agriculture to the Grover Cleveland administration while remaining as UT president. Dabney devoted a great deal of attention to the development of the Agricultural Experiment Station and its contribution to the welfare of the state.

His service in Washington convinced him that promoting the cause of advanced scientific education was essential to the economic development of the South and that a necessary step was the improvement of elementary and secondary education. He was extremely critical of the quality of southern schools, attributing much of the cause to the poverty of the people, which in turn was due to their lack of quality education. To remedy this situation, Dabney instituted a summer school for training teachers on the UT campus, which evolved into the highly successful Summer School of the South. He appointed Philander P. Claxton to head this entity and a new Department of Education at the university. More than 2,100 individuals from thirty-one states attended the first session of the Summer School of the South, although only about three hundred were expected. By 1918, more than 32,000 teachers had received instruction in advanced subjects at the school. Dabney's advocacy and his involvement in educational efforts outside the state led to discontent among the state's residents and university trustees and resulted in Dabney's resignation in 1904, when he was offered the presidency of the University of Cincinnati.

In 1904, Brown Ayres, a physicist and electrical engineer with a PhD degree from Stevens Institute of Technology in New Jersey, was appointed as UT's new president. Under Ayres, UT prospered throughout the early 1900s and World War I, beginning with construction of new buildings and reliable support from the state legislature. By 1918, the student body had grown from 729 to 1,893 and the faculty from ninety-five to 312. Along the way, the university raised academic standards and stiffened entrance requirements, and, as a result, UT was accredited by the American Association of Universities and the University of Berlin, demonstrating clearly that a UT degree had attained a level of respectability comparable to those conferred by the nation's other universities.

The U.S. stock market crash on October 29, 1929, marked the beginning of the Great Depression and had a terrible impact on families, businesses, and governments (federal and state). After a decade of increasing budgets for the university in the 1920s, the UT operating budget decreased through the early part of the next decade, and, in 1935, the state legislature cut UT's allocation in half. Throughout the decade, there was poor financial support by the state, for a number of reasons. At that time, higher education was not viewed as a responsibility of the state, and the state constitution did not make provisions for it. There was much sectarian education, and the sectionalism and competition within the three grand divisions of Tennessee (east, middle, and west) contributed to the problem. There also was class hostility to a college education.[3]

Adding to the university's financial problems was the conservative and uninspired leadership of President James Hoskins, who served as the UT president from 1934 to 1946. His administration failed to capitalize in a substantial way on obtaining federal funds for construction from the federal Works Progress Administration (WPA). This was a matching program, with 45-percent funding for a project coming from the WPA and 55 percent coming from the state. UT did achieve some success in leveraging these funds, for example, to construct a small engineering building (Berry Hall) adjacent to Neyland Stadium. But, the tight funds in Tennessee government and Hoskins' inability to cajole state legislators into supporting his cause resulted in UT tapping few of these available federal funds. Consider, for instance, that in 1938, Tennessee obtained $166,909 in WPA funds, while Louisiana received $4.7 million and North Carolina $2.4 million.[4] By 1940, the University of Tennessee was not a strong or well-led university and had yet to establish its first doctoral program.

THE RHENIUM PROJECT AT UT

The chemical element rhenium, atomic number 75, was discovered in Germany in 1925. It has the third highest melting point and second highest boiling point of any stable element,[5] and was viewed as potentially useful for industrial purposes. Today, rhenium is used as an additive to tungsten- and molybdenum-based alloys, which are used for oven filaments and x-ray machines.

Rhenium is also used as an electrical contact material, as it resists wear, withstands arc corrosion, and is added to high-temperature superalloys used to make jet engine parts. Rhenium is a rare element, comprising about 1 part per billion in the earth's crust. The element resembles manganese and technetium chemically, and its extraction from various ores is difficult. In 1928, the discoverers were able to extract only 1 gram of the element by processing 660 kilograms of molybdenite (or the weight equivalent of one-fifth teaspoon of water in almost 1,450 pounds of material).[6]

Clarence Hiskey studied rhenium for his dissertation at University of Wisconsin and came to UT as a chemistry instructor in 1939. He suggested to his colleagues that UT apply to the WPA for a grant to discover an effective method to both investigate rhenium's properties and then establish its domestic sources. The WPA provided $20,729 and UT another $6,649, along with lab space and equipment, and the project began in 1940 with fifteen staff. Initial results were promising, and the WPA provided an additional $37,204 supplemented by UT's $13,657 in 1941. Hiskey left in 1941 for a position in the Tennessee Valley Authority and was succeeded by Arthur D. Melaven, an assistant professor in the Department of Chemistry who would lead the project until its end in 1966. UT obtained its rhenium samples from Germany and, as the clouds of war began to form, UT looked for a domestic source and found it in the molybdenum and rhenium flue dust, a byproduct of processing copper ore, produced by Miami Copper Company of Arizona. Beginning in 1942 and during the next ten years, UT produced 240 pounds of rhenium from seventeen tons of flue dust. The rhenium had potential use in filaments of flash bulbs and vacuum tubes, as well as in electrical circuitry.[7]

After the war, the university continued to produce rhenium for both U.S. and foreign markets, and earned $38,000 for its efforts over the next twenty years. However, UT faculty and students also benefited from the project, gaining experience in scientific research and earning funds for grants to both students and faculty. Seven graduate students received Rhenium Research Fellowships in 1943–1957, fourteen high school seniors and UT freshmen received Rhenium Academic Scholarships during 1956–1965, and about eighty chemistry undergraduates received aid as Rhenium Scholars in 1962–1993. With other contributions added to the Rhenium Scholarship Fund, it stood at over $33,000 in 1993. Now known as the A.D. Melaven-Rhenium Scholarship, it remains a source of support for chemistry undergraduates.[8]

The originator of the Rhenium Project, Clarence Hiskey, ended up having a role in the "spy stories" addressed later in this chapter.

TENNESSEE VALLEY AUTHORITY

The establishment of the Tennessee Valley Authority (TVA) by the U.S. Congress in 1933 marked a major advancement for the region on several fronts. Indeed, Congress formed the new agency to control flooding, improve navigation, and provide affordable electrical power to the chronically impoverished Tennessee Valley. The agency marked its first major achievement in 1936 with the opening of Norris Dam (situated about twenty miles north of Knoxville) on the Clinch River in Anderson County.

TVA and UT

TVA significantly impacted UT. As an example, a chemical engineering curriculum at the university first appeared in the 1904–5 catalog. Along with other chemical engineering programs of the time—including a pioneering program at the Massachusetts Institute of Technology—UT's curriculum was heavily oriented toward chemistry rather than engineering, including detailed explorations of important industrial processes.

The impetus to establish a true chemical engineering program at UT appears to have come first from TVA and, in particular, from its chief chemical engineer, Harry A. Curtis. Shortly after his arrival at TVA headquarters in Knoxville, Curtis is reported to have approached UT President James Hoskins and advised him, "Mr. President, you offer a chemical engineering program at your university, but I would not hire one of its graduates for TVA." Having thus gotten the president's attention, Curtis explained that he would not hire UT chemical engineering graduates because there were no chemical engineers on the faculty and no chemical engineering courses were offered. Before the meeting ended, Curtis and Hoskins had reached an agreement: Curtis would pay half the salary of a chemical engineering professor who would be employed by the university and would work half-time on fertilizer development research for TVA. This position was established in UT's chemistry department and filled by Robert M. Boarts, who at the time was completing his doctoral work in chemical engineering at the University of Michigan.[9] Boarts would later become a key figure in university relations with Oak Ridge.

When Norris Dam began generating power in 1936, Oak Ridge did not yet exist. At the time, this area, twenty-five miles west of Knoxville, was agricultural land supporting one thousand deeply rooted farming families clustered in and around small communities named Wheat, New Hope, Scarboro, New Bethel, and Robertsville. But within six years, the farm fields would be transformed into some of the most sophisticated scientific facilities in the world, and its contribution to the war effort would be immeasurable.

WITH THE WORLD AT WAR, OAK RIDGE UNDERTAKES ITS CRITICAL MISSION

After the Japanese attack on Pearl Harbor on December 7, 1941, the United States was suddenly engaged in the global conflict we now call World War II. Before Pearl Harbor, much discussion occurred among scientists and even government officials in the United Kingdom and the United States about the discovery of nuclear fission and the possibility of its use to build a powerful weapon.[10 11]

The seminal experiment to demonstrate that the uranium nucleus can fission (split into two pieces) when impacted by neutrons had been performed in Berlin in December 1938. This led to an international race by scientists to understand the nature of this newly discovered physics process. Soon after, Lise Meitner, Otto Frisch, and Niels Bohr developed a theoretical basis to understand the experimental results, and fears were raised about the potential for a new and horrific weapon of mass destruction. In 1939, Leo Szilard and Eugene Wigner persuaded Albert Einstein to write a letter to President Franklin Roosevelt, urging the president to learn more about this phenomenon through the funding of research by U.S. scientific institutions. In response, Roosevelt launched an effort that would evolve into the Manhattan Project.

Experiments in Berlin, Paris, Copenhagen, England, and the United States (at Columbia University) showed that the isotope uranium-235 (^{235}U)—which represents only 0.7 percent of naturally occurring uranium—would fission upon the impact of neutrons. Experiments showed that a large amount of energy was released by this nuclear process—millions of times greater than the energy released in a chemical reaction.[12]

Physicists discovered further that two or three neutrons are emitted when ^{235}U fissions, which opened the possibility of a chain reaction (hypothesized by Szilard in 1933). And, in 1940, two Jewish physicists (Rudolf Peierls and Otto Frisch), who, respectively, left Germany and Austria to work at Birmingham University in England, conducted a crucial theoretical calculation that changed the course of science and world politics. They theorized that only a few pounds of ^{235}U, not a few tons, as previously assumed, would cause a chain reaction of this material and trigger a huge release of energy—an atomic bomb.

After Albert Einstein's letter to the president, a small research program relative to uranium fission was started at a few locations, although research was already proceeding at Columbia University led by Nobel laureate (physics, 1938) Enrico Fermi. After the Pearl Harbor attack, President Roosevelt completely endorsed the advice of Einstein and other top scientists (including E. O. Lawrence, Arthur Compton, and Vannevar Bush) to initiate the secret and expensive Manhattan Project (named after the location of the project's initial headquarters in New York City).[13] This project would learn how to possibly harness nuclear fission in order to create the world's most powerful weapon. In part, the knowledge that Nazi Germany already began to work on its own uranium project drove this move, and the Germans had a big head start since their scientists had been the first to conclude that uranium could fission. If German scientists had created an atomic bomb first, the Nazis and their allies would have most certainly won the war.

SENATOR KENNETH McKELLAR BRINGS THE PROJECT TO TENNESSEE

As former Tennessee U.S. Senator Howard Baker frequently told the story, President Roosevelt asked U.S. Senator Kenneth McKellar, chair of the all-powerful Appropriations Committee, for his support in appropriating $1 billion (and keeping it hidden from public scrutiny and all but four congressional leaders) to construct a secret laboratory, where the world's greatest weapon would be housed and would help the Allies win the war. At the time, McKellar was Tennessee's senior U.S. Senator and the only Tennessee senator ever elected to six terms. According to Baker, Senator McKellar pledged his support and then asked the president, in a classic stroke of political *quid pro quo*, "and where in Tennessee do you plan to put that 'thang?'"[14] Confirmation of this conversation has been inconclusive, as this may be just a funny story told by one good politician about another. As time goes on, good stories can take on a life of their own.

In an August 1946 interview that appeared in the *Oak Ridge Journal*, Senator McKellar explained the genesis of the Oak Ridge facility in somewhat different terms. In 1942, McKellar was invited to a meeting led by then Secretary of War Henry Stimson, who asked the Tennessee senator to pledge his support for an initial $80-million investment "because scientists are going to split the atom and win the war." McKellar wondered how he could justify this considerable expenditure to his constituents if nothing came to fruition that would directly benefit them. Sometime later, Stimson assured McKellar that a big plant would, in fact, be built in Tennessee.[15]

Col. Kenneth Nichols, the top Army official in Oak Ridge, recognized that elected officials would need more information on a regular basis to assure themselves (and their constituents) that the large projects would benefit the war effort. Senator McKellar desired details so he could answer questions about the importance of the Clinton Engineer Works (CEW) to his supporters. Nichols met with him in the Andrew Johnson Hotel with none of the Senator's staff or other visitors present. Nichols emphasized the need for secrecy as demanded by the President himself. The number of employees and town size were mentioned as well as Nichols' opinion that the end of the war would not see termination of all employment here. McKellar requested, and Nichols promised, to keep him updated periodically about the project. At the end of the meeting, McKellar apologized to his other guests that secrecy prevented him from including them in the interesting discussions.[16]

Nazi Germany surrendered on May 7, 1945, but, as the war with Japan continued into August, McKellar asked Stimson again about the project, because he said that he had heard no updates. The initial $1 billion commitment grew to be $1.89 billion by the end of the Manhattan Project, with a total of $1.19 billion spent in Oak Ridge. "We've spent $2 billion and haven't accomplished a visible thing," McKellar lamented.[17] Stimson's secretary visited McKellar the next day and informed him that a successful experiment had just occurred in New Mexico in July 1945.

When McKellar sought additional details about the cryptic project, he was told that three B-29 planes were, at that moment, in the air over the Pacific Ocean and that Stimson would soon share more information. The next day, August 6, newspapers announced the detonation of an atomic bomb over the Japanese city of Hiroshima and revealed the critical role Tennessee's "secret city" had played in the bomb's construction. Japan surrendered eight days later, after a second atomic bomb was dropped over Nagasaki on August 9, officially ending the war.

Building this crucial secret laboratory and the facilities necessary to separate ^{235}U in the predominantly rural region of East Tennessee made a great deal of sense.[18] Knoxville was not far from Washington and the Pentagon, which would operate this secret project. Meanwhile, TVA's Norris Dam provided much of the electrical power required to run the energy-intensive enterprise. And nestling a secret laboratory in the remote valleys and ridges of East Tennessee was attractive from a security perspective, because the project's location was unlikely to draw public scrutiny. Further, there was plentiful clean water from the Clinch River, barge access was possible, and the population density of the future project site was low. Indeed, only three thousand people would need relocation. One could obtain cheap land (less than $4 million for the holding of fifty-nine thousand acres), two railroads served the area, and nearby Knox County had an abundant labor force.

According to Lester Fox, who, in 1942, was a sophomore at Oliver Springs High School—a school situated in a small community located just northwest of the future town of Oak Ridge—the announcement of the coming forced relocation came rather abruptly.[19] Fox and a friend had skipped school and were walking down Main Street, past the telephone office, when the operator spotted them and, leaning out the door, hollered "Lester, go get the principal, and tell him that he has an important phone call!" Despite the fact that Lester had skipped school and was about to be caught in the act, he dutifully summoned the principal, who dashed to the telephone office and took the call.

According to Fox, after returning to the school, the principal hastily called all the students—including Fox, no longer a truant—into an assembly, where he informed his audience, "I just got a phone call from Senator McKellar. He wants me to tell you to go home and inform your parents that they are going to have to find another place to live, because the government is going to take your homes for the war effort."

Lester, who is alive at this writing, insists this is how many of the three thousand people living in what is now Oak Ridge first learned they need to relocate in order to make room for the Clinton Engineer Works. Few of the displaced residents owned trucks or automobiles to move their belongings, and many of those who did were unable to acquire tires or fuel for the vehicles, as both commodities were strictly rationed during the war. But what these families did have in abundance was patriotism. They were well aware that American GIs were being killed, and they wanted to do anything they could to win the war and end the carnage.

The federal reservation for the Manhattan Project straddled Anderson and Roane Counties. In the 1930s, Anderson was a rural and agrarian county in Appalachia, with coal mining in its western region and farms in the eastern part. As of 1930, there was only one major industry in Anderson County. Magnet Mills (hosiery) was situated in Clinton (a small town to the immediate northeast of the federal reservation) and employed 1,100 people in a county with only around 19,000 residents.[20]

After the completion of Norris Dam, the population of Anderson County grew from around nineteen thousand in 1930 to twenty-six thousand by 1940. Then the numbers truly exploded with the Manhattan Project, peaking at eighty thousand people working in Oak Ridge by 1945. This greatly impacted Anderson County beyond the dramatic increase in population: the now-booming community would require housing to accommodate the new arrivals, along with a number of enhanced services. The community's new residents, from points across the country, vastly differed—socially, economically, religiously, culturally—from existing members of a region situated in the heart of the Bible Belt. Additionally, East Tennessee's public-education system was decidedly poor compared with what many had left behind. A new K-12 public school system would soon be created in Oak Ridge that would be among the finest in the South and perhaps the nation.

General Leslie Groves was appointed to lead the Manhattan Project in September of 1942. He quickly approved (on September 19) the acquisition of a site that grew to include fifty-nine thousand acres of land in Roane and Anderson Counties. Residents of the research facility's future site were given

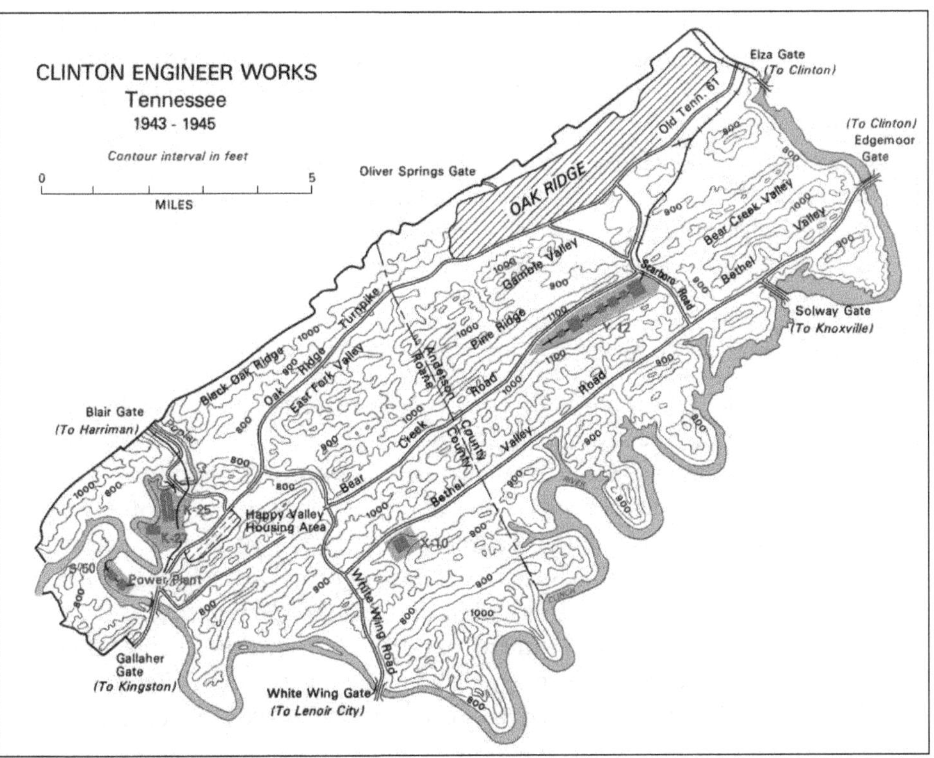

A MANHATTAN PROJECT MAP OF THE CLINTON ENGINEER WORKS RESERVATION WITH THE VARIOUS PLANTS HIGHLIGHTED: Y-12, X-10, K-25, S-50, AND THE POWER PLANT. THE CLINCH RIVER BOUNDS THE AREA ON ONE SIDE. IN POSSESSION OF THE LEAD AUTHOR.

two weeks to vacate, many after decades on their homesteads, as the U.S. government purchased the land and rapidly started building the secret facilities. Residents in Knoxville, with a 1940 population of 112,000, suddenly started seeing scores of train shipments headed for Clinton and wondered what the shipments might contain.

New arrivals quickly inundated the brand-new town site of Oak Ridge, which would number seventy-five thousand people within two years.[21] People navigated mud streets that had not been paved quickly enough to keep up with the town's growth and pass through guarded security gates to reach the government facilities and the town site. Construction workers came from far and wide to build the complex of secret facilities and laboratories.

Scientists and engineers from across the country, many in the middle of graduate-degree programs, were recruited to work on the massive project. Secrecy was paramount, and few had a clue about the project's full purpose or mission.

As farmland began to change into Manhattan Project plants and thousands of people arrived, the government facilities burgeoned. UT, likewise, soon would undergo a dramatic period of growth and become a key supporting partner in defining the future of the Oak Ridge enterprise.

THE KNOXVILLE AREA IN THE EARLY 1940S

The Knoxville area significantly contributed to the war effort even before Oak Ridge began. The Aluminum Company of America (ALCOA) produced aluminum for airplane wings and fuselages. Rohm and Haas Chemical Company's Knoxville plant opened in July 1943 and made Plexiglas canopies for almost three-hundred thousand airplanes. The Fulton Sylphon Company made many products but perhaps most important was the seamless metal bellows used to control depth charge and bomb detonation. Each American bomber used over a hundred of these bellows and two were used in precision detonation devices for the two atomic bombs being developed. Tennessee Eastman of Kingsport produced RDX explosives that were 50 percent more powerful than TNT.[22]

Although few people knew, Knoxville also became an international hub of science and technology during World War II. If a poll had been taken before the war years asking respondents to name the world's top centers of science, the Knoxville area certainly would not have made the list. However, a similar poll taken in late August 1945 might have placed the same area among the top few in the country. What a difference $1.19 billion—the cost of the Oak Ridge portion of the Manhattan Project through December 1946—can make.

The Manhattan Project's economic boost to the Knoxville area was significant. According to analysis conducted after the war, around four-hundred thousand people were hired (with much ongoing attrition) at CEW in Oak Ridge for construction and operations from 1942 through December 1946, with about forty thousand still employed at the end of that period. This large number of employees was due to the high labor turnover: estimates were 17 percent turnover per month in construction workers and almost 7 percent in operations. Construction employment peaked in March 1944 at sixty-nine thousand, and operations employment peaked at forty-seven thousand in June 1945. In August 1945, there were seventy-one thousand employed, with twenty-five thousand engaged in construction and the remainder in operations.[23]

Because of the extreme secrecy of the project, there were restrictions on who could live in the new town of Oak Ridge. If you lived closer than forty miles when hired, you had to commute. Bus service was provided from as far away as Calhoun to the south, Jellico to the north, Monterey to the west, and Bean Station to the east. In 1944, daily roundtrip bus service from these places was provided for $3.50 per week, less if you lived closer.[24] Bus service was also provided within the city's residential areas to a central terminal where employees transferred to buses bound for the individual plants. There were position openings of all types, including a large number of construction jobs and also an even bigger workforce providing food, coal, laundry, building maintenance and repair, and other services. Women played a significant role in employment, numbering around 30 percent of all workers.[25]

This huge investment of federal funds and human capital was related to four secret federal facilities in the new area of Oak Ridge, labeled by the Army as X-10 (also known as Clinton Laboratories), Y-12, K-25, and S-50. At the peak of the Manhattan Project, eighty thousand people worked at these sites, and seventy-five thousand people lived in Oak Ridge, "the city behind the fence." Ultimately, 60 percent of the total budget for the Manhattan Project was expended in East Tennessee.

Relationships between Oak Ridge and its neighbors were unusual, as almost everyone living in Oak Ridge came from outside East Tennessee. Uncertainty and unfamiliarity bred jealousy as these new folks came to Knoxville with mud on their shoes and spoke differently from others: They clearly were "not from around here."[26] Fences were erected around Oak Ridge, and armed guards were deployed. Wages in Oak Ridge were much higher than wages throughout the region. The impact of this disparity became apparent when Oak Ridge began hiring for its school system and paid wages substantially higher than those of other systems in the area. In addition, funds were available for new school buildings and textbooks. If high school sports teams wanted to play in Oak Ridge, the visitors had to declare in advance who would ride on the team bus, and all of the bus occupants would need to present identification. Meanwhile, the roster of Oak Ridge players did not include surnames; the players were identified only by their uniform numbers and first names.

Even though there were constant trials and tribulations in the relationship between this new, well-funded federal facility and the far more impoverished surrounding counties and cities, signs of goodwill and partnership soon emerged. For example, Oak Ridge residents began to participate in events in Knoxville after the war ended. A 1947 article in the Nashville *Tennessean* describes Lt. Col. Donald Williams, who served as a piano soloist for the Oak Ridge symphony and was one of a dozen musicians who played in both the

Knoxville and Oak Ridge orchestras as needed.[27] As the years progressed, increased contacts enriched both cities, Knoxville and Oak Ridge.

A hallmark of the developing partnership between UT and Oak Ridge was the strong support of Tennessee governors from the 1950s onward. However, this good relationship with the governor's office got off to a rocky start in 1943, as then Governor Prentice Cooper was indignant that he had not been forewarned as the land in East Tennessee was purchased by the federal government. Difficult communication with Governor Cooper was finally resolved when Col. Nichols, the top Army official in Oak Ridge, invited the governor for a November 1943 visit, which included a carefully orchestrated reception at Nichols' home. All was well thereafter, leading to agreements on use of state funds for road and bridge upgrades adjacent to the federal reservation.

Governor Prentice Cooper and the Army

It was early summer of 1943, and Tennessee Governor Prentice Cooper was confused and upset. Why had he not been given details about the massive land purchases in Anderson and Roane Counties by the Army in late 1942? After all, he was only the state's governor! He had recent experience with the Army coming to Tennessee and working closely with the state and local governments when buying land for ammunition plants as part of the war effort. These facilities were not small: Holston Ordnance Works in Kingsport (begun in 1941, 6,000 acres), Milan Army Ammunition Plant (1940, 22,357 acres), Volunteer Ordnance Plant in Chattanooga (1941, 6,078 acres), and Chickasaw Ordnance Plant in Millington (1940, 6,000 acres).

In Cooper's opinion, the impact on the local area and population around Black Oak Ridge was extremely significant. The Army had taken one-sixth of Anderson County and one-seventh of Roane County for this new secret project. He had heard reports that the landowners were not adequately compensated for their property and in some cases were forcibly evicted from their homes. The county governments had not been compensated for the roads and bridges built that were now used solely by this new federal government project. This was totally unlike other dealings with the federal government earlier in the war and even before the war started, when the Tennessee Valley Authority took land for dams and reservoirs and when Great Smoky Mountains National Park was formed.

When the U.S. Government filed the declaration in federal district court in Knoxville in advance of taking of the land in Anderson and Roane counties, the governor received no forewarning. Apparently, someone in the Army mistakenly forgot about the governor. Initially, this new East Tennessee project had been called the Kingston Demolition Range, and now it was known as Clinton Engineer Works (CEW). A notice, dated March 1943 and published in the June 30, 1943, in the *Federal Register*[28] announced that the area, almost sixty thousand acres (50 percent larger than all Tennessee ammunition plants combined), would now be a military exclusion area under the control of the Army, with the state having no control over the land. There was still no mention of its intended purpose.

An initial meeting on July 12, 1943, with Army representative Captain George Leonard, resulted in Cooper tearing up the letter given to him and declaring the project an experiment in socialism.[29] A letter to the governor two days later from Lt. Col. Thomas Crenshaw chastised Cooper for destroying the initial letter, rejected the charge that this was a socialist ploy, and stressed that this was a crucial part of the war effort. Several weeks later, the governor's meeting with Col. Kenneth Nichols went better, as Nichols was able to describe the size and scope of the Oak Ridge project and the traffic problems and road conditions that would require the governor's attention. Nichols also invited the governor to tour CEW later that year and to be his guest at a reception in his home.

Governor Cooper visited Oak Ridge on Wednesday, November 3, 1943, with a police escort whose sirens startled an earth-mover operator so much

COL. KENNETH NICHOLS. COURTESY OF OAK RIDGE NATIONAL LABORATORY.

that the machine's blade left a narrow roadway and rolled over several times before coming to rest at the bottom of a steep bank. The motorcade stopped to assess the condition of the operator, who was unhurt, and the man met the governor, who agreed the roads were in terrible condition and asked if the rollover was staged. It was not, assured Nichols.[30]

The reception at Colonel Nichols' home utilized a punch bowl carved out of a block of ice. Cooper remarked that he had never seen bourbon flow so freely (Anderson County was dry at the time), which literally started as the block of ice cracked and the whiskey spilled forth as the governor entered the room. This event also effectively broke the ice for the rest of his visit, according to Nichols. The working relationship between the Army leaders in Oak Ridge and the governor improved thereafter. Over the coming weeks, mutually agreeable solutions were found to resolve the county road and bridge problems, and a new state-funded access road was soon under construction.

A son of Governor Cooper is U.S. Representative Jim Cooper, who served the Nashville area from 1983 to 2022 and was supportive of science and technology throughout his long career. His brother John became mayor of Nashville in 2019.

THE FEDERAL FACILITIES IN OAK RIDGE AND ELSEWHERE

The Oak Ridge facilities were constructed primarily for two reasons: first, to supply ^{235}U for an atomic bomb utilizing uranium (Y-12 and K-25) and, second, to build an experimental reactor to learn how to breed plutonium enriched in isotope 239 (^{239}Pu) and separate it from the natural uranium fuel elements (X-10). Both of these missions were crucial for the work at two other Manhattan Project sites. Hanford in southeastern Washington produced plutonium based on the pilot-scale operations of the Graphite Pile at X-10. Los Alamos in northwestern New Mexico focused on the design and construction of nuclear weapons, utilizing the enriched uranium sent from Oak Ridge and the reactor-produced plutonium from Hanford. All sites were distant from population centers (Knoxville was the largest city proximate to any of the three sites). Relative isolation was important, in terms of both security and safety concerns.

The town and the laboratory at Los Alamos developed rapidly in parallel to Oak Ridge. Robert Neyland, the most famous football coach in the history of the University of Tennessee, played a role in Los Alamos development. Neyland was a 1916 West Point graduate, was then commissioned as an officer in the U.S. Army Corps of Engineers, and served in France during World War I. An excellent athlete at West Point, he later moved toward coaching football when he was appointed as a UT Professor of Military Science in 1925. He served as head football coach from 1927 until he was recalled to active duty in 1940, setting records and winning national championships. Back in the Army, Neyland was head of the Corps of Engineers Southwestern Division and received the request for acquisition of fifty-four thousand acres on and around the mesa that became Los Alamos. In late November 1942, Neyland issued a contract for dwellings and laboratories to be built and occupied by mid-May 1943.[31]

Compared to Los Alamos and Hanford, CEW had the largest number of incoming workers—between three to ten times the workforces at the other sites. All sites relied on the benevolence of government-funded housing and government-supplied infrastructure. The University of California managed Los Alamos, and DuPont managed Hanford. DuPont was also the contractor employing the X-10 Graphite Pile to train workers in producing and processing plutonium. Tennessee had experience with government-funded development through TVA, particularly with Norris Dam construction and operation fifteen miles from CEW. Near Hanford, the Columbia River basin dams were constructed by the Bureau of Reclamation and focused on hydroelectricity production. The three sites together accounted for almost 90 percent of Manhattan Project costs.

Work in Oak Ridge depended on earlier and concurrent research and development at two other Manhattan Project sites: the Metallurgical Laboratory (Met Lab) at the University of Chicago and the Substitute Alloy Materials (SAM) project at Columbia University in New York. At the former, the team led by Nobel Laureate Enrico Fermi built the first pile of uranium to demonstrate that a nuclear fission chain reaction is possible.

The Met Lab team led by later Nobel Laureate Eugene Wigner worked to understand the physics of neutron multiplication in the pile and to design the follow-up nuclear reactor to be built in Oak Ridge at X-10. UT physics faculty member Katharine (Kay) Way was part of Wigner's group, as was future Oak Ridge National Laboratory director Alvin Weinberg. People from both of these groups at the Met Lab came to Oak Ridge to help build and operate the Graphite Reactor. The work at the Columbia SAM project was crucial for the development of K-25. UT physics faculty William Pollard worked at Columbia during the Manhattan Project, as did future UT physics faculty members Richard Present and Jack Craven and future ORNL associate director Don Trauger.

Two sites in Oak Ridge served as production facilities for enriching the isotope ^{235}U from the minuscule 0.7 percent in naturally occurring ore to more than 90 percent, termed "bomb-grade." Y-12 employed electromagnetic mass separation for enrichment, and K-25 primarily used a gaseous-diffusion process. Both facilities required multiple processing stages, with significant needs for construction and operation staff, along with huge electrical demand. By the end of the war, Oak Ridge's electricity usage accounted for more than 12 percent of TVA's overall electricity production and 1 percent of the nation's generating capacity.[32]

For both facilities, the end product was the same—enriched uranium—and both techniques were used because, in the initial design, both gaseous diffusion and electromagnetic separation were unproven technologies, and the urgency of the war did not allow the luxury of serially trying one and then the other. No pilot facilities were built, as there was an indication that both techniques would work when scaled up from a laboratory environment. Enriched uranium was needed for the atomic bomb, and the United States needed it to win the war. About $990 million was spent on these two efforts, as part of the total of $1.2 billion allocated to Oak Ridge in the Manhattan project.

Y-12 used 1,152 calutrons (massive electromagnets) of two stages called Alpha (864 units) and Beta (288 units).[33] The Alpha first stage could separate isotopes with a maximum of 15 percent efficiency, while the Beta second stage could separate with a much higher efficiency to obtain bomb-grade ^{235}U enrichment of 80 to 90 percent. Each of these large magnets was operated independently in batch mode, to purify and process the uranium compounds before and after every separation step. The devices could be operated for only a few days at a time before needing to be opened; taken apart and the ^{235}U harvested; cleaned; and put back into service. This process took several days.[34]

Initially, the purification process did not go according to expectations, and Berkeley's E. O. Lawrence (the inventor of the calutron) proposed solutions to several problems. When the calutrons at Y-12 began to overheat, he had the plant's fire department use its hoses on the cooling towers to boost the cooling efficiency. Another remedial effort involved Frank Oppenheimer (brother of J. Robert Oppenheimer, Los Alamos director), who was invited to assist with resolving corrosion involving a chlorine-nickel reaction in stainless steel piping. A solution involved installing copper liners in the calutrons.[35] More than five thousand chemical operators were required at Y-12, representing about 25 percent of the labor force.[36] In May 1945, more than twenty-two thousand people worked at Y-12 to provide the enriched uranium for the Little Boy atomic bomb dropped on Hiroshima, Japan, on August 6, 1945.

Because of a wartime shortage of copper, Y-12's massive electromagnetic coils were made from 14,700 tons of coinage silver from U.S. government vaults at West Point. In August 1942, Col. Kenneth Nichols, CEW district engineer, met with Under Secretary of the Treasury Daniel Bell and requested between five and ten thousand tons of silver.[37] Bell's stunned reply: "Colonel, in the Treasury we do not speak of tons of silver; our unit is the Troy ounce." The Manhattan Engineer District requested 175 million troy ounces initially (6,000 tons),[38] and the request was approved; with two subsequent agreements, this request was raised to 395 million troy ounces of silver from the West Point Depository ("the Fort Knox of silver") for the duration of the Manhattan Project.

The silver was first transported to the Defense Plant Corporation in Carteret, New Jersey, where four-hundred thousand silver bars were melted and cast into cylindrical billets of about four hundred pounds each. The resulting seventy-five thousand billets were further processed by the Phelps Dodge Copper Products Company in Baywater, New Jersey and were rolled into strips, 3-inches wide by 5/8-inch thick and 40 to 50 feet long. These strips, if laid end-to-end, would stretch from Washington, D.C. almost to Chicago. More than seventy thousand strips were then sent to Allis Chalmers in Milwaukee, Wisconsin, where they were wound around steel magnet casings and sent to Oak Ridge for installation at Y-12. Another 250,000 pounds of silver were sent directly to Oak Ridge's Y-12 Plant to form huge busbars, almost a foot square in cross-section, that carried electricity to the magnets.[39]

Special guards and accountants were assigned to monitor the silver every step of the way to ensure that the precious metal was not wasted; silver shavings, turnings, scrap, and dust were collected and re-used. Machines were taken apart and cleaned to recover as much silver as possible. Each magnet, weighing ten to seventeen tons, used a great deal of electricity to generate the powerful fields needed to separate isotopes. Each month, Col. Nichols had to sign a statement attesting that all of that silver, specified to one-hundredth of an ounce, was accounted for. By the end of the war, less than 0.036 percent of more than $300 million worth of silver had been lost. Much of the silver was returned to the U.S. Treasury as the Y-12 calutrons began their shutdown in late 1945 to 1947, with the last of the silver, a seventy-ton shipment, returned in 1970.

K-25 operated as a continuous process, located in what was then the world's largest building, comprising forty-four acres under one roof. The gaseous-diffusion process forced uranium molecules (in the form of uranium hexaflouride) with different molecular weights through tiny holes, billions per square inch, in a barrier material. The lighter molecules (UF_6 with ^{235}U) moved slightly faster than the heavier ones (UF_6 with ^{238}U) and therefore preferentially passed through the barrier, enriching the uranium in the desired 235 mass after passing through several thousand stages. Developing and testing the barrier was a major challenge in the Manhattan Project and was led by Columbia's SAM lab.[40]

Don Trauger was a Nebraskan farm boy who received an undergraduate physics degree from Nebraska Wesleyan University in June 1942. He was recruited to join the SAM laboratory and chose to do that rather than start graduate school in physics at the Illinois Institute of Technology, beginning work at Columbia on September 1, 1942.[41]

Trauger was initially assigned to the barrier development group led by Francis Slack,[42] on leave from Vanderbilt University where he was the chair of the physics department. Soon thereafter, Trauger was moved to the barrier testing group led by Robert Lagemann, another Vanderbilt physics faculty on leave. Developing a barrier that had a high enough density of sufficiently small holes and had the strength to withstand the harsh uranium hexafluoride environment was a huge challenge. Many barrier fabrication strategies rapidly developed, and Trauger was put in charge of a laboratory to test the developed barriers.

In January 1943, Carbide and Carbon Chemicals Corporation signed a contract to operate the Oak Ridge gaseous diffusion plant (K-25), to be built by the Kellex Corporation. Construction of K-25 proceeded rapidly, and, at the same time, the Columbia group was under intense pressure to have a barrier ready for the huge gaseous diffusion facility, and the barrier development was the most difficult part of the process. After many attempts of different strategies, in January 1944, scientists agreed on using a sintered nickel powder as the barrier material. The barrier was fabricated as tubes, and these tubes were placed in more than three thousand barrel-shaped cylinders, used for transition of uranium hexafluoride gas from one gaseous-diffusion stage to the next. A huge number of stages was needed to achieve sufficient enrichment of uranium; ultimately, the total length of nickel powder tubes employed in the process exceeded 6,500 miles.[43]

Trauger shared an office at Columbia with William Pollard (on leave from the UT physics faculty), who worked on the physics and chemistry theory of barrier performance, along with Richard Present, who later moved to Oak Ridge and then became a long-time member of the UT physics faculty.[44] Pollard returned to the UT faculty at the end of the Manhattan Project and led the formation of the Oak Ridge Institute of Nuclear Studies. Pollard and Present made important contributions to understanding barrier performance, helping to guide those fabricating and testing barriers.[45]

As the Columbia group worked feverishly to develop and test many barrier possibilities, two construction projects were rapidly proceeding, both dependent on success in finding the right barrier strategy. One was the construction of the K-25 facility for enriching uranium by gaseous diffusion and the other was the barrier production plant to make the large number of barriers needed at K-25, once the SAM lab decided on the best barrier strategy. This barrier production facility was built in Decatur, Illinois, by the Houdaille-Hershey Corporation.[46]

Besides the delays resulting from the tests of various barrier strategies, the project also faced the lack of available hydrogen gas needed for barrier production at the Houdaille plant. The Girdler Company of Louisville, Kentucky, was the leading manufacturer of equipment that converts natural gas into hydrogen. However, Girdler had just finished filling a sizeable order of this large hydrogen-producing equipment, which was to be shipped by rail to Seattle and then to Vladivostok, Russia, as part of the Lend-Lease Act enacted by Congress in 1941. A new order of this refinery-size equipment would take six months to fill, much too long for K-25's critical need. Lt. Col. James C. Stowers of the K-25 procurement team was ordered to write a memo detailing what would happen if this vital equipment was not obtained. Stowers complied, and, two days later, the natural-gas-to-hydrogen equipment destined for Russia became the property of the Manhattan District and was re-routed to Decatur, Illinois, for assembly near the barrier plant. The U.S. Army wrote a letter to the Soviet Army Purchasing Mission promising to replace the "borrowed" equipment.[47] By the end of the war, enough progress was made on the barrier materials and the gaseous-diffusion process to establish the promise of this uranium-enrichment strategy for the next thirty years.

S-50 was a rapidly built liquid thermal diffusion plant, as another way to enrich uranium in the fissionable ^{235}U. It was located near the K-25 facility because of the availability of water from the nearby Clinch River and steam from the K-25 powerhouse. S-50 was used as the first step in the enrichment process, boosting uranium from the natural 0.71 percent ^{235}U component to 0.89 percent, which was then further enriched in the K-25 gaseous diffusion process before the final step in the Y-12 calutrons to weapons-grade material. Operation of S-50 ceased at the end of the Manhattan Project.

Names of the Oak Ridge Plants

Oak Ridge observers often wonder about the origin of the terms K-25, Y-12, X-10, and S-50 as names of the four plants built on three sites during the Manhattan Project. Some have speculated that they represent coordinates on an Army map of this area as the building commenced in 1942. However, no such map was produced to logically explain such coordinates. Even General Leslie Groves, the Manhattan Project leader, admitted in his book that he did not know the origin of these names.[48] The most convincing argument for the naming pertains to K-25.[49] The Kellex Corporation was formed as a subsidiary of M. W. Kellogg Company early in the Manhattan Project to take the lead on the construction of a gaseous diffusion plant as part of the Clinton Engineer Works. In March 1943, correspondence indicated that the output of this plant was called K-25 as to keep a veil of secrecy over the production of this core of the atomic bomb that the Manhattan Project built. The "K" of K-25 evidently referred to the Kellex Corporation that manufactured the gaseous diffusion process equipment. The "25" was derived from taking the "2" from the right side of the atomic number for uranium, 92, and the "5" from the right side of the isotopic number, 235. The facility officially became known as the K-25 Plant and, in 1965, the Oak Ridge Gaseous Diffusion Plant.

The origins of Y-12 and X-10 are less clear. Prince and Stanley wrote that the letter Y was used as code for uranium in the early 1940s, for secrecy purposes. However, Oak Ridge Historian Ray Smith (an author of this book) has not heard this code for uranium and doubts if it is true, especially since Prince and Stanley present no reference for this.

The purpose of the X-10 site was to build the first continuously operating industrial-size nuclear reactor to learn how to make and purify fissionable ^{239}Pu, which does not occur in nature. Prince and Stanley suggest that the X in the name relates to this being an *experimental* reactor, but once again Smith has never heard this theory and doubts that it is true. No one knows the origin of the "10" in the name of the X-10 plant, the "12" in Y-12, or the name S-50 for the plant to enrich uranium based on a thermal-diffusion process.

There were multiple sites for the Manhattan Project across the country, including Oak Ridge (designated as X), Los Alamos (Y), and Hanford (W). The need for secrecy produced this patchwork of names for laboratories and sites during this extremely important time. Historian Smith often tells tourists the following: "Of the code names for Manhattan Project sites in the Clinton Engineer Works, only K-25 likely has any meaning attached. All the other code names, X-10, Y-12, and S-50, have no known meaning."

X-10 (Clinton Laboratories) was the site for constructing the world's first continuously operating nuclear reactor, devised to breed small amounts of ^{239}Pu, a fissionable isotope more potent than ^{235}U. X-10 served as a pilot-scale plant for the reactors later built in Hanford, Washington, which supplied the material for the Fat Man atomic bomb dropped over Nagasaki on August 9, 1945, precipitating Japan's surrender and the end of the war.[50] Earlier, in 1942,

Enrico Fermi and his colleagues built a successful pile (the term "reactor" did not come into use until the early 1950s) of uranium and graphite in a squash court under the football stands at the University of Chicago, with close monitoring as the pile went critical on December 2 of that year.[51] This pile was built with enough uranium and graphite (the latter, a crystalline form of the element carbon, served as a moderator of neutrons emitted by fissioning ^{235}U) and with a compact enough geometry that Fermi observed rapid doubling of the emitted neutron flux and a chain reaction.

That pile (Chicago Pile 1; CP-1) had no provisions for shielding or cooling and would have melted down and killed the fifty people present if left running for an hour or more. But Fermi understood this, knew then that he could trigger a chain reaction, and put in cadmium control rods to absorb neutrons and stop the chain reaction after four minutes. This was a low-power experiment (a half of a watt) and was scaled up ten days later to two hundred watts. That ended the CP-1 experiment. Fermi now knew how to build a nuclear reactor.[52]

The next step was to scale up this pile, add radiation shielding and cooling, and build a reactor that could safely run with a chain reaction in a steady-state way, capturing neutrons on ^{238}U to breed the desired ^{239}Pu. Ground was broken for construction of the concrete-encased X-10 Pile in Oak Ridge in February 1943. The device essentially was a cube of graphite weighing seven hundred tons with slots to receive uranium fuel slugs. This first continuously operating 1-megawatt (MW) reactor of uranium and graphite went critical on November 4, 1943. In 1944, once the techniques for extracting plutonium from the fuel were established, the larger (250-MW) versions of the X-10 Pile started to turn out significant quantities of ^{239}Pu at Hanford, Washington.[53] From there, the generated plutonium was sent to Los Alamos in New Mexico for incorporation into an atomic bomb.

Sam Beall

Over the decades, many UT students and alumni played prominent roles in programs at the federal facilities in Oak Ridge. Among the first was Sam Beall, from Plains, Georgia, a UT graduate in engineering. Following graduation in 1942, Beall was hired by DuPont and sent to work on the urgent project to build the first uranium pile at the Metallurgical (Met) Laboratory at the University of Chicago. In September 1943, Beall and his mentor, Arthur Rupp, were sent by DuPont to Oak Ridge to help construct the X-10 Pile. In 1944, once the techniques for extracting plutonium from the fuel rods were established, Beall and others were sent to Hanford, Washington.[54][55]

SAM BEALL (FAR LEFT) AT OAK RIDGE WITH PAUL HAUBENREICH, SEN. JOHN KENNEDY, ORNL DIRECTOR ALVIN WEINBERG, AND SEN. ALBERT GORE SR. (RIGHT FRONT), 1959. COURTESY OF SAM BEALL.

At the end of the war, Beall returned to Oak Ridge and spent much of his career building nuclear reactors of different types for different research purposes, and he directed the ORNL Reactor Division from 1963 to 1971. Later, Beall was the first director of the Energy Division at ORNL, as the original X-10 facility is known today, before leaving the laboratory to enter the private sector in the early 1970s. Beall's son, Sandy, had just opened a restaurant called Ruby Tuesday, and the father helped his son develop the brand into a national chain. In 2022, Sam Beall died at age 102 and was one of the few living survivors of the Manhattan Project.

EDUCATION WAS A PRIORITY IN THE FOUNDING OF OAK RIDGE

Just as Oak Ridge formed in wartime through the innovative development of facilities that intertwined industry, universities, and research science under generous government funding, the leaders of the Manhattan Project also recognized that its success depended on developing a forefront educational system in the community. General Leslie Groves, the head of the Manhattan Project, quickly understood this and moved aggressively to establish a top-notch school system in Oak Ridge.

General Groves won praise for his leadership in the construction of the Pentagon in Washington, D.C., and then was appointed in September 1942 to lead the Manhattan Project. Groves was a no-nonsense man who demanded action, hard work, and attention to detail. He faced a significant challenge in attracting high-level scientists and engineers to Oak Ridge for the war effort—and in making them happy and, thus, retaining them in Oak Ridge for the duration of the Manhattan Project. Since many of those brought to Oak Ridge had families and school-aged children, Groves realized that he needed to develop a school system equivalent to what the in-coming professionals were leaving when they moved to East Tennessee from university environments in Chicago, New York, and elsewhere. Their satisfaction with the education system was essential to a successful Manhattan Project, just as when later many wished to pursue advanced degrees but needed access to graduate programs to continue their stay in Oak Ridge.[56] UT provided these graduate courses starting in late 1945, but the Oak Ridge school system was born two years earlier.

In July 1943, an Army captain interviewed Alden H. Blankenship, 33, of Columbia University Teachers College, about establishing and directing a non-existent school system in the hills of East Tennessee. There were two questions posed to Blankenship: Could he do the job, and, by the way, could he decide by 4 p.m. that day to accept or reject the offer? Blankenship accepted and arrived in Oak Ridge on July 12, 1943, and was given complete responsibility for the operation of the school system.[57] Over the next three years, Blankenship hired experienced teachers and staff whose backgrounds represented 160 colleges and universities. The system operated as a unit independent of the Anderson County and Roane County school systems, and all funds and direction of operation were provided by the Army. Oak Ridge residents began moving into their homes in July 1943, and the school year began on October 4, with an enrollment of more than six hundred students. By the end of the 1943–44 school year, five thousand children were enrolled in Oak Ridge schools. School construction continued over the next two years, and by 1945, school enrollment was close to eight thousand students, when the town population peaked at seventy-five thousand.

Teacher wages for Oak Ridge were far above the local scale—almost double that of Anderson County—and were on par with salaries paid in New York City.[58] This was not surprising since all of the educators were fully qualified to teach their subjects and had at least four years of college education (the high school teachers held graduate degrees). In October 1946, the Educational Policies Commission chose the Oak Ridge School System as one of the forty most forward-looking school systems in the nation.[59]

While Oak Ridge was able to entice the qualified teachers it wanted, local school systems were not so fortunate. Knoxville, which had a population 50 percent greater than Oak Ridge in 1945, increased its beginning teacher pay by 33 percent to $1,200 per year to avoid losing more teachers to Oak Ridge. Clinton and Anderson County faced two additional problems beyond an inadequate teacher pay-scale. Tax revenue declined, as CEW took about one-seventh of Anderson County's land, and the now government-owned property generated no tax.[60] In addition, the number of students in these systems increased because there was not enough housing in Oak Ridge, and some newcomers rented units in Anderson County, providing little additional tax funds for the schools. General Groves approached Senator McKellar and suggested federal funds be provided to assist the nearby counties that were severely impacted by the arrival of CEW, but no funds were made available to resolve these problems.

UT DURING THE MANHATTAN PROJECT

During the war, the defense industries needed extensive training programs, and UT responded by enrolling two thousand students in more than fifty courses by January 1942. All told, between 1942 and 1944, UT's Engineering, Science, and Management War Training Program prepared more than eight thousand people for entry into defense-industry fields ranging from ordnance inspection to accounting, from engineering safety to pre-flight training and airport management.[61]

UT provided its first educational program in Oak Ridge already in September of 1944, offering a three-month course on safety engineering to adult Oak Ridge residents, free of charge. These safety courses were taught in the high school auditorium and led to a certificate for those that completed the training.[62]

Several UT faculty were involved directly in the Manhattan Project. Julian Fleming and Clayton Plummer, professors of hydraulic engineering, worked for Tennessee Eastman at Y-12. Elwood Shipley, another member of the engineering faculty, worked with the calutrons at Y-12. In the 1950s, he became an assistant research director for ORNL, overseeing the laboratory's operations at Y-12 and exploring nuclear fusion.

Another UT engineering instructor, Charles L. Segaser, was active at CEW and designed reactors at ORNL after the war. Wiley Thomas, who directed

UT's engineering co-op program, helped train three hundred calutron operators. Physics instructor E. T. Jurney served at Los Alamos during the war, and Kay Way of the physics department joined the Met Laboratory at the University of Chicago. Fellow physics professor William Pollard, who received his bachelor of arts degree from UT in 1934, joined the SAM Project at Columbia University in early 1944 to focus on enrichment of uranium through gaseous diffusion. Way and Pollard played critical roles in post-Manhattan Project thinking about the future of the Oak Ridge facilities and the role of universities.

THE END OF WORLD WAR II

Germany surrendered on May 7, 1945, eleven months after the successful (but bloody) Allied Forces' invasion of Normandy in June 1944. On the Pacific front, an even bloodier invasion of Japan was planned for October 1945, but, fortunately, would not be necessary.

The Little Boy atomic bomb was dropped on Hiroshima, Japan on August 6, 1945, fueled by highly enriched ^{235}U fabricated at the Y-12 facility in Oak Ridge. A second atomic bomb, Fat Man, fueled by ^{239}Pu and produced at the Manhattan Project reactors in Hanford, Washington, was dropped on Nagasaki on August 9, 1945. The damage to both cities was horrendous and prompted the Japanese to surrender on August 14, even though many of the nation's military leaders wanted to fight on. These bombs ended World War II, and only then did the workers in Oak Ridge understand what they had feverishly worked on for the previous three years. President Truman said in his announcement of the bomb on August 6, 1945: "The battles of the laboratories held fateful risks for us as well as the battles of air, land, and sea, and we have now won the battle of the laboratories as we have won the other battles."[63]

Premonition of the Atomic Bomb

While the role of Oak Ridge in building the atomic bomb surprised most of its inhabitants, the subject of nuclear fission and atomic energy was a pervasive theme featured in comic books in the late 1930s and early 1940s, before the Manhattan project had even begun. Many of the books' fanciful heroes employed atomic power to provide propulsion, deadly weapons, and protection. William Bugg, one of this book's authors, recalls, as a high school freshman, spending the summer on his grandparent's farm and hearing on the radio the Truman announcement of a super bomb. He told his grandmother, "I'll bet that is ^{235}U." Bugg distinctly remembers the science center section of a comic book describing a mysterious material of that name with the explosive power of the Empire State Building filled with TNT. He also confesses that, at the time, he had no concept of what ^{235}U actually was. He would learn soon enough as his family moved to Oak Ridge after the end of the war and he attended Oak Ridge High School.

SECURITY AN OVERWHELMING CONCERN

Secrecy was a paramount concern of the Manhattan Project. This was emphasized by the decision of President Roosevelt to put the Army in charge of the Manhattan Project, as long as "adequate provision had been made for absolute secrecy."[64]

General Groves created a special counterintelligence group in the Manhattan Engineer District, reporting to him, to use the resources of Army counterintelligence and the FBI without disclosing the nature of the work involved. After Groves began his assignment, he learned that the only known espionage was by Russian agents focusing on the Berkeley research group using American Communist sympathizers. According to Groves, compartmentalization was key to the project security—everyone should know what one needs to know in order to do his job and nothing else. This philosophy not only limited the flow of information but also focused attention on the project's goal: building the atomic bomb. Curiosity and growth of scientific knowledge were not relevant.[65] Unfortunately, these efforts to achieve absolute secrecy were not completely successful.

Oak Ridge was a largely peaceful community during the days of the Manhattan Project, in part because of a large and conspicuous security presence. The community became a military area under control of the Army in March 1943. Security concerns for Oak Ridge—a highly classified installation that eventually covered fifty-nine thousand acres—included prevention of physical trespassing, smuggling, sabotage, and the more common physical and property crimes, combined with extensive visual surveillance. By early 1945, a total of 6,315 security personnel safeguarded the large site.[66] Army counterintelligence officers regularly gave sacks of sugar—normally in short supply due to war-time rationing—to residents and asked, in return, that they keep the security force informed of anything suspicious.[67] Some of the scientists at the three facilities were enlisted and became part of a Special Engineer

Detachment of 1,200 people. Among the Manhattan Project sites, only Los Alamos (with 1,800 people) had a larger number of these scientifically literate and highly educated enlisted men.

Informers in all aspects of Oak Ridge life monitored behavior and reported suspicious activities. People were approached by a representative of the security force, given a stack of stamped envelopes (pre-addressed to the Acme Credit Company at a Knoxville address), and asked to send a note, if he or she heard or saw something that might interest the Army.[68]

This came to light many years later at a meeting of the '43 Club, a group of residents who were among the first to live in Oak Ridge in 1943. A question arose at the meeting, when one person said he had just found some envelopes addressed to Acme Credit at his home and asked if anyone else had participated in this activity almost fifty years earlier. A good number of folks raised their hands, some surprising their spouses, who previously had no idea that their husbands or wives had participated in a clandestine security force.

When the Army was given responsibility for the Manhattan Project, agreement was reached with the FBI that the Army would manage the project's personnel clearance process. The process involved intensive investigations of prospective employees' backgrounds and associations, often causing interesting speculation and gossip back in their hometowns. Everyone was required to wear a badge that not only identified the person but also indicated where within the area the person was permitted to go. This requirement applied to town residents as well. Indeed, all persons 12 years and older were required to have badges, and they had to have these badges in their possession any time they left home. Roving patrols enforced compliance and maintained security.

If one did not have a job within the fenced-in laboratories or city, he or she could not live in the community. If someone lost employment, the person and family were summarily evicted. Folks from nearby towns were not permitted to enter the fenced-in city on their own. Instead, visitors could gain access only upon the invitation of a resident. This was one provision that many Oak Ridge residents welcomed, because, as a result, they were not bothered by salespersons or nosy (uninvited) relatives. If a bus transited the area, an onboard guard confirmed that only those with badges were permitted to depart the bus, while all others remained in their seats until the bus left the area. These badge restrictions remained in force until the gates of the area opened to all on March 19, 1949.

Repercussions for site workers who had forgotten their badges were swift, embarrassing, and, often time, consuming. Residents were required to present their badges upon entering the community, and the same requirement applied if they hoped to leave. On at least one occasion, William Bugg managed to evade these strictly enforced rules. After his junior prom at Oak Ridge High School, Bugg and some friends decided to head into nearby Knoxville for an after-dance snack. As they approached the gate, Bugg realized that he had forgotten his badge. Determined to beat the system, the group successfully sequestered Bugg in the back seat of the car under the skirts of the young ladies. They employed the same strategy in re-entering Oak Ridge later in the evening. Teenagers can be resourceful.

When the Atomic Energy Commission (AEC) was created in 1946, all employees and all of the contractor staff had to receive security clearances approved by the FBI, a massive undertaking amid an environment of fear mongering and blind accusations of Communist association. There was a perception that Communists were infiltrating the atomic weapons establishment and sending secrets overseas to the Soviets; unfortunately, there were some cases of this happening.

In the U.S. Congress, the House Un-American Activities Committee (HUAC) was created in 1938 to investigate alleged disloyalty and subversive activities on the part of private citizens, public employees, and those organizations suspected of having Communist ties. It sent investigators to Oak Ridge in June 1946 and held hearings in July 1946, after claiming an investigation had discovered problems with security and instances of insubordination by scientists.

All claims were stridently refuted by security officers and scientists, who were incensed that their patriotism had been questioned. Nevertheless, the June 1947 issue of *Liberty*, a popular magazine of the time, featured the article "Reds in Our Atom Bomb Plants: The Full, Documented Story," displaying a hammer and sickle imposed over an aerial photo of the maternity ward of the Oak Ridge hospital.[69] Eight Oak Ridge residents were identified as security risks in 1947 by AEC or the Joint Committee on Atomic Energy, but all were eventually cleared.

THERE WERE SPIES IN OAK RIDGE

A few individuals were tied to spying at the facilities in Oak Ridge, and several were convicted of espionage-related activities. One person, George Koval, was identified as a spy only after his death. Koval was born in Iowa to parents who had emigrated from the Soviet Union. Following his high school graduation and during the Great Depression, he and his parents traveled to the Soviet Union. He graduated from college, received Soviet citizenship, and was recruited into Soviet military intelligence due to his facility with English, familiarization with the American lifestyle, and knowledge of chemical engineering.[70 71]

In 1940, Koval returned to the United States on a false passport and was directed to gather information on chemical warfare toxins, while working in New York City for Raven Electric Company, which supplied parts to large companies such as General Electric and had Soviet connections. Meanwhile, his wife and mother were given an apartment in the Soviet Union during his absence. While Koval lived in New York, he enrolled in Columbia University's adult education courses in chemistry in 1941. Conspicuously, one of his professors had a reserve appointment in the Army's Chemical Warfare Service.[72] Koval registered for the draft in January 1942 but received two draft deferments because he worked for a company, albeit a Soviet-managed one, doing essential war work in New York.

The occupational deferments ended, and Koval was inducted into the U.S. Army on February 4, 1943.[73] He was assigned to the Army Specialized Training Program and studied electrical engineering at the City College of New York (CCNY). Koval then joined the Special Engineer Detachment and was assigned to CEW as a health physics officer; he monitored radiation levels at all of the sites with a security clearance that allowed access to all facilities.[74] His Soviet handlers had him focus on the use of polonium as a chain reaction initiator in atomic weapons, and Koval soon accepted a transfer to Dayton, Ohio, where polonium was fabricated.[75]

Throughout this time, he sent documents and technical details to the Soviets. After the war, he returned to CCNY and graduated with an electrical engineering degree in 1948. In October 1948, Koval told friends he was taking a European vacation, left by sea for Russia, and never returned to the United States. He died in his Moscow apartment in 2006 at age 92. In 2007, he received the posthumous title of Hero of the Russian Federation, bestowed by Russian President Putin with the following proclamation: "Mr. Koval, who operated under the pseudonym Delmar, provided information that helped speed up considerably the time it took for the Soviet Union to develop an atomic bomb of its own."[76]

Alfred Dean "Al" Slack was a Y-12 Manhattan Project shift supervisor who, in 1943, came to Tennessee Eastman in Oak Ridge from Eastman Kodak at the Holston Ordnance Works in Kingsport, Tennessee.[77] Before coming to Tennessee, he worked at the Eastman Kodak Company in Rochester, New York, where he sold proprietary commercial information to Richard Briggs, a former associate at the Eastman company, thinking the information went to a Kodak competitor. Slack soon learned it went to the Soviet Union. Later, after Briggs' sudden death, Slack was handed off by Soviet agents to Harry Gold, who then became the courier to transfer the information to the Soviet Union.

While Slack worked at the Holston Ordinance Works in Kingsport, Gold pressured him to deliver additional information or he would expose the fact that Slack had previously provided information that had been shared with the Soviets. In response to this pressure, Slack provided Gold with information on the explosive RDX, manufactured as "Compound B," that was used in World War II and was more powerful than TNT. This information helped spread knowledge about the most powerful explosive that existed at the time. Later in 1943, Slack came to Oak Ridge to work at Y-12, but there is no evidence that he passed secret Manhattan Project information to the Soviets. He, however, received a fifteen-year sentence for providing information on RDX to Gold.

Yet another person was linked to spying activities in Oak Ridge but never convicted. Russell McNutt was born in Kansas to a father who had founded the Communist Labor Party in that state. McNutt, following in his father's footsteps, later became a secret member of the Communist Party.[78] Initially, McNutt attended Kansas State University, where he pursued a degree in civil engineering, but he later transferred to Brooklyn Polytechnic Institute. Studying nights and working for the Works Progress Administration during the day, McNutt graduated in 1940. After several years of engineering jobs, he transferred to the Manhattan office of the Kellex Corporation, the organization responsible for the design of the K-25 gaseous diffusion plant in Oak Ridge.

In February 1944, Julius Rosenberg connected McNutt to the Soviet state security force, the KGB, which, at that time, had no agents involved in the Manhattan Project. Mainly based in Manhattan, New York, McNutt traveled to Oak Ridge during K-25 construction. He did not have access to experimental data, although he was able to pass design schemes along to his handlers. The KGB wanted McNutt to move to Oak Ridge, but he refused, for two reasons. First, the site was too remote and primitive for his family. Second, he wanted to remain close to his investment property in a nearby upstate New York resort.

He was transferred to another agent in March 1945, and then contact with the KGB ceased, due to the collapse of his new handler's network of spies. After the war, McNutt met with David Greenglass and Julius Rosenberg to discuss McNutt's moving to Caracas to take on a relative's business. After the arrest of Julius Rosenberg, his wife Ethel, and Greenglass on espionage charges, McNutt was interviewed, and, although he was thought to be a Communist sympathizer and party member, no connection could be made to espionage and the Rosenberg spy network. McNutt lived the rest of his life without further investigation. The Rosenbergs were the first American civilians executed for espionage. They died in the electric chair in 1953.

A SPY ONCE ASSOCIATED WITH UT AND TVA

As discussed earlier in the Rhenium Project section of this chapter, UT faculty member Clarence Hiskey created the long-standing program focused on the element rhenium but left UT for TVA in 1941. After a short time, Hiskey left TVA and joined the SAM Laboratory at Columbia University where he was assigned to Harold Urey, who led the development of the gaseous diffusion process. In September of 1941, Hiskey met with Arthur Adams, probably introduced by Koval. Adams was a Soviet military intelligence officer (code name Achilles) who recruited Hiskey into Soviet espionage.[79] Hiskey and this research team subsequently moved to the Met Laboratory at the University of Chicago in September 1943. Several years earlier, while a graduate student at the University of Wisconsin, Hiskey had joined the ROTC program and received a commission in the Army reserves.[80]

In April 1944, the FBI observed Hiskey meeting with Adams, and FBI agents raided Adams' hotel room in New York City, finding "sophisticated camera equipment, materials for constructing microdots, and notes on experiments being conducted at atomic laboratories at Oak Ridge, Tennessee."[81] No proof existed of Hiskey passing information to Adams, but the fact that they had met was irrefutable and raised numerous questions about Hiskey's loyalty. The FBI informed the Army, and this led to Hiskey's mobilization from reserve status into the Army on April 27, 1944, and (punitively) sent to Canada near the Arctic Circle and later to the South Pacific for the duration of the war.

On his journey north, Hiskey's luggage was secretly searched, and classified notes from the Met Lab were found. Upon his arrival in Canada, however, a subsequent search of the luggage failed to locate the documents. Hiskey was the subject of several Congressional hearings beginning in 1948 while he was employed as associate professor of analytical chemistry at Brooklyn Polytechnic Institute.

In May 1949, at a hearing held by the HUAC, Paul Crouch, who had been a Communist party organizer in Tennessee, told that he and Hiskey had failed to persuade ten former members of the TVA branch of the party to resume their activities. Crouch also told HUAC that around six UT professors and twenty to twenty-five TVA employees were Communists. Additional testimony was provided suggesting that Hiskey had asked a former UT Rhenium Project worker, Edward Manning, with whom he had discussed Communism, to join him as a laboratory technician on the SAM Project and then at the Met Lab when it moved to Chicago.

Manning recalled meeting Arthur Adams several times at Hiskey's home, including the day Hiskey was taken into the Army, and then several times in New York City. After Manning's suspension from the Met Lab in August 1944, he felt it was based on association with Hiskey and possibly Adams, so he wanted to ask Adams about this. He did so on his way overseas in July 1945 when he stopped in New York City to meet Adams and Hiskey's wife. After that meeting, Manning felt he was under surveillance and that Adams knew too much not to have received information about the Manhattan Project from someone.[82]

AEC's Personnel Security Board investigated Robert Oppenheimer, director of the Manhattan Project at Los Alamos, in 1954. Its report stated that Oppenheimer had been found in the company of "Joseph W. Weinberg and Clarence Hiskey, who were alleged to be members of the Communist Party and to have engaged in espionage on behalf of the Soviet Union." Oppenheimer's security clearance was revoked the following month, June 1954.[83] When the KGB archives were opened after the Soviet Union's collapse, documents revealed that Hiskey was a Soviet agent with the cover name of RAMSAY (the name occurs in decrypted Soviet message transmissions). No information has emerged on UT and TVA employees who were purportedly said to be members of the Communist Party. Oppenheimer's security clearance was never restored, but, in 1963, President Lyndon Johnson presented him with the AEC's Enrico Fermi Award. Three years later, the physicist retired from Princeton University and died of throat cancer in 1967.

SUMMARY AND LOOK FORWARD

The period from 1940 to 1945 was an extraordinary one, certainly for the United States and especially for East Tennessee. A secret set of federal facilities was built on farmland, after resident families were required to sell their property (for less than they wanted) and leave within a few weeks. In part as a result of the buildup of scientific and technological resources in the region, rural farming counties in East Tennessee changed rapidly in population, culture, and technical assets. And Knoxville and UT were also swept up by these significant changes.

In many ways, the birth of Oak Ridge provided the impetus for the university to evolve from primarily a teaching college to a competitive research institution. However, none of this was clear in 1945. The dropping of two atomic bombs on Japan ended World War II, but no one could foresee what would happen to the Oak Ridge facilities following Japan's surrender. Indeed, no one knew if the rapidly evolving relationship between UT and Oak Ridge would continue or sputter and die, after the Manhattan Project facilities had achieved their primary mission.

2

THE LATE 1940S

A Symbiotic Relationship Forms between UT and Oak Ridge

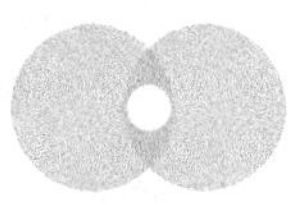

THE REALITY THAT the conflict had cost the nation 420,000 lives tempered the United States' celebration of the war's end. World War II's close also brought great uncertainty to East Tennessee in terms of the fate of the federal facilities in Oak Ridge and the thousands of men and women employed by them. The facilities' mission was accomplished, and many in government assumed that the facilities would soon close. Some of the scientists and engineers who were brought suddenly to Oak Ridge for the Manhattan Project were just as suddenly packing their bags to return to their previous lives, many returning to graduate schools around the country.

However, some scientists and engineers wanted to stay in Oak Ridge and continue conducting the research and development (R&D) that had engaged them through the war years. They actually came to like "Dogpatch," as the area was derisively called when the Manhattan Project brought many of them to the hills of East Tennessee from larger metropolitan areas.[1] Some who wanted to remain needed graduate course work to complete master's or doctoral degrees begun prior to the war.

At UT, William Pollard returned to the physics department after his stint at Columbia University. Kay Way chose to join the staff of Clinton Laboratories in Oak Ridge instead of returning to the UT physics department.[2] In September 1945, Kenneth Hertel sat in his office in UT's Ayres Hall, wrestling with a challenge. As head of the UT physics department, Hertel was anticipating heavy demand for instruction in his department and throughout the university after the end of the war. The benefits afforded returning veterans by the GI Bill, signed in 1944, included payments for tuition and living expenses to attend college. Fellow physics professionals, including some in Hertel's department, had relocated to other regions of the country because of the wartime need for their expertise, and it was not known if they would return to Knoxville.

Hertel was a highly respected UT faculty member whose service to the university extended over several decades. He joined the physics faculty in 1926

KENNETH HERTEL. COURTESY OF BETSEY B. CREEKMORE SPECIAL COLLECTIONS AND UNIVERSITY ARCHIVES, UNIVERSITY OF TENNESSEE, KNOXVILLE.

and headed the department from 1930 to 1956. With the arrival of the Great Depression of the 1930s, Tennessee state government had to be convinced of the practical value of the university—particularly its research enterprise—to the state's citizens during the dire economic downturn. One morning, Hertel read in the local paper that the state's appropriation to the university was cut in half, reflecting the then-prevailing popular opinion about institutes of higher learning.

Hertel and other faculty responded by emphasizing the importance and practical value of university research, and, at the instigation of a local cotton breeder, he began a study into the physical properties of the plant's fiber, a mainstay of the textile industry.[3] Hertel's research led to the invention of several instruments for assessing cotton's strength and other qualities, including the Fibrograph, a photoelectric scanning device, and the Arealometer, which assesses the linear density of cotton fibers. Those tools, with improvements and modifications, remain in use today. In 1934, Hertel became a founder of the UT Research Corporation, an independent nonprofit organization whose primary goal is to promote UT research and patenting of the results.[4] [5] Hertel, who served as president of the corporation from 1954 to 1969, was awarded the newly formed UT Research Corporation's first U.S. patent in 1939 for a device that measured the relative humidity of a gas. Over the course of his career, Hertel was awarded fourteen additional patents.

In the early 1940s, Hertel was aware that something momentous was happening in the high-security installation in Oak Ridge, but he did not know what. One evening, while walking in downtown Knoxville, he happened upon a clue. There, sitting on a bench in front of the Andrew Johnson Hotel, were three men Hertel immediately recognized, though their presence in Knoxville seemed glaringly out of place. All three were Nobel laureates: Niels Bohr (University of Copenhagen, Nobel Prize in physics in 1922), Arthur Compton (Washington University, Nobel Prize in physics in 1927), and Harold Urey (Columbia University, Nobel Prize in chemistry in 1934). Hertel sensed that the presence of these three eminent scientists must have something to do with the happenings at the secret facilities near Clinton. And, of course, he was right.[6]

Hertel learned the full story of the momentous endeavor following the announcement of the atomic bomb detonations over Japan in August 1945. The Smyth Report, issued by the Army Corps of Engineers soon after Japan's surrender, provided an unclassified technical summary of the Manhattan Project and the development of the atomic weapons.[7] Hertel hoped his department, its teachers, and future students could take advantage from the amazing resources in Oak Ridge, a mere twenty-seven miles from the university campus.

In September 1945, UT Graduate School Dean Fred C. Smith arranged a dinner meeting with Martin Whitaker, director of Clinton Laboratories, and Warren Johnson, director of Clinton's Chemistry Division. Whitaker requested the meeting because he faced challenges with his staff. Whitaker had about six hundred technical staff, many of whom wanted to pursue additional education, but there were no advanced-degree programs in sciences or engineering anywhere nearby.[8]

At this point, the Clinton Pile (later called the Graphite Reactor) at X-10 had fulfilled its role in demonstrating production of plutonium and training Hanford reactor operators. Even during the war, the pile began to be used to produce radioisotopes on demand for a range of scientific purposes, while also enabling staff to learn more about the health effects of radiation. After the war, the study of the effects of radiation on various materials—including human tissues infiltrated by cancer cells—became a priority, and research in neutron scattering was underway. Clearly the discoveries achieved during the Manhattan Project presented a range of exciting opportunities ahead in physics, chemistry, biology, and health studies. Whitaker knew that significant research challenges loomed and that resolving them would require an educated workforce. His goal was to enable his workers to get the education they wanted locally, instead of having to relocate or commute long distances.

This dinner meeting importantly led to the first official critical connection between UT and Oak Ridge, relating to graduate education and an eventual university consortium. From the dinner meeting, a two-pronged response emerged.[9] One dealt immediately with Whitaker's request for graduate-level instruction for Clinton Laboratories personnel, and the second, by far the more ambitious of the two, involved development of a comprehensive plan for the use of Oak Ridge facilities by a consortium of southeastern universities, including UT.

Whitaker and the Manhattan Engineer District asked UT to conduct graduate classes on-site at Clinton Laboratories. The university responded immediately and asked Hertel to supervise development of the course offerings, and, by October 1945, the UT physics department had facilitated the offering of two courses: Theoretical Physics 511 (taught by visiting professor F.C. Hoyt) and Atomic Physics 411 (instructed by lecturer Harry Soodak, a Clinton Laboratories employee who later worked with Eugene Wigner on the design of a sodium-cooled breeder reactor). The two courses enrolled a total of forty-eight students.

In winter and spring quarters, additional classes, taught by UT physics professors William Pollard, Richard Present, and John Trimmer, followed the initial course offerings. Soon, UT offered courses in chemistry, chemical engineering, and mathematics for employees of the Oak Ridge contractors and the Army. A total of 166 graduate students received instruction during spring quarter 1946, and seventy more enrolled over the summer.[10 11] There is no record of organizational approval by CEW or the U.S. Army of these courses, other than provision of classroom space and permission for employees to enroll. Thus began UT's Oak Ridge Resident Graduate Program, and, with it, the first official phase of the UT-Oak Ridge partnership, which, over coming decades, would continue to evolve and prosper as the partnering institutions matured.

At the time of the initial course offerings, UT faculty and administrators began to explore a broader future relationship with Oak Ridge. While some still believed that Oak Ridge was primarily a wartime production facility that likely would disappear now that the war was over, many recognized the tremendous potential of proximity to a large pool of research expertise, scientific facilities, and educational opportunities. In his 1945–46 report to the dean of the College of Liberal Arts, Hertel emphasized the importance of Oak Ridge in terms of UT faculty research opportunities.[12]

The report also stressed the value to UT graduate students of conducting research at Clinton Laboratories. Hertel recommended that the university support these efforts with the addition of PhD programs, not only in the sciences but also in allied fields of knowledge. He further encouraged maximum support for joint programs already underway involving other southern universities. In anticipation of the approval of Hertel's recommendations, the UT physics department prepared a course structure for a full PhD program. The chemistry department developed a similar program the previous year when it became the first UT department to offer a PhD.

BOARTS THROWS A PARTY FOR KAY WAY

In September 1945—a month after the public was made aware of Oak Ridge's crucial role in building the bombs and winning the war—Robert Boarts, UT professor of chemical engineering, hosted a party at his Knoxville home to celebrate the return of Kay Way. Way, a distinguished nuclear physicist, earned a bachelor's degree in physics from Columbia University in 1932 and a PhD in nuclear theory from the University of North Carolina (UNC) in 1938. At UNC, she worked with and studied under John Wheeler, one of the nation's more widely renowned nuclear theorists. After a year teaching at Bryn Mawr College, Way joined the UT physics faculty in 1939. She departed the university early in the war to accept a position with the University of Chicago's Metallurgical Laboratory, where she worked with future Nobel laureate Eugene Wigner (physics, 1963) on reactor design. After the war, Way joined the staff at Clinton Laboratories.

In 1939, while still a graduate student at UNC, Way presented a research paper at a conference in New York about her work on calculating the deformation of a spinning atomic nucleus. Way's calculations were based on a new concept developed by Niels Bohr, who contended that one could regard the nucleus as a liquid drop as one tried to understand how it rotates and vibrates. In preparing for her conference presentation, Way noticed that a theoretical instability resulted when one added excitation energy in the form of rotation to a highly deformed (football-shaped) nucleus.[13] [14] At the time, neither she nor Wheeler understood the implications of this theoretical result that the rotating nucleus becomes more deformed and could even become unstable. However, a few months later, it began to make sense when

KAY WAY. COURTESY OF BETSEY B. CREEKMORE SPECIAL COLLECTIONS AND UNIVERSITY ARCHIVES, UNIVERSITY OF TENNESSEE, KNOXVILLE.

they learned about the now-famous experiment led by Berliners Otto Hahn and Fritz Strassmann. Hahn and Strassmann's experiment showed that the uranium nucleus elongates and fissions when slightly excited by an incident neutron.[15] Way and Wheeler noted this instability in their calculations. This discovery marked the beginning of the race to see if a bomb based on nuclear fission could be built. The pity is that Way and Wheeler missed a great opportunity to predict nuclear fission before it was observed. It is possible that such a theoretical prediction could have earned them a Nobel Prize.

At Boarts' party, attended by various UT faculty members, including several from the physics department, Way mentioned that several midwestern universities were exploring the formation of a consortium with the University of Chicago's Metallurgical Laboratory, which a year later, in 1946, would be rechristened Argonne National Laboratory and become the first named national laboratory.[16] [17] Way wondered why a similar consortium of southern universities could not coalesce around the Oak Ridge science facilities. The multi-institutional and multidisciplinary response to that query changed the scientific landscape in Oak Ridge and the Southeast. Boarts and Pollard drafted proposals for such a collaboration, which were presented to UT President James Hoskins, who then appointed Hertel to chair a committee to explore further action. Although it was initially proposed that UT manage the research institute, the decision was made to involve other southern universities, following further discussions at UT. Also, Hertel received a note from Vannevar Bush, head of the U.S. Office of Scientific Research and Development through which wartime R&D—including the Manhattan Project—was carried out, suggesting a broadening of involvement.

Then a sequence of events quickly unfolded that changed southern science forever.[18] On November 8, 1945, Dean Fred Smith had UT host an exploratory dinner meeting at the Andrew Johnson Hotel, at which the Hertel Committee met six prominent members of Clinton Laboratories to discuss this approach.[19] This successful meeting led to an event on December 5, 1945, that was preceded by visits to Vanderbilt, Duke and North Carolina Universities, and the Universities of Virginia and Kentucky. UT sponsored the "Conference on Research Opportunities in the Southeastern United States," organized by Pollard, with goals to consider the potential of CEW for academic research training, identification of mutual benefits to be derived from collaboration between CEW and universities, and recommendations for future actions.

This early December meeting gathered an interested audience, including the District Engineer of the Army Corps of Engineers and representative research staff, scientists from all Oak Ridge facilities, and faculty from ten

WILLIAM POLLARD. COURTESY OF BETSEY B. CREEKMORE SPECIAL COLLECTIONS AND UNIVERSITY ARCHIVES, UNIVERSITY OF TENNESSEE, KNOXVILLE.

universities. In addition to presentations by heads of each of the Oak Ridge facilities, there were science talks that described in detail where collaborations might occur at Clinton Laboratories. An interim committee formed under chairman Pollard to draw up a plan for action. Pollard arranged a meeting during December 27–29, 1945, at Oak Ridge High School to formulate a specific action plan, with fifty scientists attending.[20] They were divided into teams, mainly focusing on the business and legal aspects of the proposed new organization, along with a few research areas. Results of the meeting were to elect an executive committee (with Pollard as chairman) and to name the organization Oak Ridge Institute of Nuclear Studies. The goal of this organization would be to facilitate access of university faculty to CEW resources.

It was already apparent that Oak Ridge needed educational opportunities to help recruit and retain technical staff at Clinton Laboratories. Hertel led a special committee on graduate education to institute courses under UT's auspices, because of its proximity and the difficulty of arranging credits from a multiple-institution arrangement. At its meeting on January 5, 1946, the committee proposed to conduct courses at the three operating facilities in Oak Ridge, during both working and non-working hours.

The Army Corps of Engineers approved this plan on January 31, 1946, subject to contract approval. Contractors at K-25 (Carbide and Carbon Chemicals Corporation, a division of Union Carbide Corporation) and Y-12 (Tennessee Eastman) decided that courses would no longer be offered on site. The alternative was to look for a location in the city of Oak Ridge. The outcome was that graduate courses in chemistry, mathematics, physics, and engineering would be held at Oak Ridge High School, beginning in the spring quarter of 1946. These courses were designated collectively as the UT Resident Graduate Program.[21]

An unknown for both UT and the Oak Ridge parties was the shift of oversight and priorities from the Army Corps of Engineers to a new organization now that the war was over. None of the Oak Ridge facilities had more than a caretaker budget, few funds were available for building repair, and certainly none were available for new construction. At the time, Congress mulled priorities for research, which was difficult because of the classified nature of many of the projects. Still, there was a growing interest in fundamental research that might be performed at universities, outside of these Manhattan Project facilities. This was definitely a change from the blank-check and mission-driven-at-all-costs focus that had prevailed during the war. Any agreement between UT and the Oak Ridge organizations had to happen before new bureaucratic procedures could be put into effect. Indeed, the established procedures and managers one knows are definitely preferable over those that one does not know or that have yet to be created.

Clinton Laboratories in 1945 and 1946

Japan officially surrendered on September 2, 1945, ending World War II and starting intense speculation about the future of K-25, Y-12, and X-10 (Clinton Laboratories). The K-25 facility rapidly evolved to enjoy a long future, as the gaseous-diffusion process became the preferred method for enriching uranium for the coming fleet of nuclear power reactors and naval propulsion reactors. Y-12 thus ended its role as a uranium-separation facility and became a key part of the country's nuclear weapons program. However, the future existence of Clinton Laboratories was in question for more than two years, and its role in the country's research portfolio would not be fully defined for the rest of the decade.

Successful construction and operation of the Clinton Pile in the early 1940s had demonstrated how one could build a 1-MW reactor in sustained operation for the purpose of producing ^{239}Pu in the natural uranium fuel rods and then separating microscopic amounts of plutonium from the irradiated uranium material. Lessons learned in Oak Ridge enabled the rapid construction of larger graphite-moderated reactors at Hanford, Washington and the production and separation of kilogram quantities of ^{239}Pu for use in one of the atomic bombs dropped on Japan in August of 1945.[22]

In a two-year period of uncertainty in Oak Ridge after the end of WWII, scientists used the Graphite Reactor as a wonderful research tool for a variety of programs, as it was then the world's best source of neutrons. Among other applications, physicists and chemists learned how to produce and export radioisotopes for medical uses, which had already started at a low level during the war. Between 1943 and 1950, Clinton Laboratories (renamed Oak Ridge National Laboratory in March 1948) made nearly twenty thousand shipments of radioisotopes, as applications expanded to include treating cancer, detecting childhood diseases, battling insect pests, and aiding in the identification and diversion of oil and gasoline products flowing through common-carrier pipelines.[23] Researchers started working on the design of a "high-flux" reactor, a larger reactor that would produce higher power and a much greater flux of neutrons for research purposes.

In August 1946, Eugene Wigner and Alvin Weinberg decided to shift to a very different reactor design that depended on the development of efficient aluminum-clad water-cooled fuel elements (circulating air was the cooling medium in the Graphite Reactor).[24] In experiments using the reactor, physicists demonstrated that an appropriate arrangement of ordinary water (as a coolant and a neutron moderator) and natural uranium would become critical in a larger reactor if the natural uranium were enriched by only a few percent in ^{235}U. This meant that compact and powerful reactors could be built using forced-water cooling and slightly enriched uranium, and this became the workhorse of naval propulsion and electrical power generation for decades to come. In 1946, Clinton Laboratories leaders assumed that Clinton would be the United States' leading facility for development and implementation of new and novel nuclear reactor concepts for the foreseeable future.

The Discovery of the Element Promethium

Charles Coryell was a California Institute of Technology chemistry PhD who joined the Manhattan Project in 1942 and worked at the Metallurgical Laboratory at the University of Chicago until moving to Oak Ridge in late summer of 1943. He led a research group (the Fission Products Section) charged with characterizing radioactive isotopes formed in the fission of uranium. He oversaw the X-10 construction of a "hot-lab" building that was essential for chemical separations of highly radioactive fission products. A crucial project was the chemical separation of a large amount of radioactive barium-140 (^{140}Ba), with a 12.7-day half-life, which decays to shorter-lived lanthanum-140 (^{140}La), with a forty-hour half-life. Weapons designers at Los Alamos needed ^{140}La to test and perfect the difficult process of implosion to be used in the Fat Man atomic bomb, which employed fission of ^{239}Pu manufactured at Hanford.

The idea was that a symmetrical spherical shell of conventional explosives would detonate and rapidly compress the plutonium core so that a critical mass would be achieved and a chain reaction (atomic explosion) would occur.[25] Plutonium could not be used in these compression tests because one did not want to have a chain reaction occur during testing, but radioactive ^{140}La worked well, as it emits a gamma ray that could be detected around the sphere to see if uniform implosion occurred. It was a huge challenge in Coryell's group to extract, separate, and ship hundreds of curies[26] of radioactive barium for this crucial project.

At Los Alamos the urgency was great to solve the problem of uniform compression to enable the atomic bomb utilizing plutonium. This translated to high stress at Clinton Laboratories and especially Coryell's group to produce significant quantities of RaLa (radioactive lanthanum) needed at Los Alamos for the compression tests. When the stress of that essential project subsided a bit, two young chemists in the group, Larry Glendenin and Jacob Marinsky, worked on another fission-product experiment under the guidance of Coryell. At the time, element 61 was a "hole" in the periodic chart, as it was the last element yet to be discovered up through lead (element 82). Elements 43 (technetium) and 61 have no stable isotopes, so they were difficult to isolate and study.

The experimenters bombarded samples of uranium in the Graphite Reactor and extracted the sample for the analysis of the many uranium fission products. Chemical separations were used to isolate together elements 59 (praseodymium, Pr), 60 (neodymium, Nd), and hopefully 61. However, the challenge was to separate these three elements from each other and thus isolate element 61 for the first time, since these three elements share similar chemical properties. To resolve the challenge, Glendenin and Marinsky employed the new method of ion exchange chromatography, which was pioneered in Oak Ridge during this time. The order of elution (i.e., extraction) from the ion exchange column depended on the atomic number (the number of protons, which determines the element). They were able to separate and identify a three-year activity of element 61 and thus discovered the formerly missing element as it came off the ion exchange column after known elements praseodymium (59) and neodymium (60).

In the summer of 1946, Charles Coryell left Oak Ridge to assume a professorship at the Massachusetts Institute of Technology (MIT), and Glendenin and Marinsky followed him there to be his graduate students. They finally had time to write a paper[27] on their findings, and, in September 1947, they announced discovery of element 61 at a meeting of the American Chemical Society in New York City.[28] They debated which name to assign to the new element, and one strong possibility was clintonium (after Clinton Laboratories). But, instead, they went with a suggestion made by Coryell's wife, Grace Mary: promethium (Pm), referring to Prometheus, the Greek mythological Titan who stole fire from heaven for the use of humankind. It is ironic that a Tennessee-related name narrowly missed being used for this new element, but this would be rectified in 2017, when element 117 was named tennessine, due to the contributions of ORNL, UT, and Vanderbilt to its discovery in experiments conducted at the large accelerator facility in Dubna, Russia.

Charles Coryell was a giant in the field of science and died far too young at age fifty-eight in 1971. His daughter edited the results of 1960 interviews conducted with her father.[29]

FORMATION OF OAK RIDGE INSTITUTE OF NUCLEAR STUDIES

The crisis of World War II forced, perhaps for the first time, institutional partnerships that are common today. The war was an extreme military threat to the United States, and the scope of the threat required new partnerships to successfully develop a brand-new weapon, the atomic bomb. The Army needed the help of universities, academic researchers, and industry, and, with huge government funding, success was achieved, the atomic bomb was rapidly developed based on a physics phenomenon (nuclear fission) that was unknown five years earlier, and World War II ended.

At the conclusion of the war, it was not clear what the future would bring relative to such partnerships born during the Manhattan Project. The evolution of time and the development of new threats (the Soviet Union and the Cold War) did result in a gradual maturation of such partnerships between government, industry, universities, and national laboratories. In fact, the development of these liaisons was so strong that President Dwight Eisenhower used his 1961 Farewell Address to warn the public about a perhaps too extreme reliance on the military-industrial complex.

Even though universities and their faculty and graduate students were essential to the success of the Manhattan Project, many questions emerged at the end of the war. Would the Manhattan Project facilities continue? If so, what would their new mission be? Would universities play a role in laboratories that might continue? How would the emergent laboratories and other Manhattan Project functions be funded and managed since the management and oversight role performed by the Army would certainly be altered?

These questions of post-Manhattan Project facilities certainly affected UT. The university received a much-needed boost in its research awareness and partnership development during the first half of the 1940s. However, some at UT worried about how to continue that development and whether the Oak Ridge facilities would receive continuing funding as research laboratories that could benefit universities. Physics professor William Pollard had the clearest vision of what the future could hold and he took the lead on the formation of the Oak Ridge Institute of Nuclear Studies (ORINS), bringing other universities into the mix.

Forming an organization of universities proved difficult, as there were few examples of an existing collaboration like that envisioned by the founders of ORINS, and any arrangement would have to be approved by the Army. Time was of the essence, because legislation was then being proposed for a new organization to administer the Oak Ridge facilities. Uncertainty existed about a legal structure, funding mechanism, and actual role for ORINS.

Argonne Universities Association developed bylaws that served as an initial model for consideration, although there were differences. For example, Monsanto (a for-profit company and operator of Clinton Laboratories) was not the University of Chicago (managing contractor of Argonne National Laboratory), an educational institution accustomed to working with other universities. The idea evolved into ORINS becoming a nonprofit organization with each member university paying an annual membership fee. The principles behind the founding of ORINS were given a more formal status by an article about the organization, "Nuclear Research Institute at Oak Ridge," appearing in *Science* on June 14, 1946. The piece on ORINS appeared in the same issue and on the pages immediately following an article describing "the availability of radioisotopes" issued by the headquarters of the Manhattan Project, about radioisotopes from the Oak Ridge Graphite Pile.[30]

There was impetus for formal agreements with the Army on the roles and responsibilities for ORINS. This new institute was fully incorporated under Tennessee law on October 15, 1946, and had its first meeting with leading educators of the fourteen member institutions on October 17. The papers of incorporation were filed prior to the signing of contracts with the Manhattan District.

The six purposes of ORINS were:

To stimulate cooperation between the federal government and participating universities in undertaking fundamental research in the field of atomic energy;
To act as a medium through which universities and other agencies may have contact with the facilities and techniques available at CEW;
To utilize the unique facilities at CEW for graduate research and instruction to contribute to the national need for scientific and technical personnel with advanced training in nuclear and related studies;
To develop the potentialities of the region for fundamental research and discovery in the field of nuclear science;
To aid in the acquisition of scientific personnel for CEW facilities; and
To foster increased opportunities and improved programs of graduate education and studies in nuclear and other sciences in the educational institutions of the region.

Charter member institutions of ORINS were: University of Alabama, Alabama Polytechnic Institute (now Auburn University), Catholic University of America, Duke University, Emory University, Georgia Institute of Technology, University of Kentucky, Louisiana State University, University of North Carolina, University of Tennessee, University of Texas at Austin, Tulane University, Vanderbilt University, and University of Virginia.[31] This was quite an impressive group of universities, and most of these same schools would come together several other times in succeeding decades to conduct special programs in Oak Ridge.

Professor Pollard was heavily involved behind the scenes working out the kinks of establishing the organization and the contracts with the Army Corps of Engineers to arrange for funding and set up mechanisms for operations. He was identified informally as ORINS executive director, but his official designation as such was pending because there was no contract between ORINS and the Army that specified any role for ORINS. At an ORINS board meeting in February 1947, the board members re-examined the proposal and believed the original program outlined in the letter was "meager"; Pollard and Paul Gross of Duke were asked to redraft the proposal overnight for consideration the next day. The important addition was that ORINS provide training courses for research professionals, particularly in using radioisotopes for research.

With board agreement, Pollard transmitted the request to the Atomic Energy Commission's (AEC) Oak Ridge Operations Office for approval of the ORINS charter.[32] It was forwarded to AEC General Manager Carroll Wilson for approval after a brief background discussion. Then, there was a month of silence from Washington, and Pollard decided to investigate the situation personally. Because AEC was just being organized, Pollard learned that there were many documents awaiting approval. Wilson's administrative assistant found that the ORINS document was referred to a new person of the Office of General Counsel, Clark Vogel. Pollard visited Vogel as he sat behind a tall pile of documents, and together they found the ORINS proposal. Following a discussion of ORINS, its purpose, and the general picture of Oak Ridge, additions and changes were made and approved by Pollard and the General Counsel, with James Fisk, AEC research director, also consulted. Soon after, Pollard returned to Oak Ridge, a signed contract appeared in the mail, and a suite of two offices was assigned to him in the AEC Oak Ridge administrative building.[33] Being aggressive and persistent were clearly strong points of Pollard's character and success.

Pollard was finally named ORINS executive director on November 6, 1947. The press release announcing his appointment noted that ORINS participated in arrangements for the new operations at CEW in cooperation with the recently formed AEC and the University of Chicago. Under a four-year contract with AEC, the University of Chicago was expected to take over operations of Clinton National Laboratory in Oak Ridge beginning January 1, 1948.

UT EXPANDS OAK RIDGE EDUCATION CONNECTION

Graduate education was crucial to the survival of Clinton Laboratories, and UT would prove to be the linchpin. The university already began the UT Residence Graduate Program, starting with graduate courses in fall of 1945 at the request of the Oak Ridge facilities. After a meeting of ORINS during December 27–29, 1945, the participating southern universities were asked to identify potential instructors for graduate courses offered in Oak Ridge. Only Duke University offered to provide a teacher for the program, Douglas Hill, a chemist, who would also use the nearby research facilities. Clinton Laboratories asked its staff member Henri Levy to teach chemistry.[34] Levy was a physicist and crystallographer who made contributions in the field of neutron scattering by crystalline materials. He pioneered automated methodology for neutron-diffraction studies and computer programs for analysis of crystallographic data. Levy Island in Crystal Sound in Antarctica was named after him for his contributions to this field.

During the 1945 fall term, UT enrolled 1,328 students in its Oak Ridge adult education courses. This enrollment level was almost twice as large as any other program in East Tennessee. Some courses were offered for college credit, while others were taught on a noncredit (certificate) basis. For the 1946 fall quarter, more than 1,100 students were enrolled in UT classes in Oak Ridge, according to statistics released in December 1946.[35] Noncredit courses were the most popular, with 566 enrollees; 370 registered for extension courses; and enrollment in high-school credit courses numbered 34. An additional 145 students took courses offered through the UT graduate program. There was significant interest in the course offerings among contractor personnel, with 225 Carbide (K-25) staff topping the list; Tennessee Eastman (Y-12) followed with 208. Enrollees from the Army Corps of Engineers staff totaled 114, Monsanto (Clinton Labs) totaled 112, and all others added 321. UT instructors in the fall term included Pollard, who taught advanced thermodynamics, and Robert Boarts, who taught heat-transfer design.[36] This followed a successful 1946 summer term—the first summer in which graduate classes were offered—with seventy graduate students. By December 31, 1946, most Oak Ridge graduate courses were taught at Oak Ridge High School.

After enrolling more than two hundred employees from the Oak Ridge facilities in the fall quarter of 1947–48, UT announced in January 1948 that it was expanding the course offerings of the Oak Ridge branch. The university would now offer seminars and laboratory courses on the graduate level during the day, in addition to evening and Saturday classes. Courses offered were in chemistry, physics, chemical engineering, and mathematics. In March of that year, UT began offering PhD programs in botany and psychology. Dean Fred Smith noted that offering the doctorate in botany would strengthen UT's program in Oak Ridge, as only three schools in the South (none in Tennessee) offered such a program at that time. This would provide unusual resources, given UT's proximity to the Great Smoky Mountains and to Oak Ridge.[37]

The initial contract provisions between ORINS and UT needed adjustment after a few years. Initially, UT was responsible for all costs associated with instruction, administration, and record keeping, and ORINS provided the space, equipment, and supplies. While the UT portion initially was modest, the program's growth resulted in a need to change the funding profiles. Henceforth, ORINS would reimburse UT for any instructional costs, including the appropriate amount related to faculty time spent in Oak Ridge, plus the cost of clerical university employees not recovered through the assessment of student instructional fees. Other contract modifications allowed a full-time person to direct the program, and the scope of the instructional subject matter was broadly defined as the sciences, mathematics, library science, management, languages, and areas related to nuclear energy. Further, the classrooms opened their doors to other students, including residents of Oak Ridge and individuals lacking an AEC contractor relationship.[38]

THE UT-CLINTON LAB ROADSHOW: "FACING ATOMIC ENERGY"

Throughout the nation, there was post-war concern over atomic energy R&D. Clearly, the atomic bomb was a terrifying weapon, but there was a need both to build weapons and to perform classified and unclassified research on this new energy source. At the time, political parties did not have a uniform position on a management approach to atomic energy. Indeed, opinions were widespread on such issues as who should be in charge, the role of the military, full- or part-time administrators, or information secrecy.

The release of details of the February 1946 defection of Igor Gouzenko, a cipher clerk in Russia's embassy in Canada, complicated the situation.[39] Gouzenko exposed Joseph Stalin's use of sleeper agents to steal information from many parts of the Manhattan Project. Consensus was finally reached on August 1, 1946, when President Truman signed the Atomic Energy Act, transferring the control of atomic energy from military to civilian hands, effective January 1, 1947. This shift gave the members of AEC complete control of the plants, laboratories, equipment, and personnel assembled during the war to produce the atomic bomb. At this time, there was no role for industry or universities in anything atomic other than through contract with AEC.

During this period, Manhattan Project scientists and engineers voiced their opinions and concerns about nuclear weapons. Advocacy organizations formed on all Manhattan Project sites, including Oak Ridge, and among them were the Atomic Engineers of Oak Ridge, Atomic Production Scientists of Oak Ridge, and the Association of Oak Ridge Scientists. These later merged to form the Association of Oak Ridge Engineers and Scientists (AORES). Eventually these local and regional associations combined with organizations from other facilities such as Argonne and Los Alamos to form the Federation of American Scientists.[40]

Eight Oak Ridge scientists (including Lyle Borst and Karl Morgan) and five UT faculty (including Kenneth Hertel and William Pollard) participated in conferences throughout Tennessee in August-September 1946. The conferences were organized around the general theme of "Facing Atomic Energy," and topics discussed included: "Can There Be a Monopoly on Atomic Energy?" "Atomic Energy and the Atomic Bomb," and "Plans for International Control." The day-long sessions also included an eyewitness account of recent nuclear tests on Bikini Atoll in the Marshall Islands of the Pacific Ocean. The conferences were held in Chattanooga, Cookeville, Kingsport, Knoxville, Martin, Memphis, Nashville, and Pulaski.[41]

OTHER TRAINING AND EDUCATIONAL PROGRAMS IN OAK RIDGE

Clinton Laboratories participated in several internal educational activities that bolstered its image, beginning in the 1940s and continuing for many years. The laboratory's Clinton Training School offered a course in nuclear physics that Frederick Seitz developed in fall 1946. Seitz was a graduate student with Eugene Wigner at Princeton, and they worked together during the war in the theoretical physics division of Chicago's Metallurgical Laboratory, where Wigner served as division director. Seitz was on leave from the Carnegie Institute of Technology when he came to Oak Ridge.

Attendees of this course were mostly highly-trained professionals from university and industrial laboratories, though the program was not directly affiliated with any specific southeastern universities. Other attendees included six naval officers, under the leadership of Captain Hyman Rickover, and several representatives from industry and academia who were engaged in development of an atomic energy-based power plant. UT Professor Robert Boarts was among the course's thirty-five attendees and also an instructor who presented a lecture on "heat transfer mechanisms for energy removal from piles." Other contributors to this lecture series included Kay Way and Captain Rickover.[42] [43] [44] [45] Seitz returned to Carnegie in June 1947 and had a long and distinguished career. He was the fourth president of Rockefeller University (1968–78), the seventeenth president of the United States National Academy of Sciences (1962–69), and the recipient of many awards, including the National Medal of Science.

Another internal education program was the semester-long MIT Chemical Engineering Practice School, which began in Oak Ridge in 1948 and continued through 1962. Initially, Carbide and Carbon Chemicals Corporation (renamed Union Carbide Nuclear Company, a division of Union Carbide Corporation, in 1957) was a sponsor, serving all three Oak Ridge sites and hosting the school at the Oak Ridge Gaseous Diffusion Plant, K-25.[46] The school was reactivated at ORNL in 1966. This practical program for students in the MIT Department of Chemical Engineering traces its origins to the mid-1920s. As part of the graduate program, students spent five months in an industrial setting, developing engineering and personal skills while tackling relevant problems by working in small groups. The Practice School provided an alternative to writing a master's thesis, with students paying full tuition during their internships and receiving academic credit. In Oak Ridge, the MIT students focused on diverse areas of study, including isotope diffusion, waste management, and movement of mercury through the environment.

The Oak Ridge School of Reactor Technology (ORSORT), funded by AEC, launched in 1950 as a successor to the Clinton Training School, following discussions between Rickover and ORNL Research Director Alvin Weinberg about preparing nuclear engineers to meet the future needs of both industry and government. A one-year instructional curriculum was prepared along with accompanying laboratory experiments and research projects involving new reactor designs. The design of the current High Flux Isotope Reactor (at ORNL) originated as an ORSORT project. Attendees came from Naval reactors, reactor vendors, manufacturers, shipyards, electric utilities, and universities. James Watkins, later Chief of Naval Operations and Secretary of Energy from 1989 to 1993, was an early attendee. One of the Navy officer students was Edward (Ned) Beach, who later commanded the USS Triton on its unprecedented submerged cruise around the world in 1960. Beach also authored the classic submarine novel *Run Silent, Run Deep.*[47] International students began attending in 1959. The graduates were awarded the degree "doctor of pile engineering," signified by the initials D.O.P.E.[48]

With a total of 976 graduates, ORSORT ceased operation in 1965 for reasons both international and domestic. Other countries began forming similar schools, eliminating the need for international travel, and, as a consequence, it became increasingly difficult for foreign students to receive financial support to attend. Within the United States, colleges and universities developed research and instructional capabilities in nuclear engineering to meet the academic demand for master's and PhD-level programs.

OAK RIDGE WON THE WAR; CAN IT WIN THE PEACE?

During the war, each of the three major facilities in Oak Ridge was managed by a different contractor: K-25 by Carbide and Carbon Chemicals Corporation, Y-12 by Tennessee Eastman, and Clinton Laboratories by the University of Chicago.[49] Monsanto took over management of Clinton from the University of Chicago in July 1945. The programmatic role of Y-12 changed after the war because its uranium-enrichment technology using electromagnetic separation was more labor intensive, more expensive, and less efficient than gaseous diffusion.

With the end of the war, the period of unlimited spending that culminated in the defeat of Japan was over, and constricted budgets became a reality for the Army and later AEC. Highly enriched uranium was delivered from K-25 in appropriate quantities for the country's increasing military needs, beginning in December 1946. K-25's role as an enriched-uranium production facility continued for almost forty years. With K-25's primacy in producing enriched uranium and the mothballing of Y-12's electromagnetic separators, employment levels at Y-12 declined precipitously. Indeed, only about two thousand employees remained on the payroll as of January 1947, a 90-percent decrease from twenty months earlier.[50] In that same year, Carbide and Carbon Chemicals Corporation became the new Y-12 contractor, and soon AEC assigned Y-12 a new mission: the manufacture of nuclear weapons components, which remains the mainstay of the Y-12 program to this day.

Meanwhile, K-25 remained focused almost entirely on gaseous diffusion of uranium. After a period of expansion in Oak Ridge, other gaseous diffusion facilities opened in Paducah, Kentucky (1952), and Portsmouth, Ohio (1955), responding not only to the need for highly enriched uranium for defense purposes but also to produce uranium as fuel for the world's growing number of nuclear reactors used to produce electricity. Following military intervention by China in the Korean conflict in October 1950, national security concerns in AEC led to construction of other uranium enrichment plants more than five hundred miles from the Atlantic coast—including the facilities in Paducah and Portsmouth—to limit concerns of submarine-launched missiles.[51]

The Graphite Reactor at Clinton produced radioisotopes in previously unheard-of quantities. Although radioisotopes was widely known before the war, specific radioisotopes previously were not available for study in terms of applications and effects. The Graphite Reactor's prodigious output of a range of purified isotopes changed all that, as it could produce gram quantities of isotopes instead of the microgram quantities available earlier. Indeed, among other contributions, Clinton Laboratories' chemical processing and purification of radioisotopes marked the culmination of the wartime production of plutonium, radioactive lanthanum (to aid in design of the plutonium weapon), and bismuth from which polonium was extracted for the scientists at Los Alamos.[52] Other radioisotopes would serve more peaceful purposes in diagnosing and treating disease.

The first medical radioisotopes produced in the Graphite Pile after World

War II were presented to the director of the Barnard Free Skin and Cancer Center of St. Louis at a ceremony at the Graphite Pile on August 2, 1946, the day after President Truman signed the act establishing AEC.[53 54] News reports indicated requests for radioisotopes had been received from many universities, including UT, as well as industries and government institutions. However, St. Louis was the headquarters of Monsanto, the managing contractor of Clinton Laboratories at that time. The importance of radioisotope production in Oak Ridge cannot be overstated. Alvin Weinberg wrote in 1976: "If at some time a heavenly angel should ask what the Laboratory in the hills of East Tennessee did to enlarge man's life and make it better, I daresay the production of radioisotopes for scientific research and medical treatment will surely rate as a candidate for the very first place."[55]

Following the war, there was uncertainty about the future role of Clinton Laboratories. A February 1947 report to Robert Oppenheimer, the Manhattan Project scientific leader who had been stationed at Los Alamos during the war, lamented the poor living and working conditions in Oak Ridge—conditions greatly complicated by poor management. The report described three groups often working at cross purposes: scientists brought to Oak Ridge by Eugene Wigner from Chicago, those brought by Monsanto, and the contract administrators from the Army Corps of Engineers and the new AEC (both providing local direction in the absence of input from Washington).

A month later, a local AEC manager wondered if Clinton Laboratories would cease to exist. As a reflection of the laboratory's contentious and fractured management structure, the Army intervened to stop Wigner, then the laboratory's research director, from launching an experiment using a critical mass of ^{235}U.[56] General Groves had previously issued an order requiring safety reviews of all proposed criticality experiments, following the death of Los Alamos scientist Louis Slotin, who previously worked at Clinton Laboratories.[57 58] Slotin was a Manhattan Project researcher first in Chicago, then in Oak Ridge, and finally in Los Alamos. On May 30, 1946, Slotin and colleagues experimented with a sub-critical sphere of plutonium. In this test, reflecting material was placed around the sphere to measure how many neutrons would be reflected and how this would affect the plutonium sphere. Slotin used a screwdriver to keep the hemispheres apart instead of shims as called for in the experimental protocol. The result was that a larger number of neutrons (from plutonium fission) than expected were directed back to the sphere, which went critical causing a chain reaction and a large emission of fission neutrons. Two experimenters, including Slotin, died within weeks due to acute radiation injury.[59 60]

In Oak Ridge, a bitter disagreement concerning safety review of criticality experiments ensued, as Wigner believed that Groves' order requiring safety reviews had been revoked the previous month at a meeting of various laboratory directors. When the Army insisted that the order had not been rescinded, and that its rules were to be followed, Wigner angrily responded that he took orders from Monsanto, not the Army. While Wigner was permitted to move forward with the experiment following a two-week delay to conduct safety reviews and implement corrective actions, this event sent a message both to him and to Monsanto. Clearly, AEC in Washington needed to take responsibility and establish rules and directions for programmatic activities at its laboratories.

At a meeting of the AEC General Advisory Committee in March 1947, there was much discussion of the future of AEC laboratories.[61] Plans for a new Brookhaven National Laboratory were reviewed, and the budget for the ongoing construction of Argonne National Lab was discussed. Iowa State University requested funds for a new laboratory to replace its wartime facilities. Concerning Oak Ridge, the General Advisory Committee felt that the Clinton Laboratories were not worth saving, and committee member Oppenheimer said, "Most of us think that the evidence is in that Clinton will not live even if it is built up," and suggested that Clinton Laboratories be limited to the research and production of radioisotopes.[62]

However, Carroll Wilson, AEC general manager, thought the main problem with Clinton was not its geographic location but its management by Monsanto. This view, along with the AEC decision not to build a proposed high-flux nuclear reactor in Oak Ridge, played a role in Monsanto's decision to withdraw from its management contract for the laboratory. AEC hoped that Monsanto would continue in its management role, but Monsanto named its price for doing so, and it was a steep one: build the high-flux reactor and fund development of another reactor then under discussion. Later, Monsanto said it wanted the high-flux reactor built not in Oak Ridge but at another Monsanto facility, either in Dayton, Ohio, or St. Louis. Ultimately, before its demands could be met, Monsanto announced its decision to withdraw from management of Clinton Laboratories.

Towards the end of the Manhattan Project, there were changes to and movement of its funded programs. For example, the Substitute Alloy Materials (SAM) Laboratory at Columbia University focused on research and development of barrier technology crucial for the gaseous diffusion process for enriching uranium. Control of this program was transferred on March 1, 1945, to Carbide and Carbon Chemical Corporation, which operated the K-25 gaseous diffusion site in Oak Ridge. At its peak the SAM Laboratory employed around 2,500 people,[63] but, at the time of the transfer, the workforce had shrunk to 837,[64] since the successful barrier process was developed.

Subsequently, the program and many of its people started transferring from New York to the K-25 site in Oak Ridge.[65] Don Trauger, who played an important role in barrier testing at Columbia, moved to K-25 in February of 1946, along with his friend and colleague Jack Craven, who had taken over leadership of the barrier testing group when Robert Lagemann returned to Vanderbilt.[66] Craven continued as Trauger's boss in Oak Ridge until Craven decided to move to UT to direct an expanded research facility for improving cotton textiles,[67] a laboratory that had been started a decade earlier by Physics Department head Kenneth Hertel. Succeeding Craven as head of the K-25 Barrier Testing Department, Trauger started his rise in importance in Oak Ridge, moving later to ORNL where he eventually served as associate laboratory director for nuclear and engineering technologies from 1970 to 1984.

Craven then had a long career at UT, eventually moving to a teaching role in the physics department and becoming a popular astronomy teacher. In fact, he was crucial to the astronomy education of new faculty member Lee Riedinger, who was hired by William Bugg as a young assistant professor in physics in 1971. Riedinger knew nothing about astronomy when Bugg assigned him to teach honors astronomy in his first semester, but Craven mentored him through that challenging process, and Riedinger eventually became a good astronomy instructor.

In Oak Ridge, it took intervention by General Groves to stop the attempted poaching of pieces and parts of Clinton Laboratories—not to mention its personnel—by representatives of northeastern universities who appeared in Oak Ridge soon after the formation of ORINS was announced. The northeast research community tried to persuade Oak Ridge scientists to relocate to its universities.[68] [69] Several scientists went so far as to propose moving Clinton Laboratories, even if only the staff, to the Northeast.

While clearly impossible, this proposal showed the determination of the northeastern university community to move programs from Oak Ridge to their campuses and, in the process, to allow them to build on Oak Ridge scientific successes that resulted from the contributions of the universities' faculty, who had been on leave to participate in the Manhattan Project. Oak Ridge was the largest Manhattan Project site in terms of money and staff, and Clinton Laboratories, with its reactor, radioisotope production, and active research agenda, was destined for a prominent post-war role. However, some people wanted to move that expertise and future programs out of Oak Ridge. Shortly after Columbia University's Isidor Rabi (1944 Nobel Prize in physics) visited Oak Ridge to promote the benefits of moving to the northeastern intellectual center of physics, Groves' displeasure was forcefully and definitively communicated to Rabi, and no more was heard of this proposal.[70]

Operation of Clinton Laboratories

Early in the Manhattan Project, DuPont managed engineering and construction at Clinton Laboratories (the X-10 site in Oak Ridge), while the University of Chicago managed research due to the close coupling between the Metallurgical Laboratory there and the R&D programs at Clinton. In fact, the design and construction of the Clinton Pile evolved directly from Enrico Fermi's CP-1 uranium pile at the University of Chicago. There was a migration of talent from the Chicago-based laboratory to X-10 to take the lead on operating the Graphite Pile and developing programs based on its use.

This flow of talent included Charles Coryell (who left Oak Ridge for the Massachusetts Institute of Technology after the war in 1946), Gale Young (who would cofound the first privately owned nuclear firm, Nuclear Development Associates, in 1948), Eugene Wigner (who returned to Princeton in 1948), and Alvin Weinberg (who stayed at ORNL for most of his career and served as the laboratory's director for eighteen years). In 1962, Weinberg brought Young back to ORNL, in part to work on desalination.

Near the end of World War II, the Manhattan Engineer District contracted with Monsanto Chemical Company of St. Louis to operate Clinton Laboratories. However, this period of management was rocky, and, in May 1947, the newly formed AEC needed to replace Monsanto, which did not want to continue its role in Oak Ridge. This led to a year of management uncertainty until AEC announced issuance of a contract to Carbide and Carbon Chemicals Corporation to assume responsibility for operating the newly christened Clinton National Laboratory, starting March 1, 1948. (Carbide had already been operating two other Oak Ridge installations: the Y-12 Electromagnetic Separations Plant and the K-25 Gaseous Diffusion Plant.)

In the 1940s, the concept of universities having a role in managing national laboratories became popular. In Tennessee, the Oak Ridge Institute of Nuclear Studies (ORINS) formed in the fall of 1946 under the leadership of UT physics professor William Pollard. Some Oak Ridge researchers even wondered in 1947 if ORINS should be appointed to operate Clinton National Laboratory, as the facility was briefly named before becoming Oak Ridge National Laboratory in 1948. Indeed, a meeting during that time period was described by Charles Moak, who had joined the Metallurgical Laboratory in 1944 and moved to Clinton in 1945. He spent his career in Oak Ridge and rose to be the director of ORNL's Van de Graaff Laboratory before retiring from the Physics Division in 1985. The meeting took place in UT's old physics building, and Moak noted that researchers in attendance urged Pollard to propose that ORINS should operate Clinton Laboratories. Pollard replied that it was too early for this, as ORINS was just starting up and did not have the expertise and reputation to seriously propose assuming that weighty role.

One year later, in 1948, Carbide and Carbon Chemicals Corporation was appointed to manage Clinton National Laboratory. Moak speculated that if Monsanto had stayed on the job for one more year, ORINS possibly could have successfully competed for the operator role and become the long-term manager of the laboratory.[71] But timing is everything. A similar opportunity came and went in 1982 when Union Carbide Nuclear Company (formerly Carbide and Carbon Chemicals Corporation) decided to relinquish its Oak Ridge management role. At that time, UT mounted an effort to study possible competition for the management contract but decided that the university was not ready, as William Pollard had concluded in 1947. However, in 1998, UT *was* ready to compete—and win. It took a half century for UT to evolve, mature, and prepare itself for the right time and the right opportunity.

MISSION CHANGE FOR CLINTON LABORATORIES?

In May 1947, it was announced that Clinton Laboratories' Biology Division was moving from the X-10 site to Y-12, where some of the large buildings formerly used by the uranium separation efforts were being vacated, cleaned up, and made available. At the time, no funds were budgeted for new construction at the X-10 site. More than seventy staff members were relocated to the Y-12 area within a few weeks. Monsanto and AEC agreed in the same month that Monsanto would leave as Clinton Laboratories' managing contractor. By 1947, employment at Clinton swelled to 2,130, up from 1,234 during the World War II peak.

The search began for a new contractor at the time, while AEC was determining its budget and research priorities and awaiting action by Congress. By fall 1947, apparently, associations of universities would operate laboratories at Argonne and Brookhaven. The fourteen ORINS member universities then asked themselves: Why should we not run Clinton Laboratories? Instead, an article appearing on the front page of the *Oak Ridge Journal* in September 1947 announced that AEC would negotiate with the University of Chicago to manage the laboratory.[72] [73] A group of directors of Clinton's research groups issued a statement soon thereafter saying they welcomed the University of Chicago and expressed confidence that fundamental research would continue in the physical and biological sciences, with an emphasis on nuclear reactor R&D. The statement concluded by saying that training of scientists by ORINS using Clinton Laboratories' facilities would continue in collaboration with Clinton personnel.

About a month later, the facility was renamed Clinton National Laboratory. The AEC announcement said that arrangements for the operation of the laboratory had been worked out with the staff and industrial participants at Clinton, representatives of ORINS, the University of Chicago, and AEC. All national laboratories were expected to be strong centers of atomic research. In terms of research focus, Clinton and Argonne were two parallel but separate and independent institutions, and reactor R&D was to be performed at both facilities. That said, the major projects designated for Clinton were radioisotope production and performing applied and fundamental research on AEC topics, including nuclear-reactor research.

Specific reactor projects that were also included in this scope were the design and construction of a new high-flux reactor, one that would expand greatly on the original Graphite Reactor and produce higher neutron fluxes and higher operating temperature that would allow testing of materials needed for the next generation of reactors. The U.S. Navy was interested in converting its submarine fleet to nuclear power rather than diesel fuel so that the subs could stay submerged for longer periods before needing to refuel. Also, industry was looking into the development of a commercial nuclear power plant for generating electricity. Clinton National Laboratory was also the site of the ORINS operation of a school for academic and industrial personnel and a program for on-the-job training of graduate students for credit toward advanced degrees. Dozens of participating industrial companies were named in the announcement.

As it would turn out, the University of Chicago's anticipated role as manager of Clinton National Laboratory never happened. In December 1947, while in Oak Ridge as part of an AEC industrial advisory board meeting, AEC chairman David Lilienthal met Clark Center, president of Carbide and Carbon Chemicals Corporation, which managed K-25 and Y-12 for AEC. The two shared ideas about how management of Clinton National Laboratory by Carbide might answer the needs of both AEC and Carbide. AEC wanted an established management presence, and Carbide wanted to simplify labor issues, as different Oak Ridge plants had different union affiliations (K-25 was a CIO union, and Clinton was AFL). Some of the labor problems could be abated if these plants were managed by one company.[74]

One of the more traumatic episodes in Oak Ridge history began in the days after Christmas 1947 (Black Christmas according to Oak Ridgers), when AEC commissioners made two significant decisions. The first was to name Carbide and Carbon Chemicals Corporation as the managing contractor of Clinton National Laboratory. The second was to move reactor development work from Clinton to Argonne. The combined announcement came as a major blow to the nuclear scientists in Oak Ridge. The announcement explained that Carbide had been asked to take over Clinton's management because "Clinton Laboratory will develop a strong project for research in chemical and chemical engineering problems involved in both the production operations carried on by the Commission (AEC) and in the solution of fundamental chemical problems in the broad field of atomic energy, with emphasis on industrial applications."[75] Among other impacts, AEC's relocation of the reactor research to Chicago would take a significant bite out of employment in Oak Ridge.

The staff of Clinton Laboratory went apoplectic on receipt of the news. The world's first operating nuclear reactor (Graphite Pile) was built in Oak Ridge, and the R&D staff were following that with new ideas and new projects for reactors with higher neutron fluxes and reactors that could be used on submarines. Clinton National Laboratory became the premier facility in the country for designing new reactors, but now that reactor program would be shipped north to Argonne. Promises made a few months earlier had been broken. Was Clinton Laboratory to be reduced to a radioisotope production facility with some ancillary chemical processing operations? ORINS members were not happy, because their prospects for future research at Clinton now appeared greatly diminished. Stridently worded letters and telegrams were sent to AEC commissioners and President Truman, protesting the decision and its effect on southern universities.[76] [77]

TRANQUILITY RETURNS AS CLINTON BECOMES OAK RIDGE NATIONAL LABORATORY

Soon after 1948 began, Lilienthal and others managed to communicate their ideas more clearly and effectively than they had been able to over the holidays and, in the process, hoped to ease the concerns of the people of Clinton National Laboratory and the laboratory's affiliated universities. In a letter sent to assuage the Clinton managers, AEC said the new contract ensured the permanent operation of Clinton as a national facility. The letter specified that "basic research programs will be maintained and developed in the fields of biology, physics, chemistry, and health physics." Two weeks into the new year, AEC Commissioner Robert Bacher spoke to the Oak Ridge Rotary Club, affirming that "research work will continue vigorously at Clinton Laboratory." He justified centralizing reactor development at Argonne and said he hoped that the movement of the reactor work would in no way impair the well-being of Clinton or impair the work of ORINS to carry forward work on basic research and training.[78] [79]

On Sunday, February 1, 1948, AEC Chairman Lilienthal, in a radio broadcast to Oak Ridge, said Clinton would henceforth be known as Oak Ridge National Laboratory,[80] although, at the time, there was no precise definition of what constituted a national laboratory relative to its range of research programs, extent of impressive staff and facilities, and links to educational institutions. The only other national laboratories at the time were Argonne and Brookhaven, which would settle into their permanent locations later in that same year. The AEC chairman assured that ORNL would maintain its position as a major center for basic and applied research into the challenges of atomic energy. He also reiterated that basic research programs would continue. He further pledged a future major construction and renovation of facilities and utilities in the town site as well as at the plants. And Lilienthal wanted ORINS to play a major role in ORNL and noted that he had been involved in the institute's incorporation (he was one of the members of the early board of directors of ORINS).

Just a month later, in March 1948, Oliver C. Carmichael, president of the Carnegie Foundation for the Advancement of Teaching, spoke at a regional planning conference in Gainesville, Florida. He urged southern states to band together to take over Oak Ridge in an effort to produce a cadre of outstanding scientists for the future. Carmichael noted that Oak Ridge had exceptional resources in the physical sciences and urged southern states to support development of a research center utilizing these resources.[81] However, it would take another fifty-two years before universities assumed a role in managing ORNL.

UT RESEARCH GOES ATOMIC

Following World War II, UT initiated a number of PhD programs, and graduate students became involved in research at Oak Ridge for their theses and dissertations. The Department of Biological Sciences and the College of Medicine at the UT Medical Units in Memphis began using radioisotopes for research following receipt of a shipment of radioactive sulfur from Clinton in December 1946. Researchers in UT's Departments of Chemistry and Physiology in Knoxville studied the flow of the isotopes through the bodies of white rats. New instructors were added to the botany, bacteriology, and zoology departments in June 1947 to keep pace. The three departments planned to use the radioisotopes from the Oak Ridge facilities to mark atoms as an adjunct to their courses studying vegetables and animals. Among other applications for the radioisotopes, they hoped to use them to tag bacteria and other organisms to study their behavior.[82]

UT's Department of Physics established a "Radiations Laboratory" to provide equipment and offer advice to UT departments eager to work with radioisotopes.[83] Among its goals and responsibilities, this laboratory would help researchers make full use of the radioisotopes, build the needed experimental equipment, and furnish devices designed to protect researchers from harmful radiation. Many UT faculty in the physical and biological sciences began to appreciate the importance of the study of radioisotopes. UT Graduate School Dean Fred Smith, recognizing the need for an orderly process of radioisotope research among staff and departments, appointed a faculty committee in May 1947 to facilitate the use of radioisotopes and to act as a clearinghouse for UT's requests for radioactive materials from Oak Ridge.[84]

"FALLOUT" CATTLE COME TO OAK RIDGE AND BECOME UT'S HERD

Oak Ridge welcomed some unusual visitors in December 1945. At the time, with the United States still struggling to overcome food shortages that resulted from wartime rationing, fifty cattle arrived from New Mexico. They differed from other livestock of the time—they had been exposed to radioactive fallout from the Trinity atomic explosion in New Mexico on July 16, 1945. Trinity, the first test of an atomic bomb, was detonated in the Jornada del Muerto desert about thirty-five miles southeast of Socorro, New Mexico. While Los Alamos leaders were confident that the enriched-uranium bomb (Little Boy) would work, the plutonium-based Fat Man relied on a complicated implosion mechanism that definitely required confirmatory testing. The successful Trinity test provided President Harry Truman with great confidence at the Allies' Potsdam Conference concerning the eventual

NORRIS BRADBURY STANDS BESIDE THE TEST BOMB AT TRINITY. COURTESY OF OAK RIDGE NATIONAL LABORATORY.

terms of surrender for Japan. Norris Bradbury organized the Trinity test and became director of Los Alamos after Robert Oppenheimer left the position at the end of World War II.

While the Army tried to clear the test site of cattle, fallout from the first atomic explosion rained down on cows grazing in the vicinity. Following their exposure to the radiation, the affected Herefords began to grow gray hair in irregular patterns on their skin—certainly not a common trait of the otherwise mottled reddish-brown and white animals. They were purchased by the Army after their owner could not sell them because of their blemished hides. The animals more severely afflicted by the fallout were kept in New Mexico at Los Alamos, and the others were sent by rail to a pasture in Oak Ridge. Several years later, the remaining Los Alamos cattle and their progeny were also sent to Oak Ridge.[85 86 87]

Over the next few years, the cattle were studied carefully, and radiation levels were monitored frequently, along with the animals' health and behavior. They were normal in all respects except for the gray hair that did not go away. The Trinity cattle arrived scrawny after having grazed on arid New Mexico ranges, but, less than six months later, a newspaper article described their appearance as healthy and well nourished, plus their milk was not radioactive. The gray hair returned after shedding, although, over time, some natural red hair also came back.[88] [89]

AEC realized it had a tremendous resource in these cattle and, in 1948, asked UT to study them at the newly named UT-AEC Agricultural Research Laboratory (the name changed to the Comparative Animal Research Laboratory [CARL] in the 1970s)[90] in Oak Ridge. This was part of a larger UT-AEC contract that facilitated the study of the effects of radioactive isotopes and radiation in agriculture. The contract also sought to create a learning environment for graduate students and scientists in the fields of nuclear energy and agriculture. The facility occupied about five thousand acres on the southeast boundary of the Oak Ridge Reservation.[91] Cyril Comar was hired to direct the UT-AEC Ag Research Lab in 1948, after he earned a PhD from Purdue University and held positions at Michigan State and the University of Florida.[92]

At UT, Comar built a highly regarded program in research on the use of radioactive tracers in biological research. He departed Oak Ridge in 1957 to move to Cornell University and direct the Laboratory of Radiation Biology as part of the New York State Veterinary College. Ultimately, Comar became a national leader in the effects of radiation on biological systems. The UT-AEC Agricultural Research Laboratory opened another critical connection between UT and Oak Ridge and represented the first large AEC contract to UT in the area of radiation effects in animals.

By July 1949, forty-three of the fifty cows and heifers originally sent to Oak Ridge were still living, with seven having died of causes not attributable to their exposure to radiation. One of the exposed bulls (nicknamed Ferdinand after a popular children's book) mated with several of the exposed cows, and all thirty-three calves produced were normal. There was no evidence of adverse effects on the fertility of the herd. Administration of CARL transferred[93] on October 1, 1981, from UT to Oak Ridge Associated Universities (ORAU), which was the new name for ORINS. William Bugg, one of the authors of this book, actually spent high-school summers in the late 1940s working with these special cattle. Perhaps because of this, he became a physics professor rather than a farmer.

Interest in the cattle reflected the growth in international research on the effects of radiation on humans and other animals that followed in the wake of the Manhattan Project. This area of research was high on the agenda of Clinton Laboratories' Biology Division after it was formed in 1946.[94] Health physics became a focus of the safety practices at each of the national laboratories then coming into existence.

The field of nuclear medicine developed in part due to research on animals at the UT-AEC Agricultural Research Laboratory. As described by Marshall Brucer, a pioneer in this field, work at the Bethesda Naval Hospital in 1949 indicated that the element gallium seems to concentrate in cancerous tumors on bone.[95] A way to detect the gallium concentration and even to kill the tumor would be to inject the radioisotope gallium-72 (^{72}Ga). However, the ^{72}Ga short half-life (fourteen hours) and the difficulty of producing the isotope for these experiments made it challenging to pursue this research in Bethesda. The solution was to move this work to Oak Ridge, to utilize the Graphite Reactor for making this radioisotope and the nearby UT-AEC Ag Research Lab for performing experiments on animals, and to test this idea that gallium concentrates in bone tumors.

The researchers from Bethesda, working with Comar's team in Oak Ridge, decided to utilize a large animal for these experiments, to better approximate a human. Comar was asked to procure a piglet with the goal of administering a human-sized therapy dose of the gallium isotope when the pig got to approximately one hundred pounds. The experiment was a success, as the autopsy showed that gallium did concentrate in the bone, which meant that a therapeutic dose of radioactive gallium could be delivered to a bone tumor for the purpose of killing the tumor.

Many future developments were needed before this technique could be used on human subjects. For example, a more sensitive set of detectors—essentially a whole-body scanner—was needed for external measure of the radioactivity in the body. And, over time, other radioisotopes—other isotopes of gallium or strontium or technetium—were found to work better for detection and treatment of tumors in the human body. Today, nuclear medicine is a well-developed and important field of human therapeutics, and the experiment in Oak Ridge with "Comar's pig" played an important role in the early development of radiation therapy. Comar left the UT-AEC Ag Research Lab in 1954 and moved to ORINS to become the head of biological research, focused in part on early experiments on use of radioisotopes in humans. In 1957, he moved to Cornell University.

Comar's Pig and Chanel No. 5

Cyril Comar and his staff at the UT-AEC Agricultural Research Laboratory selected a small piglet to use in an experiment to see if radioisotope ^{72}Ga would concentrate in bone, as a future way to diagnose and treat bone cancer in humans. But first, the small pig would need to be fed sufficiently to increase its weight to around one hundred pounds to better approximate a human body. Neutron irradiation in the Graphite Reactor at ORNL would produce the radioisotope. The produced ^{72}Ga, with a fourteen-hour half-life, would be rushed seven miles to the nearby UT-AEC Ag Research Lab, where it would be injected into the pig. While the reactor began to produce the desired isotope, Comar and staff started feeding the pig to get it up to the desired weight.[96]

Schedules were made to accommodate isotope production and safety, and purification trials were conducted. Various delays in the schedule ensued. Time was taken by scientists to prepare for and attend scientific meetings, as the pig's weight continued to increase well beyond the desired one hundred pounds. The pig was allowed to wander in the nearby woods of the farm and feed itself, combined with frequent observations and generous feedings by Comar's staff. The fall university term began and the team of faculty and students returned to Oak Ridge. Preparations began in earnest, and duties were assigned for each step of the experiment, from isotope production and purification to anesthesia application and administering the gallium to the pig, and finally to data collection. As the team approached readiness, Comar continued to feed the pig.

The big day finally arrived and the scientific cast assembled included a physiologist, health physicist, anesthesiologist, and various technicians. The irradiated gallium came out of the Graphite Reactor at 8 a.m., and soon many unexpected events began to transpire. The pig was lost and had to be found roaming around the farm, barely fitting through the door of the operating room due to the extensive weight gain. Six men could barely get the huge pig onto the autopsy table, and the facemask to be used for delivering anesthesia to the thrashing pig was too small.

A white laboratory coat was quickly improvised to serve as a mask. Due to the size of the pig, the ether dose was greatly increased, which meant that the room ventilation had to be quickly improved so that staff would not pass out. As the ether began to calm the pig, the surgeon realized that the syringe to deliver barbiturates was not big enough. Eventually a vein was found that allowed the barbiturate delivery. Now that the pig was mellow, everyone waited for the radioactive gallium delivery from the Graphite Reactor, which was late because the delivery truck ran out of gas.

Finally, the gallium arrived and was injected, setting off radiation detectors as the radioisotope was far "hotter" than expected. The pig relaxed and eventually succumbed, releasing its sphincters of fluids and feces over the nearby area, causing a quick cleanup effort in the sterile environment. Staff assembled outside to escape the smell and the ether. To ameliorate the foul odor, a technician sponged the pig with Chanel No. 5 perfume.

In the end, accurate external radiation counts of the gallium in different parts of the pig body could not be made, due to the amount of radioactivity and the lack of sufficient collimation to define the points from which the radiation emanated. Conclusions on the gallium absorption throughout the body had to wait until after the autopsy. Though a difficult experiment, the results of the autopsy marked an important step forward in the treatment of bone disease by delivery of a radioisotope to a region of bone building activity. Nuclear medicine progressed greatly thereafter.

ORINS CONTINUES TO GROW, AND OAK RIDGE OPENS ITS GATES

At the request of AEC, ORINS convened a meeting of twenty southern medical schools (including the UT, Memphis, School of Medicine) on March 1–2, 1948, to explore the feasibility of establishing a clinical research program to study cancer and other malignant diseases by using special facilities to be developed at the Oak Ridge hospital.[97] An advisory medical board, proposed to be a part of ORINS, was elected to establish policies for treatments using the unique types of radioisotopes produced at Oak Ridge National Laboratory, which had just been renamed from Clinton National Laboratory.

The medical schools wanted to emulate the Rockefeller Foundation Research Hospital (in New York City), which, besides being a top-flight hospital and research organization, was also a leading medical radioisotope research center. Two weeks later, at its regular meeting, the ORINS board approved the recommendation to administer the program. Initially, the Oak Ridge hospital had eight to ten beds devoted to this new line of research; the goal was to expand to twenty or thirty beds, with patients brought from throughout the South for treatment. In its January 1949 semiannual report, AEC noted that a clinical cancer research facility was being established at ORNL under the direction of southern medical schools through ORINS.[98] Earlier, in May 1948, representatives of these institutions met with AEC staff members in Washington for preliminary discussions. Seven months later, in November, AEC designated a clinical director of the program. This clinic would receive full financial support from AEC and would utilize existing Oak Ridge facilities. The participating medical schools would refer their cancer patients to the new facility for research, diagnosis, and treatment.

After studying the use of radioisotopes produced in Oak Ridge during the first few years following the war, AEC determined that lack of education and training in safe handling of radioactive materials inhibited growth in use of this promising resource.[99] In response, AEC announced a broad training program in the medical and biological use of atomic energy, to begin in summer 1949. The program provided fellowships for students to attend more than a dozen schools in six states. The instructors selected for the program took courses at the University of California, ORINS, and Brookhaven in summer of 1948. ORINS also hosted a training course for physicians and health physicists in the use of radioisotopes.

OPENING OF CITY OF OAK RIDGE ON MARCH 19, 1949.
COURTESY OF OAK RIDGE NATIONAL LABORATORY.

After the war, discussions were held between AEC and Oak Ridge community leaders about relaxing some of the security restrictions that continued to limit access to the town. At first, residents opposed this idea, as they did not want door-to-door salespersons, preachers, confidence men, and beggars (among others) to disrupt their way of life. However, AEC persisted in efforts to make Oak Ridge just like other towns in America. On March 19, 1949, in ceremonies that included Vice President Alben W. Barkley, AEC Chairman David Lilienthal, and movie star Marie McDonald, AEC guards took down the barriers, and Oak Ridge became open, with unrestricted access. There would be no further checks of badges for those entering or leaving the area. Although not yet a municipality, Oak Ridge had taken the first step toward the goal of self-government. It would take another decade before there would be an official city of Oak Ridge.[100] [101]

ORINS also began operation of the American Museum of Atomic Energy in Oak Ridge as a public service for visitors eager to learn more about nuclear energy. The facility was open to the public at a nominal charge. The museum's displays were designed by AEC industrial contractors for the "Man and the Atom" exhibit, featured at the New York City Golden Jubilee Exposition in 1948. After approval by AEC in February 1949, the displays were erected in an abandoned cafeteria building in Oak Ridge, which had been remodeled in time to host eight hundred people on opening day.[102] This began ORINS' national role in public education in science and technology. Henceforth, the institute created atomic energy displays and other educational materials, which were provided to universities, academies of science, and other public or educational organizations.

SUMMARY AND LOOK FORWARD

The 1940s saw enormous change in the institutions associated with the Manhattan Project in Tennessee. At the beginning of the decade, UT had no doctoral programs and sustained little faculty research. But following the war, UT quickly mobilized to begin offering graduate courses in science and engineering in Oak Ridge, in addition to a host of new courses on campus relating to nuclear science. This was the first critical connection that evolved in the fall of 1945 to begin a formal partnership between UT and Oak Ridge institutions. The university's decision to offer this specialized instruction in Oak Ridge had come in response to appeals from the leaders of the federal facilities, who saw the need to offer continuing education for their employees. Absent such educational opportunities, these leaders realized, many of their employees would depart East Tennessee to resume their graduate studies where they had begun them before the war.

By the end of the decade, UT had established five doctoral programs in science or engineering, building on the proximate federal research facilities in Oak Ridge. ORINS was launched primarily under the leadership of William Pollard, a UT physics professor. While the future of the federal facilities in Oak Ridge had looked highly uncertain at the end of the war, facility leaders, in collaboration with UT, devised a pathway to survival, based in part on the keen interest by UT and other southeastern universities in using the Oak Ridge facilities to bolster faculty and student research endeavors. Without ORINS and UT's off-campus education programs, Clinton Laboratories would have had difficulty in maintaining the scientific workforce needed to avoid being shut down after Japan's surrender in September of 1945, and even Y-12 and K-25 might have been endangered. Losing these plants would have sounded a death knell for the community of Oak Ridge.

Clinton Laboratories survived amid intense uncertainty to become ORNL. Y-12's primary wartime mission, to enrich uranium, was taken over by the K-25 gaseous diffusion facility, but Y-12 adapted quickly in undertaking a new mission that would contribute to its survival: machining components for nuclear weapons. In addition, Y-12 modified a small number of its calutrons to separate a broad range of stable isotopes, which provided the feedstocks necessary to produce radioisotopes in the Graphite Reactor for the benefit of medical research programs and patient treatment.

In 1949, Alvin Weinberg, then ORNL research director, summarized the relationship between ORNL and the southern educational institutions this way:

> Oak Ridge National Laboratory represents a unique experiment in scientific and governmental administration. It is a national institution operated by a private corporation for the purpose of furthering nuclear chemical technology on the one hand, and basic research, in conjunction with the Southern universities, on the other. It is thus a microcosm in which are projected many elements of our modern American and Southern society. . . . But it may be that the laboratory draws its essential strength from its position as the largest scientific institution in the South. It is commonplace to observe that the Southland is undergoing a modern industrial revolution, that living standards are increasing, and that, as a concomitant, a cultural rebirth is in the making. But the South has a long way to go, especially in the sciences. In making its influence felt throughout the scientific departments of the Southern universities, Oak Ridge National Laboratory, through the agency of the Oak Ridge

> Institute of Nuclear Studies, has a worthy educational mission to perform. Should it fulfill this mission then this fulfillment, this curious by-product of the atomic bomb, will almost surely rank in importance with any future technical advances which Oak Ridge National Laboratory or any laboratory can hope to achieve.[103]

The decade began in great uncertainty, with a raging world war and a formidable alliance of enemies, but ended with the belief—shared among leaders and staff at the various UT and Oak Ridge institutions—that the future was bright. And indeed, it was, as the following chapters will reveal.

3

THE DECADE OF THE 1950S

Oak Ridge Transitions to a City and Deepens Relations with UT

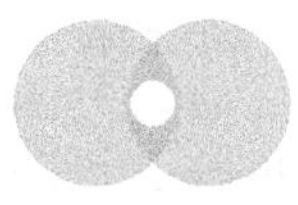

BY 1950, the relationship between the Oak Ridge facilities and UT was firmly established. A measure of a new critical connection was headlined in an article appearing in the 1950 edition of *ORNL Laboratory News*, listing the more than thirty UT faculty, representing ten departments, who served as consultants in all three plants in Oak Ridge.[1] This marked another crucial step in the partnership that eventually grew to include a joint faculty program between UT and ORNL. The Oak Ridge Resident Graduate Program, discussed in the previous chapter, not only sent faculty to teach in Oak Ridge but also recruited scientists from the Oak Ridge facilities as lecturers for graduate and undergraduate courses taught in the community.

It is difficult to overemphasize the importance of these efforts in nurturing the close relationship between UT and the facilities in Oak Ridge. UT, due to its proximity, was able to interact with ORNL in ways not possible for other universities. In physics, chemistry, chemical engineering, and later nuclear engineering, approvals of PhD programs was accelerated by the needs of Oak

Ridge employees who had their graduate studies interrupted by World War II and the Manhattan Project. Equally important, senior academic and scientific personnel at ORNL were eager to resume teaching and research-based interaction with students. Both factors were instrumental in the postwar retention of key scientists and administrators at Oak Ridge and the numerical increase of UT faculty. From the UT viewpoint, the research opportunities were unparalleled. Large increases in science and engineering faculties and their resident research expertise occurred in the 1950s on the UT campus due in part to the partnership established with Oak Ridge facilities and the research programs that resulted. The city of Oak Ridge also experienced remarkable changes during this decade.

OAK RIDGE EVOLVES FROM HIGH-SECURITY COMPOUND TO CITY

There was not a blueprint for the way forward for Oak Ridge, from the time of site selection for the Clinton Engineer Works in 1942 until Oak Ridge became a city. Never before had a small isolated farming community been so rapidly transformed into a prosperous city with a post-war population of about thirty thousand and ninety square miles of real estate. Though initially funded and constructed entirely by the federal government, within a decade, Oak Ridge would progress toward self-sufficiency, while the organization that had founded it still controlled the overwhelming portion of its economy as well as most of the land.

During the war, Oak Ridge expanded from an original projection of thirteen thousand residents to seventy-five thousand. The federal government had to rapidly build everything, not only the facilities where the secret work happened, but also housing for residents, utilities, and all infrastructure, including schools, hospitals, stores, cafeterias, etc. It was a massive and successful job. After the war, uncertainties about Oak Ridge's future only grew. Employment decreased, and less than half of the peak wartime population (around fifteen thousand people) remained as the Oak Ridge plants continued operating with redefined missions.

At the close of the Manhattan Project, AEC wanted to reduce the funds provided to the formerly secret city and get out of the community-management business. Facilities built to last only through the war—among them, hospital, schools, and roads—were starting to deteriorate by the late 1940s. Through the war years, the cost of provided water, sewer, and electricity was included in the rent, and these utilities were not metered for individual households. In 1950, the change to metered electricity at the standard TVA residential rate resulted in large reduction in the use of electrical power (residents, now paying their way, began to conserve energy). Water began to be metered in 1953, and water usage subsequently also decreased.

While it was planning to somehow sell the community's houses to their current occupants, AEC was not interested in selling land to large industries (which would have generated tax revenue) or making a commitment to support the future municipality, even though the agency paid no taxes on its developed property. While the federal government totally funded the plants (K-25, Y-12, and ORNL), residents were pressed to pay more for rent, utilities, and services. AEC still controlled and paid for fixes and upgrades to the aging physical infrastructure and supported trash pickup and the upkeep of streets and roads.[2]

One important element of Oak Ridge was the school system, ranked as the best compared to those in eleven southeastern states. During the Manhattan Project, the Army Corps of Engineers built (at a cost of $3.9 million)[3] and maintained the physical plant of the schools while the operational costs were funded by Anderson County—costs that were reimbursed under the Lanham Act by the Federal Works Agency. This agency later reported that the Oak Ridge school expenditures were the largest they made for education anywhere during the war.[4] After the war, the Oak Ridge schools continued to be part of the Anderson County school system, with expenses for the schools reimbursed by AEC. In 1947–48, AEC supported 14,720 school-age children on three reservations (1,155 at Los Alamos, 9,165 in Oak Ridge, and 4,400 at Hanford); the rest of the government's installations had about 45,000 students combined.[5]

AEC costs in operating the community decreased significantly from 1948 to 1951, partially due to elimination of substandard housing. New areas of the community received permanent homes, apartments, and schools. Older homes were upgraded to extend their useful life by twenty years.[6] The bus service subsidy was reduced as more cars were used and the service area reduced. Dormitory and housing rates were increased in line with neighboring communities. As a result of these measures, the total cost for the maintenance and operation of the community was lowered from $10 million in FY 1948 to $3 million in FY 1950, a reduction of 70 percent. Production efficiencies reduced these costs by another million and a half in FY 1951.[7] These funds for Oak Ridge came from the Congressionally appropriated AEC budget. In FY 1952, AEC received $1.6 billion in three appropriation bills.[8] This was up from $622 million received in FY 1949.

An initial vote on Oak Ridge incorporation failed in 1953 by a four-to-one margin, as residents remained uncertain about the future. Would AEC continue its financial commitments to the community, and how would it handle sales of government-owned homes to their current occupants?[9] In 1954, the Chamber of Commerce sponsored studies by UT political science professor Lee Greene on possible revenues that could develop following incorporation and the anticipated expenditure pattern based on other Tennessee cities.[10] Through the efforts of Senator Al Gore, Sr., AEC agreed to smooth the transition to self-government with additional payments for at least ten years.

In 1955, Knoxville industrialist Guilford Glazier built a new central business district later called the Downtown Shopping Center. J. C. Penney himself was there for his store's opening.[11] Also in 1955, the Community Disposition Act was signed into law providing for the sale of federally constructed Oak Ridge homes, with the first home transactions occurring the following year. According to local legend, in 1957 Oak Ridge went from having the fewest number of homeowners to having the highest percentage of single-family home ownership in the nation.

Active participation by Oak Ridge residents in these changes was viewed as essential by AEC, as the community was moving from being government owned and operated to a municipality with a continued strong government presence and influence. The community would soon establish a school board, build a $3-million state-of-the-art hospital to replace the wartime structure, and construct a new municipal building near the Downtown Shopping Center. These buildings were funded by the federal government.[12]

By the end of 1958, Oak Ridge's 3,526 single-family homes was sold to residents, and AEC soon announced a sizable financial-aid package to the city. On May 5, 1959, the citizens voted fourteen-to-one to incorporate (5,552 to 395). The first city council and school board were elected on June 2, and the AEC financial-assistance document was signed December 5, 1959. In 1960, the total cost of municipal, educational, and utility infrastructure given to the city exceeded $28 million, and the hospital, valued at $3.6 million, was handed over to the Methodist Church.[13] The original cost of acquisition, with no depreciation for these assets, was $84.5 million—worth almost $750 million in 2020.

Especially in view of the unique origin of Oak Ridge as initially a government-owned town with no taxable industrial base, it has been important in later decades for managing contractors of the federal facilities to aid the city through various forms of investment. UT would provide graduate educational opportunities for residents, the state of Tennessee would create a two-year college campus, and UT-Battelle LLC would make sizeable community investments.

The Oak Ridge Boys

Oak Ridge's reputation had some interesting side effects. The gospel country singing group known now as The Oak Ridge Boys began in Knoxville in 1943 as Wally Fowler and the Georgia Clodhoppers. Playing gigs throughout the area, they frequently appeared in Oak Ridge during the war years, entertaining on some Saturdays at the Grove Theater. One often-told story that accounts for the change in name contends that, after the war, the group was on its way from Knoxville to Nashville to perform at the Grand Ole Opry and passed by Oak Ridge. Wally suggested the quartet portion of the band, known as the Harmony Quartet, be renamed the Oak Ridge Quartet, to take advantage of the Oak Ridge name, at the time arguably the most famous city in southeastern America.

Wally insisted there was magic in the Oak Ridge moniker. The group became a member of the Grand Ole Opry in the late 1940s. Group members came and went, and the group officially became The Oak Ridge Boys in 1961. In 2007, The Oak Ridge Boys were the featured entertainers at Oak Ridge's annual Secret City Festival, had a road in the city named for them, and were given a tour of the Y-12 National Security Complex by Ray Smith, former Y-12 historian and an author of this book. In addition to other honors, The Oak Ridge Boys were inducted into the Country Music Hall of Fame in 2015.[14] Often in their concerts they will now refer to the origin of their name and fondly recall the time they were taken to see the "secrets" of Oak Ridge.

OAK RIDGE'S FIRST TECHNOLOGY TRANSFER

The first semi-industrial business came to Oak Ridge in 1951, when AEC concurred in the assignment of a lease for a defunct mortuary building in downtown Oak Ridge to Abbott Laboratories.[15] Abbott used this space as an office and laboratory facility to manage sales and distribution of isotopes created at the Graphite Reactor. AEC sold isotopes to Abbott, which processed and marketed them for medical and pharmaceutical research. This additional processing was necessary because AEC did not guarantee the safety of the original isotopes for human use, while Abbott did.

In response to increasing demand by the medical community for iodine and phosphorous radioisotopes, Abbott proposed that it establish a facility proximate to the Graphite Reactor, where the isotopes were made. In 1951, Abbott built a plant at ORNL near the reactor to develop a process for delivering accurate and reproducible quantities of radioactive iodine used to treat thyroid diseases. By 1954, Abbott was shipping more radioisotopes than AEC, soon making thirty thousand shipments per year to medical facilities around the world. In 1956, AEC estimated that more than five-hundred thousand people annually were receiving diagnoses or treatments with these products.

The operation of the Graphite Reactor ceased in 1963, after which industry fully assumed this profitable business for producing and shipping these popular radioisotopes.[16] This was the first technology transfer from the federal facilities in Oak Ridge, but many more would follow in later decades. Industry took on the production of the profitable isotopes, but ORNL continues to produce those that are not profitable. The High Flux Isotope Reactor (HFIR, opened in 1965) still makes radioisotopes that are deemed critical or are in short supply for research, industry, and medical use.

OAK RIDGE BUILDS A RELATIONSHIP WITH THE REGION

The Oak Ridge community continued to develop, as did its expectations for additional supporting services. Many regional scientists and engineers had to travel to Oak Ridge for their work and to attend meetings, and their colleagues visited Oak Ridge from around the world. The seventy-five minutes spent driving the thirty-five highway miles to and from the Knoxville airport were neither reasonable nor convenient. In August 1954, a shuttle helicopter service between Oak Ridge and the airport was initiated by the local AEC office for travelers on official business.

The flight engaged a loaned Air Force Sikorsky H-19 that covered the nineteen air miles in fifteen minutes. Operated by a two-man civilian crew for up to ten passengers, the shuttle service met both departing and arriving flights, and reservations could be scheduled. More than a thousand passengers flew to and from the Knoxville airport monthly, with future increases expected. The helicopters were based at McGhee-Tyson Airport with the Oak Ridge landing site situated near the AEC Administration Building. This service, which continued into the early 1960s,[17] marked Oak Ridge's initial foray into establishing an airport, something that would be revisited more than once in coming years.

In 1957, U.S. Representative Howard Baker, Sr. suggested that a two-year community college be established on a 370-acre site near the AEC Administration Building in Oak Ridge.[18] That evolved into interest in establishing a local branch of UT. The community college idea did not engender significant support, but the concept of a UT satellite campus somewhere in Oak Ridge gained enthusiasm. Backers of this idea toured unsold properties in the area and recommended that two recently constructed apartment complexes (with almost six hundred units) would be most suitable for university use as classroom buildings, even though the complexes were the highest-grade multi-family units in Oak Ridge and were 70-percent occupied.

When AEC stated its opposition to the transfer of these housing units to the university, the community appreciated the potential loss of tax base and the fact that the university already had a strong presence in Oak Ridge. In March 1958, UT president Cloide E. "Charlie" Brehm and the Oak Ridge Office of AEC agreed on a new potential site for a UT campus—a plot of land near AEC headquarters (as suggested by Representative Baker). However, this did not move forward, and no progress came on establishment of an Oak Ridge campus of higher education for several decades. In the 1960s, there was an attempt to start the College of Oak Ridge. Prominent Oak Ridgers pushed this concept, including Floyd Culler (later ORNL deputy director and then interim director) who chaired the board trying to establish it. Over one thousand citizens contributed a total of $107,300 and a president (Sumner Hayward) took office in January 1967, but he left suddenly in December and all momentum was lost and this idea died.[19] Success did not come until the Roane State Community College opened its permanent home on that site in 1999.[20]

Soon after Oak Ridge became a city, the newly elected city council met for the first time on June 17, 1959, and an item on the agenda was UT's interest in acquiring 2,260 acres of land in Oak Ridge for educational (forestry and agricultural) purposes.[21] The U.S. Department of Health, Education, and Welfare transferred the land with the approval of the General Services Administration. This soon became the UT Arboretum. Some of the deeded land, on top of Chestnut Ridge, became the subject of a potential swap with UT in the 1970s in support of Oak Ridge's next attempt to build a general aviation airport in the city.[22] This attempt failed, as would several future attempts.[23]

With discussion about Oak Ridge's incorporation, Anderson County and the state of Tennessee became worried about the financial implications, since it appeared direct federal support of city infrastructure and services would end. For several years, Tennessee had pressed AEC for its contractors to pay sales and use taxes to the state.[24] But contractors, such as Union Carbide

Corporation, maintained that they acted as agents of the U.S. government and thus did not pay local taxes. This debate continued until a 1955 U.S. Supreme Court decision that formalized the exemption of federal government contractors from paying local taxes. In 1957, it was estimated that Tennessee lost $2 million annually in revenues due to the current tax exemption afforded to government contractors. Union Carbide and several other AEC contractors agreed to settle the tax dispute in 1957 for $413,000.[25] This tax issue continued into future decades as states (such as Tennessee) sought annual payments from agencies (such as AEC) "in lieu of taxes" that would have been collected if non-government enterprises were sited there. In 1994, the Oak Ridge site (the city plus Anderson and Roane counties) received $392,272, and this grew to $3.4 million in 2017, including $1.7 million for the city of Oak Ridge (around 4 percent of the city budget) and the rest split between the two counties.[26]

The incorporation of the city of Oak Ridge had a big impact on Anderson County, which had to share with the new city beer taxes, traffic fines, and court costs assessed against Oak Ridge residents. In addition, the county had to share school taxes with the new city, even though Oak Ridge schools were funded by the federal government since the Manhattan Project. Anderson County had to almost double school taxes in order to send money to Oak Ridge and also keep county school budgets level. The state appreciated the impact of Oak Ridge's incorporation and added language to appropriation bills reducing state aid to school systems receiving federal assistance. Then, in 1959, the Tennessee General Assembly enacted the "Oak Ridge Rider" barring Oak Ridge schools from receiving any state and county funds: If the federal government was willing to pay for Oak Ridge schools, why should the state and Anderson County contribute funds?[27] Over the following several years, federal support of Oak Ridge schools ended, the collection of property taxes in Oak Ridge started to be collected, and the new city became eligible to receive state and county funds in a manner similar to other Tennessee cities. The transition from a federally supported city to one supported by state and local taxes was difficult.

DESEGREGATION AT UT AND OAK RIDGE: PARALLEL COURSES

Oak Ridge was formed and initially operated according to the rules of the region, which mandated racial segregation. UT had a long history of dealing with issues of segregation and trying to decide when and how to start admitting Black students. The pathways to desegregation in Oak Ridge and at UT are rather parallel, both taking crucial steps in the 1950s, driven in part by a presidential order relating to federal facilities and government contracts to non-federal entities. In effect, UT got a push on desegregation through its partnership with Oak Ridge.

Desegregation in Tennessee education had roots in the 1862 Morrill Act, which provided land grants to states that established agricultural and mechanical colleges.[28] The Tennessee legislature attached several additional conditions: free tuition must be provided to three students from each county, nominated by the state's senators and representatives, and the scholarships must be awarded without discrimination on the basis of race. Instead of establishing separate agricultural and mechanical colleges between races as other states had done, East Tennessee University (which became UT in 1869) offered Tennessee a cheaper alternative. It would accept all African American students who were nominated for the free scholarships and then transfer them to Black institutions—Fisk in Nashville and Knoxville College. At the time, this segregated approach, though clearly discriminatory by today's standards, was eminently satisfactory to UT, which received the proceeds of the fund created by the sale of some three-hundred thousand acres of public lands (the land grant). The African American colleges complained about the quality of the cadets[29] assigned to them and about the minimal sums they were paid by UT to educate the students.

Nevertheless, Blacks did become associated with UT under this arrangement beginning in 1881. Knoxville College became the exclusive agency for educating UT's African American students when the contract with Fisk was terminated in the mid-1880s. Under a new contract, Knoxville College was designated the "industrial department" of the university and received a lump sum annually to pay faculty assigned to the instruction of the Black cadets. There were complaints over the amount of funds furnished for the education, and many Blacks wanted a separate minority agricultural and mechanical college that could provide for their education. In response, the Tennessee Agricultural and Industrial Normal School for Negroes was established in Nashville in 1912 and received its share of Morrill Act funds, and, with this, UT's obligation to educate African Americans ended, not to be resumed until 1952.[30]

The federal land-grant funds came with a stipulation that the state had to match some of the funds. This state match did reliably come to UT but not to Tennessee Agricultural and Industrial Normal School for Negroes, which is now Tennessee State University.[31] A joint Tennessee legislative committee met on April 5, 2021, and said Tennessee State could receive between $150 million and $544 million of unappropriated state matching funds dating back to the 1950s.[32] Whether this bill will be paid remains to be seen.

The "separate but equal" stance on education in Tennessee was reinforced with a 1941 statute saying the establishment of educational facilities for Blacks equaled those for Whites, even if those facilities were not yet created. The absence of graduate programs at Nashville's Black agricultural and industrial school exacerbated the racial differences in educational opportunities.

In 1950, four Black students filed for admission to the UT law school and other graduate programs. The students were denied entry to the university and, in 1951, they filed and won a federal lawsuit. The victory upheld the students' right to admission under the equal-protection clause of the 14th Amendment. However, the presiding judge did not issue an order to the state to act on this ruling. The students probably sensed the mood of the university and filed an appeal to the U.S. Supreme Court, which, in March 1952, declined to take any action. By then, the university finally had changed its policies, granting the Black students admission to the law school and other graduate programs.

The first Black student to receive a graduate degree from UT was not a participant in the initial "group of four," but was Lillian Jenkins, who earned a master's degree in special education in August 1954. Two years later, R. B. J. Campbelle Jr. became the first African American to receive a UT law degree. At this time, Blacks were still not permitted to enroll as undergraduate students, in correspondence or extension courses, in any graduate program without the intent of completing the degree, or in any UT graduate program available at Black colleges elsewhere in Tennessee.

On May 17, 1954, in the landmark Brown v. Board of Education decision, the U.S. Supreme Court ruled that separate schools, whatever their quality, were "inherently unequal" and that separation placed a stigma of inferiority on Black students.[33] This ruling changed little at UT, as African Americans were still barred from enrolling in medical school programs in Memphis. UT's Board of Trustees delayed action for undergraduates and segments of the underserved Black educational community as long as possible—but such delays became less and less tenable. Just a week after the Brown v. Board ruling, the U.S. Supreme Court in a Florida decision ruled that, while desegregating the lower schools might require some period of adjustment and planning, no delay seemed necessary in desegregating higher educational facilities.

Then, on September 3, 1954, President Eisenhower issued Executive Order No. 10557, barring racial discrimination by private contractors receiving government funds.[34] A notice was sent to UT President Brehm on December 31, 1954, indicating that the executive order would apply to UT when it renewed its contracts with AEC.[35] This critical connection between UT and Oak Ridge put pressure on the university to begin the move toward desegregation. A subsequent Tennessee Supreme Court decision eventually stirred the trustees to action. In October 1956, the court ruled in the case of *Roy v. Brittain* that all state laws that codified segregation were invalid. In response, the university's position on segregation changed, if only slowly. Initially, African American graduate students could begin to take courses not leading to a degree and extension courses were opened to them, but there was still no relief for undergraduates hoping to enroll in the university.

In 1960, Knoxville native Theotis Robinson sought admission to UT as an undergraduate.[36] He was turned down but persisted in advancing his case through channels until he and his parents received an invitation to meet with UT President Andy Holt.[37] After expressing ignorance of the university's policies, Holt agreed to take the matter up with the trustees. At the meeting, Robinson indicated his preparation to sue the trustees and the university to gain admission. The state's attorney general indicated clearly that the *Roy v. Brittain* ruling meant Robinson could not be denied admission, and, on November 18, 1960, the trustees resolved, "That it is the policy of the board that there shall be no racial discrimination in the admission of qualified students to the University of Tennessee."[38]

THEOTIS ROBINSON. COURTESY OF BETSEY B. CREEKMORE SPECIAL COLLECTIONS AND UNIVERSITY ARCHIVES, UNIVERSITY OF TENNESSEE, KNOXVILLE.

On January 3, 1961, Robinson, joined by two other Black students, registered for classes, and his enrollment began the next day, without incident. Robinson received his undergraduate degree and began a long career of service to the city and the university. He served on the Knoxville City Council from 1970 to 1977, the first African American elected to that office in more than a half century. Robinson began his working career at UT in 1989 as a lecturer in political science, advanced through the system, and, in 2000, was appointed UT vice president for equity and diversity. He served in that position until his retirement in 2014 and received an honorary doctorate from UT in 2019. A Knoxville street underpass was named in his honor, as is a new UT residence hall.

When the Army Corps of Engineers located the Manhattan Project site at the Clinton Engineer Works, in the area that became Oak Ridge, it continued the practice of conforming to the laws and social customs of the state and local communities in which the facilities were located. Because, at the time, racial segregation was the norm in East Tennessee, it became the norm throughout the community that later became Oak Ridge. Housing was segregated, and African American men and women lived separately, even if they were married. Their children were not allowed to live in Oak Ridge. The quality of the housing for Blacks was poor, and Blacks were restricted to jobs at the low end of the pay scale. J. Ernest Wilkens, an African American scientist at the University of Chicago, worked on the Manhattan Project and was recruited to move to Oak Ridge. When told of the problems he would encounter if he moved to Tennessee in the early 1940s, he decided to remain in Chicago.[39]

In 1953, the Oak Ridge Town Council voted to desegregate the high school but had to rescind its vote after it faced public backlash. The chairman of the council (Waldo Cohn, a Manhattan Project scientist at ORNL) subsequently faced a public recall vote, which failed, garnering only 62 percent and not the two-thirds majority of those voting required for removal. Following the U.S. Supreme Court's *Brown v. Board* decision, AEC announced in January 1955 that Oak Ridge public schools would desegregate at the beginning of the following school year, September 1955.[40]

Oak Ridge elementary schools were located in the residential neighborhoods they served, according to the community school model established when the city was built by Skidmore, Owens, and Merrill in 1943. After the war, the overwhelming majority of the Black population lived in Gamble Valley, which had its own elementary school constructed in 1949. Despite some protests, on the first day of the 1955–56 school year, forty African Americans entered Oak Ridge High School and forty-five entered Robertsville Junior High School, as the two schools were integrated. Oak Ridge thus became the first school system in Tennessee and the South to integrate. The younger African American students would continue to attend the elementary school in their neighborhood, the Scarboro School. Meanwhile, the city's other junior high school and eight elementary schools remained segregated for twelve more years, until Scarboro Elementary School closed in 1967. This closure sent the 250 Black students to the eight other Oak Ridge elementary schools.[41 42 43]

Though Oak Ridge classrooms now featured an inter-racial blend, full integration did not happen swiftly, and other racial issues persisted. A 1957 Associated Press story reported that Lawrence Graham of Oak Ridge was the first Black student in Tennessee history to take to the basketball court in a high school tournament with White players.[44] He entered the district quarterfinal game against Clinton before halftime and played almost two quarters, scoring nine points, and Oak Ridge won the game. At the time, participation in interscholastic athletic activities was rare for African Americans, because the "separate but equal" provisions under which the state's education system operated prevented these activities. Several lawsuits were being adjudicated but no statewide ruling was established. For Black students to enter the athletic contests, both schools had to agree in advance to field integrated teams, or neither of them could. In Graham's case, counterparts at other schools told his Oak Ridge coaches to leave him at home if they wanted a game. Similarly, Black students could not take field trips or participate in other activities that involved a segregated facility.

Desegregation efforts continued in Oak Ridge to eliminate restrictions barring African Americans from entering stores and theaters or receiving service at cafeterias and grooming shops. While efforts to desegregate Oak Ridge schools had largely succeeded by the end of the 1950s, discriminatory practices persisted in the community's private sector, particularly in the service industry. Indeed, color-line restrictions continued into the next decade, barring African Americans from entering some privately owned businesses.

Consider, for instance, that it was not until 1966 that Ken's Barber Shop was established as the first barber shop in Oak Ridge to provide service to all customers, regardless of race,[45] and several years earlier, the popular Davis Brothers Cafeteria, located in a downtown shopping mall, became a focal point in the city's efforts to dispel the community's lingering "Jim Crow" restrictions.

At Long Last, Oak Ridge Fully Confronts a Shameful Legacy

The integration of Davis Brothers Cafeteria in Oak Ridge involved city government, a business leader, and the Reverend James Spicer, a respected pastor of the Chapel on the Hill in Oak Ridge. According to a presentation on the history of integration in Oak Ridge by Ray Smith,[46] the story began when Oak Ridge mayor Bob McNees called Spicer in 1963 to ask him to serve on a new Human Relations Committee. McNees was a chemist who joined ORNL in 1952, and, in 1963, he became a special assistant at the laboratory for programming and budgeting, in addition to being the mayor of Oak Ridge. McNees was spurred to action by pressure from the Kennedy Administration to address ongoing segregation in this town that was so heavily federal in scope. This committee was filled with city leaders (White and Black), and, in fact, McNees appointed Spicer to chair this committee that was expected to act quickly and produce solutions on how to end segregation in Oak Ridge.

In the first meeting, the committee addressed the question, "What is the most difficult issue for African Americans in our town," and the immediate answer by the Black committee members was restaurants. The dining establishments in Oak Ridge hired Blacks for cooking, cleaning, and maintenance but would not allow them to eat there as customers. The principal antagonist was Davis Brothers Cafeteria, which featured a row of African American men who stood at the end of the tray line and carried food to customers' tables for a prescribed tip of a quarter—no more, no less.

Spicer visited the local manager of Davis Brothers Cafeteria and explained calmly and professionally that the exclusion of African Americans as customers and the degrading use of men as "boys" to carry the trays could not continue. However, the manager was unprepared for this issue and explained that he had been instructed to say: "Davis Brothers would not change anything in the way they were doing business." They had multiple cafeterias across the South and they saw no need for a change.

Seeing no hope for local change, Reverend Spicer enlisted Mayor McNees to go with him to the corporate offices for Davis Brothers in Atlanta. The meeting did not go well and ended with both Davis brothers strongly stating that they would never integrate their restaurants.

On the airplane ride back to Knoxville, serendipity intervened. Spicer was upset and told the story of the disastrous meeting to a young man seated next to him. The fellow passenger asked a few questions, but nothing seemed to result from the exchange as the two departed upon landing in Knoxville. But, two days later, prominent Oak Ridge attorney Gene Joyce called Spicer to a meeting in his office. Joyce explained to Spicer that he represented the Glazer Corporation and had information that would be helpful in the work of the Human Relations Committee.

It turned out that the young man who sat next to Spicer on the airplane was Emerson Glazer, Guilford Glazer's son. The Glazer Corporation built (and owned) the Downtown Shopping Center in Oak Ridge, which opened in 1955 and included the Davis Brothers Cafeteria. The Davis brothers, in addition to leasing their restaurant space, also leased a bowling alley owned by the Glazer Corporation in Oak Ridge. They made money with their restaurant but lost money on the bowling alley. Their lease for the bowling alley was $8,000 per year and had six more years to run. The Glazer Corporation proposed to let the Davis brothers out of the lease on the bowling alley space if they would integrate their restaurant. The Davis brothers immediately accepted and integrated the cafeteria. In addition, one of the Davis brothers was president of the Southern Restaurant Association and announced that all Davis Brothers Cafeterias would be integrated—a remarkable step forward for restaurants in the South.[47]

At the time of the 1954 *Brown v. Board of Education* decision, the atmosphere in neighboring Clinton was much less congenial toward students of color. Following a January 1956 federal court order to initiate desegregation beginning the following school year, Clinton High School became the first integrated state-supported school in the South (Oak Ridge High was federally supported and desegregated a year earlier). The National Guard had to be called in to restore order a few days after school started. School resumed, but periodic episodes of violence continued until October 5, 1958, when a series of explosions tore through Clinton High School, completely destroying it. AEC and the Oak Ridge School Board offered Clinton the use of then-vacant Linden Elementary School until a replacement building was available. The Oak Ridge High School band played the Clinton High School alma mater to welcome the Clinton students, both Black and White, as they came to Oak Ridge the next school day.[48] One of the Clinton Twelve, as these first Black students at Clinton High came to be known, was Bobby Cain, who later graduated from Tennessee State University and worked at ORNL before being drafted by the Army.[49]

The processes and pathways for desegregation at UT and in Oak Ridge differed, and the forces for change varied widely. But, the university and the city arrived at the same point of integration at about the same time, and, of course, both were better off because of these long-overdue changes.

GOVERNMENT-SPONSORED RESEARCH UNDERGOES SIGNIFICANT CHANGES

The support of research and development by the U.S. federal government evolved in new directions in this decade in a number of areas. Many of these changes in R&D funding were driven not by a world war (which led to the huge investment in the Manhattan Project), but rather by the Cold War between the United States and its allies (via NATO) and the Soviet Union and its allies (the Warsaw Pact). No direct armed conflict occurred between these adversaries in the decades-long Cold War, although there were more localized wars supported by these two camps, e.g., the Korean War. The Cold War led to a huge expansion in the U.S. and USSR nuclear arsenals, now focused on hydrogen bombs working on the fusion of isotopes of hydrogen and triggered by "little" uranium or plutonium fission bombs. This led to expansion of funding for these sophisticated nuclear weapons, huge in numbers on both sides. The Cold War led to increased expenditures for espionage, propaganda, sports competitions, and embargoes, and evolved to include a Space Race, sparked by the launch of Sputnik on October 4, 1957.

The Cold War greatly impacted government-funded R&D programs in many fields of science and engineering. One new area was the development of computing, a field in its infancy at the beginning of the 1950s. AEC-funded R&D on computers was pioneered by John Von Neumann at Princeton's Institute for Advanced Study. Computing on a large scale at national laboratories began at Argonne with creation of the Argonne Version of the Institute's Digital Automatic Computer (AVIDAC).[50] AVIDAC could multiply two twelve-digit numbers in 0.001 seconds and add them in 0.000001 seconds. The Oak Ridge Automated Computer Logical Engine (ORACLE) was designed and constructed by Argonne and ORNL engineers and became operational at ORNL in the summer of 1954.[51] ORACLE cost $350,000 and could multiply two twelve-digit numbers in 0.0005 seconds and add them in 0.00000005 seconds. ORACLE also incorporated a remote-controlled magnetic-tape memory system to give ORNL the largest computer memory to date. Initially, ORACLE was used to resolve aircraft design issues and address other research challenges, but, in 1957, the device was adapted in part for budgeting and financial accounting activities and was used until it became obsolete in 1962. This was the start of a "high-speed" computing program at ORNL, which would require fifty years before it would assume a world-leading role.

In the late 1940s, considerable focus revolved around learning how to build nuclear reactors that could serve as sources of electricity for the public and as power plants for Navy submarines. The assumption in Oak Ridge was that Clinton Laboratories (becoming Oak Ridge National Laboratory in March 1948) would be the Atomic Energy Commission site to lead this reactor research and development. However, that changed suddenly with the decision in December 1947 to move the reactor program out of Oak Ridge to Argonne National Laboratory, located just outside Chicago.[52]

In 1948, there was a desire to build a reactor with a high flux of produced neutrons so that researchers could study the properties of materials in the larger reactors to be built for the Navy and for the electrical power industry. Both ORNL and Argonne vied to build this new high-flux reactor at their locations, but Argonne won the AEC decision. However, at the time, Argonne had no room for this reactor facility, which was slated to be built on Argonne's new campus outside Chicago. The problem was that the new campus was still under construction and would not be completed for at least a year. In addition, Argonne's budget would have to increase considerably to accommodate the new research areas required for this reactor program to succeed. Needed for a robust high-flux reactor program was research on (a) radiation effects on properties of materials used for the construction and operation of the reactor, (b) handling of the samples and experimental equipment, and (c) radiation health and safety from the perspective of both beneficial effects and dangers. Argonne's board of governors did not welcome this anticipated programmatic disruption—its members wanted to continue to focus on basic research, not classified reactor work.

Faced with these issues, leaders of ORNL and Argonne, Alvin Weinberg and Walter Zinn, agreed to collaborate in designing the high-flux reactor (later called the Materials Test Reactor [MTR]) that would meet AEC requirements. But building such a reactor near a populated area like Chicago or Oak Ridge would entail potential risk to human health, and, in fact, the MTR was to be substantially more powerful than any reactor proposed for an AEC site during the decade.

Soon, the collaborating national laboratories arrived at a tenable solution: the MTR would be built in sparsely populated Idaho at AEC's National Reactor Testing Station, with Oak Ridge designing the reactor and Argonne responsible for managing the physical plant that surrounded it. The MTR's principal task was to advance reactor development by testing components such as fuel assemblies and containment materials exposed to high levels of radiation. It started operating in Idaho (as part of Argonne West) in 1952 and evolved in the decade to be a 30-MW reactor with neutron fluxes ten to one hundred times greater than those in other reactors at that time.

The other main driver of reactor development in the late 1940s was designing a power plant for Navy submarines, a project inspired by Captain Hyman Rickover (promoted to admiral in 1973). An important issue for the naval and high-flux reactors was finding a new and better cladding (enclosure)[53] for the enriched uranium fuel rods. Aluminum was used in the Graphite Reactor because it was a relatively low-neutron-flux and low-temperature device, compared to these new reactors being designed. The element zirconium had the correct thermal and structural properties to serve as the new nuclear fuel cladding, but seemed to absorb too many neutrons, which would decrease the chain reaction in the core of the reactor.

However, in 1947, ORNL's Herbert Pomerance discovered during experiments at the Graphite Reactor that the element hafnium has a large neutron capture capability, and it was a small impurity of hafnium in the zirconium, not the zirconium itself, that absorbed the neutrons. Hafnium is chemically similar to zirconium and separating the two elements is difficult. Because of this, most commercial hafnium is a by-product of zirconium refining. The mineral zircon ($ZrSiO_4$) is the primary source of all hafnium, containing it at a ratio of about 50 to 1 (zirconium to hafnium).

Many laboratories funded by the Navy tried to find ways to perform this chemical separation. Finally, early in 1949, Y-12 came to the rescue and, within three months, had produced twenty-five thousand pounds of hafnium-free zirconium. The leader of this project, Y-12 chemist John Googin, received a UT doctorate in physical chemistry in 1953.[54] Y-12 also assisted reactor development by machining a beryllium reflector (starting in 1950) for future use in the MTR. The beryllium reflector is situated around the reactor's core and establishes a region of enhanced neutron flux important for testing the properties of materials.

Although Oak Ridge lost the lead role in siting the MTR, it was given an expanded role in isotope production and distribution when AEC moved Argonne's radioisotope program to ORNL. Likewise, Y-12 transferred to ORNL its isotope production divisions, which were involved in separation of stable isotopes—an undertaking that would eventually include isotopes of most elements on the periodic table. These moves centralized all radioisotope distribution at ORNL, which continued until extensive privatization occurred forty years later. As isotope separation came off its agenda, Y-12 focused on weapons-production activities, including lithium enrichment[55] and the manufacture of components of highly enriched uranium. ORNL still manages the National Isotope Development Center for DOE, supporting the distribution of radioactive and enriched stable isotopes produced across the DOE complex. ORNL still makes radioisotopes and ships them to customers. A recent addition to this capability is the U.S. Stable Isotope Production and Research Center at ORNL, announced in 2022.[56]

Sadly, on June 16, 1958, Oak Ridge's only nuclear criticality accident resulted in severe radiation exposure to eight Y-12 operators.[57] The accident occurred in an area designated for the salvage of enriched uranium. Criticality developed when a solution containing highly enriched uranium was allowed to drain into a 55-gallon drum, which was intended to contain only water. Following the original criticality burst, which did not destroy the drum, criticality seemed to oscillate for twenty minutes. The tank was finally stabilized with additional water flowing into it, stopping the chain reaction of enriched uranium. All eight workers were hospitalized at the medical facilities of the Oak Ridge Institute of Nuclear Studies. The three less seriously exposed workers were released on June 26, and the other five, who had sustained more-serious injuries, were released on June 30. All recovered and were able to return to work.[58]

ENTER THE DECADE OF REACTORS AND THE QUEST FOR AN ATOMIC AIRPLANE

At the end of 1947, the AEC announced its plan to move the Oak Ridge reactor development program to Argonne, which was perceived as a disaster for ORNL at that time. But Oak Ridge recovered from that setback, and the 1950s turned into the decade of reactor development at ORNL. The leader throughout this decade was Alvin Weinberg, who drove and guided this golden era of nuclear reactor development in Oak Ridge.

Alvin Weinberg—First Half of a Great Career

Alvin Weinberg (1915–2006) grew up in Chicago, the son of Jewish immigrant parents from Russia. He went to a public high school and excelled in math, Latin, and science, even though his physics teacher never had taken a physics course in preparation for teaching the subject. Although a good student, Weinberg's main desire then was to get picked for the high school basketball team, in part because that would make him more attractive to the girls. But he did not make the team.[59]

Following high school, Weinberg stayed at home and went to the University of Chicago as an undergraduate and then as a graduate student in physics. He produced a master's thesis in quantum mechanics, working with Carl Eckart (famous in physics circles for the Wigner-Eckart theorem), and then a PhD dissertation in the new field of mathematical biophysics. He passed his final doctoral examination in the winter of 1939, not long after the time when Otto Hahn and Fritz Strassmann announced the amazing results of their experiments in Berlin, later interpreted by Lise Meitner and Otto Frisch to be the result of uranium fission.

Weinberg was recruited by Eckart to work at the Met Lab at the University of Chicago in September 1941, who needed someone to study the diffusion of neutrons through material. This original Manhattan Project laboratory was established to figure out how to design and build a self-sustaining nuclear reactor based on uranium fission and capable of producing ^{239}Pu as fuel for an atomic bomb.

Nobel-laureate Arthur Compton was director of the Met Lab, and he brought to Chicago the best people in the country working in this totally uncharted field of science. Enrico Fermi (also a Nobel laureate) and Leo Szilard came from Columbia, and Eugene Wigner (future Nobel laureate) and John Wheeler came from Princeton. Weinberg first met Wigner when the latter arrived at the Met Lab in February 1942, and he was assigned to Wigner's group. Kay Way (her contributions are noted in the late 1940s chapter), who left the UT physics faculty to join the Met Lab team, also worked in Wigner's group.

Weinberg's initial task, as assigned by Wigner, was to calculate the diffusion of neutrons from the fission of uranium through the material in the pile (later called a reactor) that they were trying to build and control. Neutrons would be moderated (slowed down) by the graphite in the pile, migrate to cause other uranium nuclei to fission, and/or get absorbed and lost in these

ALVIN WEINBERG. COURTESY OF OAK RIDGE NATIONAL LABORATORY.

materials or others constituting the pile. The dynamics of the neutron migration were complicated but extremely important in helping Fermi decide how big the initial pile needed to be.

Under the oppressive time pressure imposed by World War II and the race against an assumed parallel project in Nazi Germany, the Met Lab's efforts succeeded in spite of many stops and starts and wrong turns. The big day was December 2, 1942. Fermi's group was methodically adding uranium and graphite (the latter a moderator used to enhance the chances of a chain reaction) in a spherical arrangement and measuring the neutron flux after each addition. They were confident—and justifiably so—that the addition of one more layer of material on the afternoon of December 2 would achieve a chain reaction for the first time—marking the dawn of the nuclear age.

This pile was squeezed into a squash court under the stands of the University of Chicago football field—the university gave up football the year before. However, space was tight, and Compton and Fermi decided that they could only invite fifty people to attend history in the making. Weinberg was number fifty-four on the list, and so he could not observe the start of the age of nuclear reactors that day, something that always chagrined him.

After that historical day, much of the action moved to the new laboratory in Oak Ridge, where a reactor equipped with cooling and shielding would be built, using knowledge gained from Fermi's pile in Chicago to construct a full-fledged continuously operating reactor. Of course, building the first reactor was not an easy task, since this was all new ground. Weinberg would be squarely in the middle of reactor development thereafter.

In 1945, Wigner accepted a position as director of research at Clinton Laboratories (later named Oak Ridge National Laboratory), and he brought along protégés Weinberg, Gale Young, and Kay Way. Weinberg became head of the Physics Division in 1947. After Wigner left Oak Ridge to return to Princeton in the summer of 1947, Weinberg accepted the position of associate director in charge of research and development. Weinberg became director of ORNL in 1955 and served in that role until 1972.

From the late 1940s through the 1950s, ORNL was a laboratory focused to a large extent on the development of nuclear reactors of different kinds, and Alvin Weinberg was the expert and the leader in these promising technical advances. In 1958, Weinberg and Wigner wrote the classic nuclear reactor textbook, *The Physical Theory of Neutron Chain Reactors.*

Weinberg's innovative work related to reactors was recognized by thirty-seven patents filed from 1944 to 1956. Many of the patents took a decade or longer between filing and issuance, probably because of classification issues following World War II. One of his important patents was #2969311, Means for

Producing Plutonium Chain Reactions, filed on October 8, 1946, and awarded on January 24, 1961. His leadership, his technical expertise, and his innovations led to many calling Alvin Weinberg "the greatest Oak Ridger."

Nine reactors achieved criticality at Oak Ridge during the 1950s, and two other reactors had substantial ORNL involvement.[60] [61] At the time, the Army wanted compact reactors that could easily be transported to remote locations to produce heat and electricity and avoid the logistical challenge of delivering thousands of barrels of more conventional fuels, like diesel, to relatively isolated zones. A study group at ORNL responded by developing the conceptual design for a 10-MW Army Package Power Reactor to be built at Fort Belvoir, Virginia. ORNL developed the fuel elements and control rods and provided other support for the reactor, which went critical in 1957. Eight other compact reactors were built, including one installed at a research station in Antarctica, which provided heat and electricity from 1962 to 1972. The *N.S. Savannah*—the world's first nuclear-powered merchant ship, equipped with a 69-MW reactor—was launched in 1959 and traveled three-hundred thousand miles around the world before it required refueling. The deactivation of the *Savannah* occurred in 1971.

The Low Intensity Test Reactor (LITR) began in 1950 as a full-scale mock-up of major components to be included in the Materials Test Reactor built in Idaho in 1952.[62] Once the mock-up was running, ORNL got AEC's permission to add fuel and reflector elements for a longer-term operation, to test reactor controls and instruments as well as conduct experimental nuclear measurements. With additional modifications, it operated at 3 MW, allowing high-flux neutron irradiation of samples. This process was (and remains) important for obtaining knowledge of how materials in the pumps, piping, and support structure of a reactor will behave under radiation conditions never before encountered. AEC closed LITR in 1968.

ORNL had a lead role in the 1950s in designing shielding for reactors in the mobile environment of a nuclear submarine or airplane. To aid these studies, the Bulk Shielding Reactor (BSR) was designed and built.[63] The reactor was suspended from one bridge into a pool of water (for cooling and moderation of reactor neutrons), while bulky equipment could be suspended from another bridge and moved near the core to measure radiation shielding. This "swimming-pool" reactor became a popular model for low-power usage at universities and research centers. BSR went critical in 1950, using fuel plates designed for the MTR. This reactor enabled studies of reactor component design, including neutron reflectors, moderators, and shielding. It shut down in 1987.

The blue glow in a swimming pool reactor is a result of Cherenkov radiation. This kind of radiation is analogous to a sonic boom from a jet aircraft traveling beyond the speed of sound. The sound waves generated by the jet travel at a speed slower than the supersonic aircraft and thus cannot propagate forward from the plane, instead forming a shock front and a loud boom. In a similar way, a charged particle can generate a light shock wave as it travels through an insulator. In the case of a swimming-pool reactor, the radioactive decay of fission products in the fuel rods generates high-energy electrons that

THE BULK SHIELDING REACTOR IN ITS "SWIMMING POOL." COURTESY OF OAK RIDGE NATIONAL LABORATORY.

FROM LEFT: SAM SHAPIRO, JACKIE KENNEDY, SEN. JOHN KENNEDY, ALVIN WEINBERG, SEN. ALBERT GORE SR. COURTESY OF OAK RIDGE NATIONAL LABORATORY.

travel through the water at a velocity greater than the velocity of light in that medium. Of course, the speed of light in a vacuum is a universal constant that cannot be exceeded. But the speed of light in water is only 75 percent of that in vacuum. Therefore, an energetic electron can exceed this speed in water, forming a shock wave that appears as a blue glow.

LITR, with its glow of Cherenkov radiation, appeared on the cover of *Scientific American* in October 1951, and the colorful radiation became an intriguing and mysterious phenomenon for visitors to the Oak Ridge facilities.[64] Senator John Kennedy and his wife, Jackie, on a February 1959 visit to ORNL, enjoyed gazing upon the glow and talked to Weinberg in the reactor control room.

Already in the late 1940s, there was interest in trying to build a liquid fuel reactor, and this became a major program at ORNL in the 1950s. An advantage of such a homogeneous reactor would be that the liquid fuel could circulate continuously from the reactor core to a processing loop where unwanted fission products could be removed, while also circulating through a heat exchanger to allow power generation. In contrast, the solid nuclear fuel rods in a heterogeneous reactor (solid fuel and liquid coolant) need to be removed in a reactor shutdown and chemically reprocessed to remove, for example, unused fissionable fuel.[65]

In the 1950s, ORNL explored creation of reactors in which the highly enriched uranium fuel was mixed with water to serve as both moderator and coolant. The result was the Homogeneous Reactor Experiments (HRE-1 and HRE-2).[66] The initial smaller reactor version (HRE-1), activated on February 24, 1953, proved that a circulating fuel reactor would operate smoothly and at a significant temperature and pressure. In fact, the reactor generated about 150 kilowatts (kW) of useful electricity. After shutdown, it was dismantled and its successor, HRE-2, was constructed in the same building and became operational in 1956. Although it ran continuously for more than twenty-one months from its criticality in 1959 to its shutdown in 1961, HRE-2 experienced several corrosion problems that affected the viability of the homogeneous reactor concept in the eyes of sponsors. ORNL director Alvin Weinberg still maintained that a homogeneous reactor could be developed for the large-scale generation of electrical power, but AEC officials clearly preferred the conventional heterogeneous reactors that became the workhorse of the nuclear power industry.

After the end of World War II, military leaders (e.g., Air Force General Curtis LeMay) were interested in developing a nuclear-powered bomber. The idea was an airplane that could fly at least twelve thousand miles at 450 miles per hour without refueling, so that it could deliver conventional or nuclear bombs anywhere on Earth. The huge technical challenge would be to design a reactor small enough to be carried on board but powerful enough to enable take-off of a heavy plane laden with nuclear bombs. Thus, in 1946, the Aircraft Nuclear Propulsion (ANP) program was born and was housed initially in an unused building at the Oak Ridge K-25 site, with Clinton Laboratories scientists providing assistance to technical staff from the Fairchild Engine and Airplane Corporation.[67] Despite skepticism, an MIT study determined that it would be expensive but feasible to develop a nuclear-powered airplane. In 1949, AEC directed ORNL to establish the ANP program, which became ORNL's largest activity throughout the decade of the 1950s. General Electric replaced Fairchild and moved part of the reactor development to its facility in Ohio, where it hoped to build a direct-cycle reactor in which air flowing through the reactor would be heated and then sent directly to the aircraft's jet engines.

Work at ORNL (in a building located at Y-12) focused on developing an indirect-cycle reactor, in which flowing helium gas would remove heat from the reactor core, which subsequently would go through a heat exchange system that would heat forced air to turn the jet engines. The Aircraft Reactor Experiment (ARE) started with blocks of beryllium oxide as moderator in a pile fueled by stainless steel tubes of uranium dioxide. However, these long thin solid fuel rods were replaced by enriched uranium fuel dissolved in a salt that would continuously flow through the reactor core, thereby avoiding structural problems of solid fuel elements at high temperature. The 2.5-MW prototype reactor ran successfully for twelve days in November 1954. The Air Force was pleased with the ARE performance and asked Pratt & Whitney to develop the full-sized power plant.

The other main effort of the ORNL work on the ANP program focused on determining the best way to protect aircraft personnel from radiation emanating from the reactor employing shielding much lighter than normally used around a stationary reactor. The Tower Shielding Reactors (TSR-1 and TSR-2)

A REACTOR AND SHIELD SUSPENDED AT THE TSF.
COURTESY OF OAK RIDGE NATIONAL LABORATORY.

were built in response to this challenge. These two reactors were suspended between the two looming structures of the 315-foot-tall Tower Shielding Facility, still visible today from the nearby Melton Hill Dam.[68] Experiments showed that weight could be reduced by having two shields, one around the on-board reactor and another around the crew cabin, compared with one robust shield around the reactor.[69]

Times—and priorities—changed as the development of long-range missiles obviated the need for a nuclear-powered bomber. Meanwhile, the nuclear-powered aircraft would be vulnerable to short-range missiles, and the prospect of a downed airplane equipped with an atomic reactor presented a troubling scenario. After a twelve-year run, the Aircraft Nuclear Propulsion program was canceled in 1961. The experience gained on use of molten salts (laden with uranium fuel) as coolant started the evolution of a large molten-salt reactor that might someday serve the nuclear-power industry.[70]

The Oak Ridge Research Reactor (ORR) was Oak Ridge's answer to Argonne's Materials Test Reactor.[71] Located next to the Bulk Shielding Reactor, Graphite Reactor, and the Low Intensity Test Reactor, ORR and the other three reactors shared a common control room. The high-flux reactor facility (a swimming-pool reactor) began operating in 1958 and provided for examination of cooling loops in which tests of various fuel elements could occur, as well as access to extremely high-radiation fields close to the fuel elements. Openings in the reactor walls (ports) allowed for neutron scattering research and irradiation of samples including isotopes, for short or long durations. ORR was a reactor widely used by many ORNL researchers for a variety of science and engineering programs, as it was the best research reactor in the world for a time in the late 1950s and early 1960s. It began operation at 15 MW, and the power was later increased to 30 MW with improved cooling capacity. ORR was still operating in 1966 when one of this book's authors, Lee Riedinger, used the reactor to irradiate samples via a pneumatic line to the core. Riedinger, a nuclear physics graduate student at Vanderbilt at the time, had to learn how to conduct radiochemistry on the highly radioactive samples for studies as part of his dissertation. Fortunately, this was done safely, and Riedinger later welcomed two healthy daughters. ORR was shut down in 1987.

EXPANSION OF PARTNERING PROGRAMS BETWEEN UT AND OAK RIDGE

In the 1950s, the UT Department of Physics flourished. Under the leadership of Kenneth Hertel, the department head since 1930, and from mid-decade under Alvin Nielsen, the department underwent a remarkable post-war upsurge in undergraduate and graduate enrollment with the return of thousands of veterans financed by the GI Bill. Faculty absented by the war returned, and additional faculty were hired to meet the increased need for instructors, particularly in graduate programs. The presence of the Oak Ridge facilities added outstanding research opportunities in fields not supported by the university's existing research programs. Among the latter were molecular spectroscopy, fiber research, nuclear theory, rocket technology, and trace element study in normal human tissue. Oak Ridge also provided UT with a ready-made pool of bright students whose graduate education was interrupted by the war and potential faculty eager to teach and interact with the university's new enrollees.

The physics PhD curriculum, approved in 1946, was in full swing by 1952, with about forty advanced students, of whom approximately half were Oak Ridge employees and the remainder full-time students based at UT.[72] Course instruction was completely integrated, with offerings at UT's Resident Graduate Program in Oak Ridge, administered by the Oak Ridge Institute of Nuclear Studies, and on campus for both student groups. Even though the number of graduate students increased, there were still not enough students or faculty to warrant two sets of course offerings, one in Knoxville and the other in Oak Ridge. As a result, many of the required advanced courses were offered in alternate years at each location.

The university provided transportation to Oak Ridge in UT automobiles, driven by students and equipped with speed governors. This course arrangement ensured extensive mixing and mingling among the UT and ORNL graduate students. To fulfill all class requirements, PhD students needed to attend about half of their classes at each location, and, thus, half of the students in a class at either location were likely to be from the other institution. William Bugg, an Oak Ridge native and one of this book's authors, was discharged from the Army in 1954 and soon entered the UT graduate physics program but found, after moving to Knoxville, that he had to drive back to Oak Ridge to take part of his course work.

The UT chemistry department, under Calvin Buehler, was also heavily involved in meeting demands by Oak Ridge for courses in chemistry, as more than four hundred chemists were employed at the federal facilities. In 1944, the department's PhD program, motivated in large part by this demand, became the first to be approved at UT and conferred its first two degrees in 1947.[73] The program was highly successful, and eighty-two graduate degrees were awarded by 1959. Three chemistry professors, Hilton Smith, W. T.

Smith, and George Schweitzer, with strong ties to Oak Ridge, supervised nearly half of these graduating students through their coursework. As of 2020, Schweitzer was still on the UT faculty. After earning his PhD from the University of Illinois in 1948, he was hired to teach graduate inorganic chemistry and radiochemistry and to supervise UT's chemistry program at ORNL. By 2020, Schweitzer, then ninety-five, had spent seventy-one years on the UT faculty and was still guiding graduate students in their research for advanced degrees. The estimate is that he has taught chemistry to more than forty-five thousand students during his career. And, in his spare time, he earned a second PhD (in philosophy) from New York University in 1964.

The chemical engineering doctoral program was approved in 1952, and the first PhDs were awarded in that same year. One of the topics included in the curriculum was nuclear processes, which became the seed of the Department of Nuclear Engineering, begun in 1957 and one of the first such academic departments in the country. The critical connection between UT and ORNL in reactor R&D led to the formation of this department.

In 1950, the physics PhD program granted its first four degrees. Alvin Nielsen served as the mentor of three of these, including Ray Murray, in molecular spectroscopy. The other dissertation was in nuclear theory directed by Richard Present, who joined the faculty in 1945 directly from the Manhattan Project in Oak Ridge. Murray had his Berkeley graduate studies interrupted by the Manhattan Project work at Y-12. He continued working at Y-12 as an ORNL employee and in 1948 resumed his graduate work through the UT Resident Graduate Program. After finishing his UT doctoral work, he took a faculty position at North Carolina State University and helped build the nuclear engineering effort there, resulting in creation of a separate department in 1962. He had a long career as a leader in the field of nuclear engineering.

The next year, 1951, saw two ORNL employees (both from the Physics Division) receive PhD degrees in physics, along with two more from UT. The following few years began a surge of ORNL dissertations, particularly in the Physics and Health Physics divisions. In 1953–54, UT conferred twelve PhDs, ten to students from Oak Ridge and two from the university. By 1960, a total of thirty-seven physics PhDs were granted, seventeen to students who were located at Oak Ridge and often mentored by laboratory personnel but academically directed by full-time UT faculty.

In 1960, M. E. Rose, a distinguished theoretician at ORNL, became the first Oak Ridge staff member to formally direct the doctoral dissertation of a UT student, C. P. Bhalla, who later became a professor at Kansas State

University. Rose was to be followed by numerous other ORNL staff members who would go on to direct UT PhD students. By 2019, the physics department had produced about six hundred PhDs, with more than fifty of them employed at Oak Ridge after graduation. At one point, this preponderance of UT graduates on the laboratory's payroll became a matter of some concern to the ORNL administration. In fact, Herman Postma, laboratory director from 1974 to 1988, led an effort to prioritize acquisition of new hires from institutions other than UT. However, the proximity of the university to Oak Ridge and the integration of UT faculty and graduate students into research taking place at the national laboratory have sustained the Oak Ridge institutions' interest in hiring UT graduates.

EARLY STANDOUTS WITH UT CONNECTIONS IN OAK RIDGE

Scores of talented people contributed to building the partnership between UT and Oak Ridge. At UT, leaders in the 1940s included faculty William Pollard, Kay Way, and Kenneth Hertel in the Department of Physics and George Schweitzer in the Department of Chemistry. In Oak Ridge, Clinton Laboratories leaders Martin Whitaker and Warren Johnson were crucial. In the decade of the 1950s, high-level leadership at UT (President C. E. Brehm) and ORNL (Director Alvin Weinberg) effectively promoted the partnership. In addition, the 1950s saw the emergence of researchers at UT and in Oak Ridge contributing to the development of joint programs of research and education. For example, the roles of six such people important for advancing the partnership are highlighted below: Ed Von Halle, Sheldon Datz, Robert Birkhoff, Rufus Ritchie, Sam Hurst, and Isabelle Tipton.[74]

Ed Von Halle was a Brooklyn native who later served in World War II and then completed his undergraduate degree in chemical engineering at Carnegie Tech. He went on to earn a master's degree in the same subject from Bucknell University before the Oak Ridge K-25 Plant hired him in 1950. Over his long career in Oak Ridge, Von Halle became an internationally renowned expert in the theory of isotope separation. After joining K-25, Von Halle did his PhD work at UT in chemical engineering. Physics professor Richard Present maintained that Von Halle was the brightest student that he ever had in his graduate course on kinetic theory. Other professors felt the same way about Von Halle.

It was customary in those days to post grades on a central bulletin board, and chemistry professor W. T. Smith posted a famous set of grades for his graduate physical chemistry class; it merely said, "Von Halle A, all others F."[75] Von Halle's UT dissertation was on the subject of separation of species by thermal diffusion, and his later contributions covered isotope separation by gaseous diffusion, gas centrifuge, and laser methods. Von Halle spent most of his career in the Operations Analysis and Planning Division at K-25, and he also taught a UT course on the theory of uranium enrichment as part of the Oak Ridge Resident Graduate Program.

Sheldon Datz's start in science came early during the Manhattan Project, when, as a New York City high school student, he worked part-time with Ellison Taylor at Columbia University's SAM Laboratory. This laboratory was partially focused on understanding the process of gaseous diffusion, later utilized to separate isotopes of uranium at K-25 in Oak Ridge.[76] After earning bachelor's and master's degrees in chemistry from Columbia, Datz was hired at ORNL in 1951 and began a long career initially focused on studying chemical reaction mechanisms using molecular beams. Ellison Taylor joined ORNL the same year and became a long-time research collaborator with Datz. Their development of the molecular beam technique was utilized by Herschbach, Lee, and Polanyi to win the 1986 Nobel Prize in chemistry.

While employed at ORNL, Datz earned a UT PhD in chemistry in 1960 under the guidance of physics professor Richard Present. He then became an expert on channeling of charged ions in crystals, where an atomic ion traveling between rows and planes of crystal atoms loses less energy than if the ion traveled randomly through the crystal. This ion channeling technique led to many breakthroughs in fundamental science and presented a way to understand the structure of crystals. In 2000, the Department of Energy presented to Datz the Enrico Fermi Award, which is a presidential award honoring scientists for lifetime achievements in the development, use, or production of energy.

Robert Birkhoff, born in Chicago, came from a family line strongly rooted in the sciences. His uncle George Birkhoff was a mathematician, and George's son Garrett excelled in his father's chosen field during a career at Harvard. Robert Birkhoff served in the Navy and later pursued his PhD at Northwestern University, while working on part of his dissertation research at Argonne National Laboratory. He was hired onto the UT physics faculty in 1949 and set up a research program devoted to the study of beta rays (a type of radioactivity) emitted by atomic nuclei. In his first year on the UT faculty, Birkhoff began a consulting relationship with ORNL and worked throughout his career to bridge what was then regarded as a significant gap between UT and Oak Ridge in terms of research prowess. He maintained that he was always amazed at how ORNL seemed to snub UT in favor of more prestigious schools, including Harvard and Yale. He once bluntly advised his Oak Ridge colleagues, "You were a cornfield 50 years ago, and a cornfield you may be again."[77]

In 1951, Birkhoff took the lead on the UT/ORNL Mobile Radiological Laboratory to measure the presence of radioisotopes in the air and water throughout Tennessee and, in the process, to recruit doctors and other scientists to become experts at measuring radiation. The urgent need for such expertise arose from public concern about a possible nuclear attack by the Soviet Union.[78] After five years on the UT physics faculty, Birkhoff moved to ORNL's Health Physics Division and assumed an adjunct professor role with the university. In so doing, Birkhoff became one of the first faculty shared jointly by UT and ORNL.

Rufus Ritchie was born in Kentucky's Blue Diamond coal-mining camp. After serving in World War II, he earned a bachelor's degree in electrical engineering from the University of Kentucky. Ritchie was hired in 1949 as part of ORNL's Health Physics Division. In the early 1950s, he worked with Robert Birkhoff on an analysis of experimental data on energy losses in the passage of charged particles through thin metal films. As part of this work on how a rapidly moving electron loses its energy in passage through a metal foil, Ritchie discovered the surface-localized collective electronic oscillation, now known as the surface plasmon. This plasmon is formed, for example, when a pulse of light shines on a metallic surface and part of the electric and magnetic fields of the light wave (light is a traveling set of oscillating electric and magnetic fields) tunnels into the metallic surface. If the wavelength of the light and the angle of incidence are just right, there is a resonance, and electrons on the surface absorb some of this electromagnetic energy and oscillate, a plasmon.

The full impact of Ritchie's discovery became clear only years later with the advent of nanotechnology in the late 1990s. Specifically, the surface plasmon can now be exploited to confine and manipulate light at the nanoscale.[79] Ritchie's 1957 paper[80] opened up the new field of nanophotonics and has inspired many practical applications in opto-electronics, photovoltaics, and solar energy conversion. In 1959, he received a PhD in physics from UT under the supervision of Richard Present, and, several years later, Ritchie took a joint faculty appointment in the UT physics department. In 1989, Ritchie was named a corporate fellow of ORNL, an honor that reflects significant career accomplishments and continued leadership in the scientific fields. Fewer than eighty fellows have been named since the program's inauguration in 1976.

Sam Hurst, another native Kentuckian, was born on a farm in Bell County. He attended Berea College and then earned a master's degree in physics at the University of Kentucky. Beginning his career at ORNL in 1948, Hurst worked as a researcher in the emerging field of health physics, to which he made significant contributions in instrumentation and field analysis. He traveled to Japan with colleagues to study latent disease effects and mortality rates among atomic bomb survivors. During his tenure at ORNL, Hurst emerged as a true entrepreneur.[81] Indeed, through his work at the laboratory, Hurst was awarded thirty-four patents, making him one of the laboratory's more prolific technology developers.

Over his career, Hurst started five companies, one based on his patent for the elograph, an electronic device with a transparent screen that allowed chart readers to easily acquire coordinates of locations on a map merely by touching the screen. In 1971, Hurst started Elographics, a small company that produced and marketed this ingenious technology. The elograph, the world's first iteration of the now-ubiquitous touchscreen, later made its debut at the 1982 Knoxville World's Fair. In 1986, Elographics was bought by the Raychem Corporation, and its name changed to Elo Touch Solutions, which today is the world's largest producer of touch-screen products and built a new operations hub in Knoxville in 2021, returning to the place where the touchscreen was invented.[82] In launching his active career, Hurst had received a PhD in physics from UT in 1959. Of the four physics doctorates awarded that year, three went to Oak Ridge employees (including Ritchie and Hurst) and the fourth went to William Bugg, who worked on campus in high-energy physics. Bugg is a co-author of this book.

Georgia-born Isabel Hanson Tipton attended the University of Georgia, receiving an undergraduate degree in physics—not her first choice of majors. In fact, Tipton wanted to be a chemist, but, at the time, the university's male-dominated chemistry department had little interest in enrolling a female student. The physics department, by contrast, accepted Tipton with open arms.[83] In 1929, she received her bachelor's degree and, in 1930, her master's, both from the University of Georgia. Tipton then went on to Duke University to continue her graduate work. While there, she met and married Samuel Tipton, a zoologist, and received her PhD in 1934. Both Tiptons later came to UT. Isabel joined the UT physics faculty in 1948, and she built a research career measuring trace element content in human tissue; such measurements were used to determine safe levels of radioactivity in the body. She was a consultant to the ORNL Health Physics Division from 1950 until her retirement in 1972.

Starting in 1952, Tipton, in an effort to protect human health, worked to determine safe exposure levels for radioactive metal present in air and water, and she acquired samples of human tissue from myriad sources with one unifying characteristic: all donors had died from causes other than disease or old age, including, for example, accident victims and criminals who were executed in the electric chair. The tissues arrived in the mail and were converted to ash in ovens in the physics building. According to William Bugg, a graduate student at the time, the smell of roasted flesh permeated the hallways when Tipton was processing a fresh batch of tissue. Bugg still marvels at the creative measures Tipton employed to acquire her tissue samples.

After the tissues were cremated, the remaining ash was treated chemically to prepare sources for spectroscopic analysis, also conducted in Tipton's laboratory. She often described her work as dealing with "instant people"—in a somewhat macabre comparison to the dried powder that was infused with boiling water to make a cup of instant coffee. Tipton and her team conducted these spectroscopy measurements to learn where various metals concentrate in the human body.[84]

EVOLUTION OF OAK RIDGE ACCELERATOR PROGRAM

In 1948, Art Snell, ORNL Physics Division director, started an accelerator program using materials readily available at ORNL and Y-12.[85] The objective was to understand more about the neutron and its interaction with various materials. In 1950, the Electromagnetic Research Division was administratively transferred from Y-12 to ORNL, and this brought substantial accelerator expertise to the laboratory. Important people came with this transfer, including Clarence Larson as the laboratory director and Elwood "Ed" Shipley as assistant director for ORNL at Y-12.

At the time, the Van de Graaff accelerator was the only known source of neutrons of precisely determined energies. ORNL's Chemistry Division acquired a 2.5-million-volt (MV) Van de Graaff accelerator from the Navy, and the Instrumentation and Controls Division managed to convert it into a 3-MV proton accelerator that could bombard lithium targets with protons to produce a stream of neutrons. This device supported research for thirty years, but perhaps its most important service to science came in the 1950s, when ORNL scientists John Gibbons, Richard Macklin, and colleagues used it to confirm a theory that elements originated through nucleosynthesis in the centers of stars.

This accelerator was used to produce neutrons of low energies that could be varied for experiments. New elements are formed in stars by the rapid capture of low energy neutrons to build up heavier elements, through the successive interplay of neutron capture followed by beta decay. Gibbons and Macklin developed the technique where the cross sections (capture rate) of low-energy neutrons on isotopes of the lightest elements (present in stars) could be measured and fed into theories successfully explaining the production of heavier elements from the lightest ones so abundant in stars (primarily hydrogen and helium).[86]

These early Van de Graaff accelerators led to ORNL's long history of increasingly large electrostatic accelerators for experiments in nuclear physics. The earliest accelerators could reach a terminal energy of 2.5 MV for acceleration of charged beams of nuclei, while the last one of the series, the Holifield accelerator, which operated from the 1970s through the 1990s, could reach 25 MV. As a result of this dramatic increase in energy, the Holifield facility, important to the UT-ORNL partnership, enabled experiments that could not have been contemplated in the 1950s.

Accelerators

To study the workings of the atomic nucleus, one needs a "hammer" to probe it and break it apart. For this purpose, physicists have built particle accelerators for decades. The accelerator boosts the energy of a beam of nuclear particles sufficiently that the beam can penetrate the nucleus to be studied, fragment it, and enable the experimenter to learn about the components and the mechanisms of the nucleus. The simplest particle to be accelerated is a proton, which is just the hydrogen atom that is ionized (i.e., with the atomic electron removed). One can accelerate beams of all the different elements as long as one is able to ionize the atom of that element.

Accelerators are generally built with either a linear or a circular geometry. A common linear accelerator was invented by American physicist Robert Van de Graaff in 1929 and is appropriately called a Van de Graaff. It is an electrostatic generator that has a belt that moves an electric charge to a hollow metal dome that sits on top of an insulated column. A tabletop version can produce an electric potential of maybe a hundred thousand volts on the dome and can store enough energy to produce a visible spark. Small Van de Graaff machines are produced for entertainment and for physics education to teach electrostatics; larger ones are displayed in some science museums. A person who stands on an insulated stool with her hand on the dome feels nothing as the electric charges flow from the dome to the ends of her hair and then repel each other, producing a wild hairdo.

Technology in building Van de Graaffs improved dramatically such that the highest potential achievable was 25 MV for the Holifield facility at ORNL.

A beam of negatively charged atoms (with an extra electron added) is accelerated in an evacuated beam tube to the high-voltage terminal, one or more electrons are removed by passage through a gas or a thin foil, and the beam is bent by 180 degrees and accelerated even more away from the terminal to ground potential and sent to the experimental gear with an acquired high energy.

A circular accelerator called a cyclotron was invented by Ernest Lawrence in 1929–30 at the University of California, Berkeley. Lawrence was awarded the 1939 Nobel Prize in physics for this invention. A cyclotron accelerates charged particles outward from the center along a spiral path. The particles are held to a spiral trajectory by a static magnetic field and accelerated by a rapidly varying (radio frequency) electric field. In general, cyclotrons can be built bigger and can generate more energetic beams of particles than a Van de Graaff. The largest cyclotron at present is the fifty-six-foot multi-magnet TRIUMF accelerator at the University of British Columbia in Vancouver, which can produce 520-million-electron-volts (MeV) protons.

ELECTRONS ON THE DOME OF A SMALL INSULATED VAN DE GRAAFF ACCELERATOR FLOW TO ENDS OF HAIR OF A GIRL STANDING ON AN INSULATED STOOL. COURTESY OF AMERICAN MUSEUM OF SCIENCE AND ENERGY.

THE 25 MV TANDEM ELECTROSTATIC ACCELERATOR AT ORNL HOLIFIELD FACILITY. COURTESY OF OAK RIDGE NATIONAL LABORATORY.

A long history of cyclotron development at Oak Ridge utilized space and assets of both ORNL and Y-12. A sixty-three-inch cyclotron was constructed in Building 9201–2 at Y-12 in the early 1950s, led by ORNL new hire Alex Zucker.[87] A decade later, the success of this research program led to construction of the Oak Ridge Isochronous Cyclotron (ORIC), which was a larger eighty-six-inch device capable of providing higher-energy beams for inducing reactions on target nuclei and studies of their properties. And, a decade after that, ORIC and the Holifield electrostatic accelerator were combined into joint operation[88] to allow the formation and acceleration of beams of radioactive nuclei, important for the study of short-lived and highly unstable nuclei. This combined accelerator operation provided a strong research opportunity for universities, and UT and Vanderbilt especially benefited. Faculty and graduate students from these two universities (plus those from many other schools) used this two-stage accelerator for producing and studying the most unstable isotopes of various elements. For example, a very light isotope of indium (^{109}I) was observed to decay to a daughter nucleus (^{105}Sb), which is important for understanding the rapid proton-capture process expected in explosive hydrogen-burning scenarios of stars.[89]

NATIONAL LABORATORIES ASSESS UNIVERSITY INTERACTIONS

In the mid-1950s, each part of the nation's atomic-energy enterprise underwent self-examination concerning its mission and research partners. During this period, Argonne National Laboratory examined its relationship with its partnering university community, and John T. Rettaliata, president of the Illinois Institute of Technology, chaired the committee to study those long-term relationships. Meanwhile, Argonne's Council of Participating Institutions went so far as to ask the University of Chicago to determine if universities should even continue to participate in Argonne's programs. At the time, Argonne's atomic-energy program was generally regarded as being solid, but its midwestern constituents generally ranked Argonne well behind ORNL as a useful partner. According to some of the surveyed university partners, all was well with Argonne as long as the universities remained fully integrated in the laboratory's research agenda. In other words, the universities were essential to Argonne's success and survival.[90]

By contrast, prior participation in ORNL's nuclear programs was regarded as an invaluable asset for the supervisory staff listed on applications for permits to construct nuclear reactors. In particular, experience in nuclear research and technology at ORSORT, ORINS, or ORNL was highly valued and viewed as an indication of technological qualification. Through support by AEC, a variety of ORSORT programs and courses provided theoretical and practical experience in the design and operation of reactors. Additionally, ORINS and other educational institutions, Vanderbilt University among them, offered instruction on the safe handling and use of radioactive materials, including radioisotopes.[91]

Such was the reputation of ORNL and its ability to collaborate with UT that, in a speech to the Tennessee Military Officers Club in March 1958, former Tennessee Governor Jim McCord proposed the creation of a state academy of science directed by UT and located in Oak Ridge. Students, all on scholarship, would be selected on merit and scientific promise for post high-school education, without regard to race, and the school would operate along the lines of the national military academies. McCord promised to introduce the measure for consideration to the 1959 Tennessee General Assembly,[92] but there is no record of any progress on this good idea. If McCord had delayed the proposed idea for a few years, by which time the UT-ORNL partnership had matured considerably, it might have gained traction.

Substantial interest at UT in participating in the growth of nuclear energy led to the establishment of the Department of Nuclear Engineering in 1957. The department initially offered bachelor's and master's degrees, later added a doctoral program, and is one of the oldest such departments in the United States. The growth of nuclear reactor R&D at ORNL contributed greatly to this new university direction. In fact, on September 5, 1958, UT faculty and students, in collaboration with ORNL reactor scientists and engineers, submitted a proposal to AEC for an on-campus 10-kW heavy-water moderated research reactor. The proposal advanced through the state's funding process, and the state building commission approved construction of a nuclear reactor building on UT's Knoxville campus.[93]

According to an article in the Nashville *Tennessean* about the project, AEC made available to UT a subcritical assembly and related equipment valued at $107,000.[94] In October, UT trustees accepted an AEC loan of 5,500 pounds of natural uranium slugs and 80 grams of polonium beryllium (a neutron source) for the new nuclear engineering facility. Pietro (Pete) Pasqua, the first head of the then-nascent nuclear engineering department, believed that the material would be received from AEC around the New Year, by which time the laboratory's completion was expected. At its December 1959 meeting, AEC evaluated the design for the reactor itself. Ultimately, though, the project stalled before the reactor was built. In fact, on May 2, 1960, UT submitted to AEC a letter withdrawing its request for approval of the proposed on-campus reactor.

The decision to scrub the project likely resulted from the university's increasing comprehension of the complexity of operating a nuclear reactor and the reality that several already operational reactors were located in nearby Oak Ridge. The decision not to move forward with construction of the reactor seems to have done little damage to the UT nuclear engineering department, which, over coming years, grew markedly and gained a national reputation. This growth and boost in prestige resulted in part from Pasqua's leadership. He continued to serve as department head until 1988, after which time the building that housed the department until 2021 was renamed in his honor.

Currently, the department's graduate program is consistently ranked as one of the top in the nation by *U.S. News and World Report*. In the years 2017 to 2019, UT ranked third in bachelor's degrees awarded in nuclear engineering (behind Texas A&M and Penn State) and first in master's degrees awarded (ahead of Penn State and Berkeley) and doctorates (ahead of Michigan and Texas A&M).[95] This is an important outcome of the partnership between UT and ORNL. In addition, there are now more than one hundred nuclear-related companies located within fifty miles of Knoxville.

CHANGING ROLE FOR "BIG SCIENCE AND ENGINEERING"

The end of World War II saw unprecedented growth in the U.S. economy. In his studies of economic growth, MIT economist Robert Solow showed that about half of economic growth cannot be accounted for by increases in capital and labor. Solow attributed this unaccounted portion of economic growth to technological innovation and development, and the sector's upward trajectory would only increase over future decades. Solow's groundbreaking research earned him the 1987 Nobel Prize in economic sciences and helped persuade governments intent on spurring economic growth to channel their funds into technological R&D.[96]

At the beginning of the 1950s, as the United States recovered from a costly world war, institutions of higher education were hard pressed, at least initially, to accommodate the flood of prospective students eager to take advantage of the G.I. Bill. However, at the time, colleges and universities were quick to appreciate the reality that the nation's economic prosperity would be predicated largely on growth in science and technology and on a highly skilled workforce. One barrier in particular stood in the way of the hoped-for technological boom and the universities' role in nurturing it: many, if not most, of the recent advances in science and technology remained cloistered and out of reach, contained within the walls of big government-funded facilities (national laboratories, atomic weapons facilities, and operations of other large government contractors).

When Dixy Lee Ray was the chair of AEC in 1974, in a discussion with Alvin Weinberg, she reportedly used the term "eunuchs of science" to describe these federal facilities, whose tight mission focus and high-security operation made it impossible for the nation at large, and universities in particular, to capitalize on and replicate these facilities' research methods and technological developments.[97] Somehow scientists and engineers at these federal facilities needed to devise ways to encourage and train student scientists in much the same way the universities did. This is, of course, almost impossible for national laboratories to accomplish on their own, since they do not have a natural source of students or the mandate to educate. But in close partnerships with universities, these facilities are perfectly positioned to achieve this goal, much as the UT-ORNL collaborative relationship would demonstrate over coming decades. ORINS began in part to build these partnerships to facilitate the use of ORNL for training of graduate students and to bring university faculty into the laboratory. The flow of UT graduate students doing thesis and dissertation research at ORNL started in this decade and accelerated in future years. Other universities worked with ORNL for the benefit of their faculty and graduate students, but, in future decades, UT and ORNL found even more intimate ways to partner in the education of a high-level workforce for science and engineering. Along with this, future decades would see an evolution in the very nature of Big Science.

SUMMARY AND LOOK FORWARD

In the 1950s, ORNL was primarily a reactor laboratory, even after AEC officially moved the leadership of the national reactor program to another lab (Argonne) in late 1947. The mission of K-25 as the premier site for uranium enrichment solidified. The role of Y-12 changed from uranium enrichment to support of the U.S. weapons program. The expertise in nuclear reactors at ORNL in this decade led to the initiation of a nuclear engineering department at UT in 1957, which resulted from the critical connection between the two institutions in this field. The need for ORNL scientists to finish their delayed doctoral studies resulted in significant growth at UT, especially in departments of physics, chemistry, and chemical engineering.

The following decade would see a decrease in reactor programs at ORNL and an increase in a broad program of collaborative research, as the UT-ORNL partnership continued to mature.

4

THE DECADE OF THE 1960S

From a Period Marked by War, Civil Unrest, and the Growth of the Environmental Movement, UT and ORNL Emerge Intently Focused on the Future

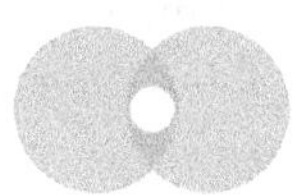

THE 1960S WAS a decade of change, for the country, for UT, and for the federal facilities in Oak Ridge. In many ways, this was also a decade of turmoil that reshaped the future. The shock of the Soviet Union's launch of Sputnik in 1957 led to creation of a large space program in the United States and a race to the moon through the 1960s. The Vietnam War raged in this decade, affecting not only budgets for science and technology research but, more important, the nation's collective psyche.

Rachel Carson published *Silent Spring* in 1962, which led to significant public interest in the deleterious effects of chemical agents, including DDT, released into the environment and, in 1970, to the first Earth Day. Three traumatic assassinations occurred in the decade: President John F. Kennedy in 1963 and presidential candidate Robert F. Kennedy and the Reverend Martin Luther King Jr. in 1968. These violent acts were accompanied by strife and turmoil associated with the nation's continuing struggle for desegregation and racial equality.

These events affected UT and the Oak Ridge facilities in profound ways. ORNL primarily was a nuclear energy laboratory in the 1950s. Its focus began shifting in the 1960s, in part due to growing environmental concerns about nuclear power amid the rapid rise in national emphasis on ecology, biology, and other "big science" topics. Already in 1961, there was a move to expand ORNL's programmatic focus and to develop it into a multi-purpose laboratory with multiple national research drivers.

The K-25 facility in Oak Ridge was affected by a declining emphasis on nuclear energy, as, in 1964, President Lyndon Johnson announced a suspension of some of the uranium-enrichment capabilities of this large gaseous-diffusion plant. Also in this decade, Y-12 responded to an increased national emphasis on nuclear weapons testing and exploited its precision machining capabilities to contribute to design of new weapons systems for the military.

This decade saw substantive changes at UT as well. Iconic president Andy Holt led a significant expansion of the university in this decade, with student enrollment increasing from 15,515 in 1959 to 40,991 in 1968 and, during this period, UT's physical plant more than doubling in size and value. By 1967, UT had become the twenty-third largest university in the United States.[1] The student protests that affected other campuses (notably, the University of California, Berkeley, and Columbia University in New York City) in the 1960s were generally absent at UT, but even this conservative campus would experience demonstrations and arrests when President Richard Nixon joined evangelist Billy Graham in a large revival-type gathering in the football stadium on May 28, 1970.[2]

In spite of these turbulent national trends, progress was made in the programs at the university and the national laboratory in the 1960s, and two leaders stand out as being responsible for much of the change and progress. Alvin Weinberg served as ORNL director from 1955 to 1972 and played a crucial role in shaping the laboratory's evolving mission. UT's Alvin Nielsen wore two hats in this decade: dean of the College of Liberal Arts and head of the Department of Physics. Separately and together, Weinberg and Nielsen brought new programs to their two institutions, including novel joint programs.

Through the early years of his career, Weinberg served as a nuclear reactor leader, but, in the 1960s, he guided ORNL through a diversification that expanded its focus on biology, ecology, and environmental science.[3] He also spoke and wrote about the importance of national laboratories teaming with universities to address the problem of the limited supply of scientists and engineers available to serve the national good. He worked toward new initiatives in tandem with UT, some of which succeeded (programs in biomedical science and ecology) and others which did not (programs designed to bolster civil defense). The UT-Oak Ridge Graduate School of Biomedical Sciences was initiated in 1967 and ran for thirty years as a UT graduate degree program based in the ORNL Biology Division. It served as a model for other academic/research partnerships between UT and ORNL established in future decades.

Nielsen took the lead on obtaining a grant from the Ford Foundation in 1963 to bring a cadre of ORNL researchers to the UT campus one day per week to engage in teaching and research with faculty and students. While UT faculty had opportunities to work as part-time consultants in Oak Ridge through the 1950s, and ORNL staff had taught courses for the UT Resident Graduate Program in Oak Ridge, this was the first opportunity for Oak Ridgers to work and teach part-time on the UT campus. This program eventually evolved into joint hiring of faculty, starting in the 1980s.

UT ARBORETUM IN OAK RIDGE

The UT-AEC Agricultural Research Laboratory, based in Oak Ridge, was home to sponsored research for Oak Ridge staff and UT faculty. It operated from 1948 to 1981 and was renamed the Comparative Animal Research Laboratory (CARL) in the middle 1970s.[4] This facility was considered such an important asset that, in the fall of 1960, UT requested more than two thousand acres of surplus property from AEC, in part to expand the scope of the UT-AEC Ag Research Lab. This parcel of land was not obtained, but former AEC land would later become the home to the UT Arboretum.

Alexander Hollaender built ORNL's Biology Division into a preeminent research organization. He liked to hike and regarded the area on a hillside of Chestnut Ridge barely a mile from his research facility as a potential site for an arboretum. Hollaender attempted to get UT to establish an arboretum there and even prepared to designate funds from the Biology Foundation to support this effort. But that would not be necessary.[5]

The State of Tennessee Educational Agency for Surplus Property in Nashville sent a letter dated June 23, 1960, to "Health and Educational Institutions near Oak Ridge, Tennessee," advising them of the availability of 2,378 acres of excess government property. UT President Andy Holt responded on July 12, 1960, expressing interest in the transfer of the entire tract and describing the potential uses of the property.[6] Eventually, on August 21, 1961, the U.S.

government, through the Department of Health, Education, and Welfare, transferred 2,260 acres to UT, following the declaration by the General Services Administration that this was surplus property.[7] A UT committee met in April, 1962, to establish an arboretum on a portion of the transferred Oak Ridge property.

The UT Arboretum continued to develop with the university's forestry department starting work in April, 1964. A nonprofit organization to support the arboretum began in February 1965 and became the University of Tennessee Arboretum Society. A funding request was made to the state legislature the following year for bridge construction and a residence for the arboretum director. Development efforts continued with a headquarters and office building dedicated on November 12, 1975. Arboretum leaders turned back a 1978 effort by the City of Oak Ridge to locate an airport on top of Chestnut Ridge, and, in the 1980s, the Arboretum was the first site in Tennessee designated as a "Watch Wildlife Area" and was also selected as the first "Tennessee Recreational Trail," according to the Tennessee Wildlife Resources Agency and the Tennessee Department of Conservation. The project reached another milestone in 1985, when the Arboretum was selected as the site for the first release of native turkey in the Oak Ridge region.

Educational partnerships expanded through collaboration with the Clinch River Environmental Studies Organization, to provide access to the Arboretum property for conducting studies by middle and high school students. Summer programs for students enabled research studies on box turtles, snakes, birds, and salamanders through a program sponsored by the Oak Ridge and Anderson County schools and funded by a DOE grant.[8] Today, the Arboretum is a 250-acre research and education facility as part of UT's Forest Resources AgResearch and Education Center, with seven miles of walking trails. It has more than 2,500 native and exotic woody plant specimens that represent 800 species, varieties, and cultivars, and hosts more than 30,000 visitors annually.

UT FORGES NEW LINKAGES WITH THE NATIONAL LABORATORY

Several approaches to improve cooperation and linkages between UT and the federal facilities in Oak Ridge were investigated during the late 1950s and early 1960s and with some success. Early in the 1960s, UT's James Montgomery (director of the Office of Institutional Research) prepared a report on the status of the UT-Oak Ridge partnership for the Board of Trustees, listing the successes of the past and making suggestions for the future. The report illustrates the progress that was made but clear in calling for the university to decide what it wanted in the future.[9] The Oak Ridge Resident Graduate Program was initiated with UT leadership in 1948, with instructors from Oak Ridge and UT (usually more from the former). Enrollment had steadily increased over the decade, with fall quarter registration of 273 students in 1960, compared with 200 in 1950. Union Carbide Corporation used the Graduate Program as an effective training and recruitment tool for prospective employees.

Various UT science and engineering departments effectively leveraged their connection with the Oak Ridge facilities in the 1950s, including physics, chemistry, chemical engineering, and nuclear engineering. In the physics department, about half of its doctoral students were employed in Oak Ridge while also pursuing graduate degrees from the university. By 1960, approximately forty UT faculty held consulting positions at the Oak Ridge facilities, led by the colleges of Engineering and Liberal Arts. UT undergraduate students took advantage of co-op programs in Oak Ridge, with forty-two participants from engineering departments and six from chemistry in 1960. For education majors, student teaching assignments in the Oak Ridge School System—regarded as one of the best in the state—were highly coveted.[10]

The Montgomery report[11] noted that the Tennessee governor had appointed a committee three years prior to investigate the uses and development of atomic energy in Tennessee, but that nothing had come from that committee. Montgomery urged UT to determine what else it wanted from its collaborations with Oak Ridge. One discussion at that time centered on an official UT presence and building in Oak Ridge. U.S. Senators from Tennessee pushed legislation to encourage AEC to build a new headquarters building in Oak Ridge, turning over the current building to UT. But no new AEC (and later DOE) administration building was ever constructed.

The Montgomery report expresses the frustration—perhaps tinged by envy—that UT leaders experienced with the successes of the Oak Ridge Institute of Nuclear Studies, led by former UT physics professor William Pollard. ORINS operated the similarly successful Oak Ridge Resident Graduate Program as a sub-contract to UT to offer courses in Oak Ridge. Still, UT desired to expand the scope of that program to operate this as a prime contractor for AEC and include research collaborations for its graduate students from many UT departments working in various parts of the Oak Ridge complex. However, the other thirty-six universities then part of ORINS did not want to have UT expand its visible presence in Oak Ridge. ORINS was also successful in winning AEC contracts to offer specialized courses and summer institutes and perform research in areas related to radiochemistry, dating of deep-ocean sediments, and preclinical studies of new isotopes related to medical treatment.[12] UT could have easily handled some of this scope, according to Montgomery. Additionally, there was concern about how much future growth in ORINS might limit growth of UT in Oak Ridge-related fields.

This report offers other advice for the future of the partnership.[13] A goal was suggested for UT to establish an advanced research institute for science and engineering in Oak Ridge, operated as a prime contract with AEC, since winning the contract to manage the Oak Ridge facilities was not seen as possible at that time. This would help expand Oak Ridge research ties for UT graduate students and provide the financial resources to support this. To achieve stronger research ties, many more UT-ORNL joint appointments were proposed, as was moving the Oak Ridge School of Reactor Technology to the UT Department of Nuclear Engineering. And, finally, the report calls for strengthening on-campus research capabilities, to bring UT and its faculty closer to the level of expertise found in Oak Ridge.

The Montgomery report effectively reflects a prevailing concern by UT leaders over decades: how to improve as a university by leveraging the Oak Ridge facilities and capabilities, thereby becoming a more robust partner. This concern and these issues persisted in future decades and progress was made step-by-step. Success in hiring joint faculty began in the 1980s, and joint institutes were added. On-campus research improved markedly. And, finally, forty years after the Montgomery report, UT began its management role at ORNL with its partner Battelle.

WEINBERG'S AMBITIOUS AND EXPANSIVE VISION

In 1962, Alvin Weinberg wrote an article in the prestigious journal *Science* about supply and demand for trained PhD scientists and engineers to work in the greatly expanding programs of what he characterized as "Big Science."[14] He was among the first to use and popularize this term to refer to new directions involving large devices and reflecting leading national priorities. Federal support of science programs in the United States increased by a factor of five from 1950 to 1962, but the generation of new PhDs increased only by a factor of two. Weinberg associated Big Science with a focus on big machines—rockets, accelerators, and reactors—in a way that previous civilizations had focused on temples, roads, palaces, and coliseums. While these earlier projects revealed a shortage of craft workers, Big Science revealed a shortage of scientists and engineers.

Weinberg proposed addressing this shortage of highly trained scientists and technicians by making better use of the national laboratories as educational resources. Universities were building up their competitive research programs in science and engineering to generate more and better-qualified PhDs, and Weinberg felt that federally supported research citadels should somehow also be given an academic mission so they could directly aid the buildup of the nation's scientific expertise. His 1962 *Science* paper was extremely forward looking and actually bore fruit later in this decade with the initiation of the UT-Oak Ridge School of Biomedical Sciences.

By the early 1960s, ORNL already played a role in science education, as each summer it hosted around sixty science faculty and dozens of undergraduates, plus a number of graduate students working on research for theses and dissertations at their home universities. ORNL researchers gave about two hundred traveling lectures annually at area colleges and universities. Weinberg envisioned that these and similar initiatives should grow substantially to benefit the country. At that time, ORNL and other government laboratories employed about 5 percent of the nation's science PhDs.

To aid in the education of new doctoral students, Weinberg estimated that the three large multiprogram national laboratories (Oak Ridge, Argonne, and Brookhaven), plus the non-secret parts of the national laboratories at Livermore and Los Alamos, could host dissertation research for 1,500 to 2,000 doctoral candidates in the biological and physical sciences. He described several methods of achieving this, including establishment of federal universities, whereby some national laboratories would be expanded to include academic programs for graduate students and would be authorized to confer degrees. There was some precedent for this, as the Rockefeller Institute of Medical Research had transitioned to Rockefeller University. Rockefeller is a private school and therefore easier to transform into a degree-granting institution. It is safe to say that large universities (private or public) at that time would not have been enthusiastic about national laboratories becoming federal universities, in part because of the competition for high-achieving students they might create.

A more workable idea that Weinberg identified was to establish "joint institutes" with neighboring universities. The students would get their degrees from the university, and national laboratory researchers would serve as faculty of that university, as they guided the students in their dissertation research. Although this idea was elegant in principle, Weinberg acknowledged a host of potential difficulties, among them the coordination of payment and benefits across institutions and the agreement at the federal level to allow national laboratory staff to spend significant time teaching courses and/or guiding graduate student research. Conceptually, this idea made sense. If graduate schools at universities are affiliated with federal laboratories (as in the case of the Massachusetts Institute of Technology and Lincoln Laboratories), then why not have federal laboratories with attached graduate schools?

Weinberg viewed this hybrid approach to education as a cost-effective way to bolster both graduate education and the level of science in the United States. Two decades later, UT and ORNL would use the same term (joint institute) to describe a key and innovative element of their collaborative approach. And, the partnering entities would establish two joint doctoral programs fifty years farther on with creation of the UT/ORNL Bredesen Center, with ORNL staff serving as faculty.[15]

Weinberg's ideas were not received well by members of the Board of Directors of the university-led ORINS, since he initially focused only on UT for participation.[16] His proposal was to establish a degree-granting institution at ORNL to utilize the national laboratory's resources and, thereby, to assist in meeting the national need for more science doctorates. Weinberg was unable to get financial support from a foundation for this bold proposal as part of ORNL, so he suggested that UT, AEC (through ORNL), and foundations administer such a program through a separate board. This proposal was discussed at the meeting of the ORINS Board during February 14–15, 1962, which took an unfavorable view of the proposal. William Pollard, ORINS executive director, suggested that opinions had been formed hastily and said that UT should have a chance to present the full proposal to his board.

UT never concurred with Weinberg's original proposal because the university was not part of the initial discussion. Only after the original proposal failed did UT become involved in drafting the second proposal, which was advanced on June 18, 1962.[17] With such a plan, UT believed it could maintain full academic control over this ORNL-based graduate program in the areas of focus: biology, chemical engineering, chemistry, metallurgy, physics, and possibly mathematics.

The proposal was regarded as a document to chart the university's institutional growth in science and engineering and did not include provisions ceding any semblance of internal control over its graduate program to the proposed entity at ORNL. There was still criticism of this proposal by ORINS members in part due to the size of the contemplated graduate program. However, this was unwarranted, as the UT graduate program already had

2,500 students, and the number in the proposal did not exceed the number currently enrolled in physics or chemistry, so the number in those departments should no more than double. President Holt sent a letter clarifying the UT position on Weinberg's revised proposal to the presidents of ORINS member universities for information only and requested no opinions or replies. Nevertheless, Holt received replies, many of them favorable.

UT would continue to work with ORNL on its long-term research and graduate education objectives but still needed financial backing to support the new doctoral program at ORNL. In essence, how to provide stipends for the cadre of new grad students in this expanded graduate program? One possibility was to increase the use of ORINS Oak Ridge Graduate Fellowships. While examining this program, Pollard realized that stipends and eligibility requirements needed to be made competitive with graduate programs offered at other institutions nationwide. He reviewed the history of ORINS' Graduate Fellowship Program, which was created to offer support to qualified graduate students in their final year of doctoral study. In part because of the fellowships' fairly rigid academic restrictions, ORINS universities had made little use of the program, except for UT, whose students were awarded almost half of the available fellowships.

The discussion ended with the ORINS council agreeing that ORINS had no role in this new ORNL graduate program, as it was a private agreement between UT and ORNL. The council did, however, encourage its member universities to make better use of ORINS' Graduate Fellowship Program. To promote that, the ORINS council increased the fellowship stipend and relaxed the eligibility requirements to focus on the candidates' readiness for research as opposed to a strict emphasis on completion of coursework and exams. This would make it more attractive for graduate students from more distant ORINS universities to take advantage of the fellowships to perform dissertation research at ORNL.

Alvin Weinberg: Second Half of a Great Career

Life at ORNL changed in the 1960s as the country began to focus more on ecology and environmental science (and on the space program), and less on nuclear energy. As director, Alvin Weinberg guided the laboratory through these difficult and changing times. He was quoted as saying that "ecologists have displaced physicists and economists as high priests in this new era of environmental concern."[18] He led the formation of the Environmental Sciences Division and the Energy Division at ORNL before his active tenure as director ended at the close of 1972, although he continued a leave of absence through 1973.

This did not end Weinberg's career; it only sent him off in new directions. He served in a government role in Washington and then came back to Oak Ridge in 1975 to direct the Institute for Energy Analysis, which he established a year earlier, at Oak Ridge Associated Universities (named ORINS until 1966). His institute was assigned the responsibility of assessing the impact of the increase in carbon dioxide that would result from continuing growth in fossil fuel emissions. In 1977, Weinberg testified at a hearing of the House Subcommittee on the Environment and the Atmosphere[19] and stated that a doubling of global carbon dioxide emissions by 2025 would lead to a 2-degree Celsius increase in global average temperature.

He was correct in this prediction but fortunately somewhat off on the time scale. In 1977 the global CO_2 content in the atmosphere was 333 parts per million (ppm). In 2020 it is around 412 ppm. The global temperature has risen by around 1°C since the mid-1800s, when the CO_2 content was around 280 ppm. Many scientists predict that our atmosphere will be at 600 ppm of CO_2 by mid to late century, at which time a second degree of warming will occur.

Weinberg was a legend in many ways. As ORNL director, he tried to attend all information meetings (where recent research results were presented and future directions discussed) in every research division, sit in the front row, and ask the first and most penetrating question of the presenter.

In 1980, Weinberg received the Enrico Fermi Award, which honors scientists of international stature for their lifetime achievement in the development, use, or production of energy. This DOE award honors the memory of Enrico Fermi, leader of the group of scientists who achieved the first self-sustained, controlled nuclear reaction at the University of Chicago in December of 1942. Weinberg's mentor, Wigner, received the Fermi Award in 1958. The citation for Weinberg's award tells the story: "In recognition of his pioneering contributions to reactor theory, design, and systems; for untiring work to make nuclear energy serve the public good, both safely and economically; for inspiring leadership of the Oak Ridge National Laboratory; and for wise counsel to the executive and legislative branches of the government."[20]

DEVELOPMENTS AT THE OAK RIDGE FACILITIES THROUGH AN ACTIVE DECADE

The 1950s ended with strength in operation of the three federal facilities in Oak Ridge—K-25, Y-12, and ORNL—but the decade of the 1960s saw change in all three as the country evolved in significant ways.

Gaseous Diffusion Plant

The role of the Oak Ridge Gaseous Diffusion Plant (ORGDP), including the massive K-25 facility, was to enrich uranium, along with similar gaseous diffusion facilities at Paducah, Kentucky, and Portsmouth, Ohio, constructed in the early 1950s. The need for uranium enriched in ^{235}U (the fissionable isotope) was great through the 1950s for defense purposes and for the nascent nuclear power industry. However, this effort would be reduced by the mid-1960s, when studies by AEC and the Department of Defense (DOD) indicated that full-capacity production of enriched uranium was no longer needed. President Johnson announced in his State of the Union message on January 8, 1964, his decision to curtail the production of special nuclear materials.[21]

The K-25 and K-27 (also a gaseous diffusion facility) buildings at ORGDP were placed on standby as of June 30, 1964. Operations in the other process buildings at Oak Ridge and other enrichment sites continued, but at reduced levels. The K-25 building was enriching uranium at over 90 percent ^{235}U for seventeen years, and, in 1967, the status of this building was changed from "standby" to "shut down."[22] In response, ORGDP began processing uranium on a toll basis for U.S. and foreign commercial nuclear power operations, typically up to 3- to 4-percent enrichment levels. Because the customer paid for the enrichment services, this added to the U.S. balance of payments to the Treasury Department.

Harold Conner began working at ORGDP in the 1960s as a UT co-op student in chemical engineering. After earning his bachelor's degree in 1968, he came to K-25 as an employee and proceeded to a thirty-three-year career working at gaseous diffusion facilities in Oak Ridge, Paducah, and Portsmouth, eventually leading environmental management and enrichment facilities at all three plants. The next two decades of his career were devoted to working with companies on the cleanup of facilities in Oak Ridge once the gaseous diffusion plant shut down.

Harold Conner—Pioneer

After attending a segregated high school in Martin, TN, Harold Conner began a journey in which he broke many barriers and contributed much to facilities in Oak Ridge and to the UT-Oak Ridge story. In 1964, he was the first African American co-op student at the K-25 site in Oak Ridge and, in 1968, he was the first African American to graduate from UT in chemical engineering. Then, Conner was hired at K-25 where there were no other Black engineers or executives. During his career in Oak Ridge, he earned a master's degree in nuclear engineering from UT and a doctorate in industrial and systems engineering from the University of Alabama, Huntsville.[23]

Conner's long career in gaseous diffusion focused on increasing the performance of the process and reducing wastes.[24] He became section head of a group and rose through the system to become site manager at the Oak Ridge Gaseous Diffusion Plant. Part of Conner's duties in enrichment activities involved improving or else shutting down buildings and processes that were less efficient. This eventually involved conversion of the plant from operation to standby and then shutdown at K-25 and moving equipment to Paducah and Portsmouth.

Conner became executive vice president for environmental management and enrichment activities for Lockheed Martin at the Oak Ridge sites as well as Paducah and Portsmouth. After the cleanup contract for K-25 was awarded in 1999, Conner left Oak Ridge for positions at the Savannah River Site, Lawrence Livermore National Laboratory, and Idaho National Laboratory. He returned to Oak Ridge in 2015 with United Cleanup Oak Ridge, the facilities' prime cleanup contractor.

In his career, Conner received many awards, including the Muddy Boot, presented by the East Tennessee Economic Council. Since 2018, he has served as a member of UT's Nuclear Engineering Board of Advisors. Over the span of fifty years, his talent, knowledge, and expertise have led to a diverse career across the DOE complex and solidified his reputation as a legend in his field and community.

Inside the powerhouse building that was shut down in 1962, ORGDP reexamined centrifuge enrichment of uranium. After World War II, gaseous diffusion was the process favored by Western nations, while the Soviet Union developed effective centrifuges largely with the guidance of captured Luftwaffe scientists. One of the scientists, Gernot Zippe, came to the University of Virginia in 1958 and wrote a report for the U.S. government describing the recent engineering efficiencies achieved in centrifuges. Zippe's enthusiastic endorsement renewed interest in the technology, which largely had been abandoned in the early stages of the Manhattan project as less efficient than gaseous diffusion for enriching uranium.

Engineers at ORGDP wanted to develop the technology of the dramatically improved centrifuges, because it appeared potentially more cost effective (and energy efficient) than gaseous diffusion, and it was necessary to reduce enrichment costs if AEC were to keep its share of enriched uranium on the world market. And, there would be yet other applications for this new class of centrifuges.

Indeed, at ORNL's Biology Division, Norman Anderson worked to develop a program with ORGDP engineers to physically isolate cancer cells, using centrifuge technology. Because this biology program used viruses, the project was relocated to ORGDP in the powerhouse to eliminate risk of exposing the Biology Division's mouse colony to infectious disease agents.[25] This was a collaborative program between AEC and the National Institutes of Health (NIH), which led to a number of developments involving centrifuges.

Earlier, vaccines were prepared by inoculating thousands of eggs with a live virus, waiting two weeks, cracking open the eggs, and harvesting the debris from the mixture. The resultant vaccine was more than 98 percent egg debris and only 1 percent virus. For those with egg allergies, the reaction to the vaccine could be fatal due to the impurities, limiting inoculations only to those between ten and sixty years of age and others under medical supervision.

Working with Eli Lilly Corporation, the ORNL, ORGDP, and NIH team developed an ultracentrifuge that produced a purified influenza vaccine that reduced side effects by more than a factor of ten in clinical trials. This eliminated the age and supervision restrictions and also increased vaccine production. A few years later, the rabies vaccine was also targeted for purification. In 1969 at a National Research Council meeting, Anderson noted that a new centrifuge rotor had been used to isolate high-purity rabies virus. He maintained that this joint effort between Eli Lilly and ORNL demonstrated the feasibility of preparing a pure rabies-virus vaccine.[26]

Y-12

During this decade, a transition in mission and direction occurred at Y-12. Isotope enrichment of uranium declined, resulting in the shutdown of many calutrons. In the late 1940s and early 1950s, about 12,500 tons of silver were returned after use of the calutrons decreased when the electromagnetic enrichment of uranium isotopes was stopped at Y-12 in favor of gaseous diffusion at ORGDP. While there were still calutrons at Y-12, most of them had their silver removed and replaced with copper windings.[27] A large portion of the remaining silver borrowed from the U.S. Treasury in 1943 was returned in 1968—around 2,145 tons of silver, valued at $124 million. The last return of silver (about 70 tons) occurred in 1970. According to Leon Love, head of ORNL's electromagnetic separations program in the Y-12 plant, the Treasury

Department would each year request its silver back, but the argument was always made that the electrical coils using the silver might be needed in the future.[28] These 2.3 million troy ounces[29] returned in the final shipment were but a small part of the total of 395 million troy ounces of silver borrowed from the Treasury during the Manhattan Project. The losses of 260,300 troy ounces represented less than 0.1 percent of the original amount borrowed.[30]

A new direction at the Y-12 Plant was the expansion and steady improvement in machining capabilities, which enabled it to be recognized as a leader in precision machining and measurement. Y-12 leaders were looking for business development opportunities beyond AEC, and they responded to the call for increased nuclear weapons testing. Throughout the decade, Y-12 worked on weapons systems for the U.S. Army, Navy, and Air Force. Its reputation for high-quality machining was established in the early 1960s during a competition among weapons complex facilities to produce as many acceptable parts as possible, within a tolerance of 0.0002 inches, from a given set of blanks. Y-12 won hands down and cemented its future in support of the country's weapons systems.

Another result of Y-12's fine machining capabilities was the construction of the "rock boxes" for NASA's lunar explorations, based on an inter-agency agreement between AEC and NASA. Officially named the "Apollo Lunar Sample Return Container," each was machined from a single aluminum forging to ensure no porosity, so that residue from Earth would not contaminate the moon and the samples on reaching Earth would have their native atmosphere intact. Beginning with Apollo 11, six missions were flown between 1969 and 1972 and brought back a total of 842 pounds of lunar material.[31] Y-12 was also involved in developing NASA's systems to receive, unload, and store the rocks, in addition to monitoring the environmental conditions of the lunar samples after they had returned to Earth. UT geologist Larry Taylor was one of the U.S. academic researchers who spent years working on analysis of the lunar rocks. He established the UT Planetary Geosciences Institute. His major contributions in geology included the discovery of the oldest basalts in a lunar mare, which is a big dark lowland region where scientists believe basaltic lavas pooled in the ancient past.[32]

ORNL

ORNL continued to grow through the 1960s, but the process involved a few fits and starts. Through the 1950s, the laboratory primarily focused on nuclear reactors and various research areas supporting this mission, but, in the 1960s, ORNL changed substantially as its focus began to expand beyond reactors and new programs in biology, ecology, and environmental science emerged. This shift in research emphases would reach full realization in the 1970s, but seeds of change were sprouting in the 1960s.

DIMINISHING ROLE FOR REACTOR PROGRAMS

The 1960s saw significant changes in reactor programs at ORNL. In 1961, it was announced that the Aircraft Nuclear Propulsion program (ANP) was canceled after having spent more than $1 billion to develop a nuclear-powered plane. As a 1963 report stated, "An aircraft had never been flown on nuclear power nor had a prototype airplane been built."[33] ORNL was a major contractor for this program, with the AEC providing $508 million in project funding and DOD providing $532 million. The effort to develop a long-range nuclear bomber was made obsolete by the arrival of in-flight refueling technology for conventional aircraft and the development of intercontinental ballistic missiles.

ORNL's Graphite Reactor achieved a self-sustaining nuclear reaction (also known as "criticality") for the first time in November 3, 1943. It was shut down twenty years later to the day and was designated a National Historic Landmark by the National Park Service on September 13, 1966, in a ceremony featuring an acceptance speech by AEC Chairman Glenn Seaborg. Since then, the reactor's public areas have been decontaminated and placed in a safe shutdown condition and now serve as an attraction for visitors to ORNL. The control room logbook remains opened to entries[34] from the date the reactor first went critical.

LITR was the first operating reactor in which the blue glow of Cherenkov radiation was visible and photographed. Initially built as a mock-up to test fuel elements for the MTR located in Idaho, this reactor reached criticality in 1950 and then served as a training reactor for MTR operators. It was shut down on October 10, 1966, after its function at ORNL was superseded by more powerful reactors.[35]

In 1946, Captain Hyman Rickover and his associates came to attend the Clinton Training School at ORNL and this led to the development of a reactor for Naval submarines. Water-cooling and gas-cooling schemes were evaluated and the decision was made to build a water-cooled reactor, which became the world standard for power reactors. By contrast, the United Kingdom favored gas-cooled designs in the 1950s for its initial reactors for producing electricity. The success of that British program prompted AEC to test a gas-cooled design—the Experimental Gas-Cooled Reactor (EGCR)—at ORNL in the 1960s. Helium was chosen as the gas to circulate through the reactor core to extract the heat for making power.

The containment vessel was huge, weighing 250 tons, standing 57 feet high, and measuring 20 feet in diameter, and was erected near ORNL after it was brought by barge on the Clinch River and then trucked a short distance. Before the fuel was loaded, however, AEC decided that reactors of this type had no future in the United States and this $50-million project was canceled in January 1966.[36] In later years, the large EGCR building with its high bay and the associated offices was used as development space for other programs of particular note. Among them were the Army's Future Armor Rearm System (FARS) and the Target Test Facility for the Spallation Neutron Source in 1998. The ORNL Robotics and Process Systems Division was also located at this site beginning in the 1980s. Currently, the building containing the former reactor is being deactivated and removed with the DOE land to be returned to ORNL for re-use.[37]

The Molten-Salt Reactor Experiment (MSRE) was perhaps Weinberg's favorite reactor scheme. The MSRE was initially based on ORNL's Aircraft Reactor Experiment that had investigated the use of circulating molten fluoride fuel for aircraft propulsion reactors. Design of the MSRE began in 1960, construction started in 1962, and the reactor went critical in 1965, operating at 7 MW.[38] The primary circulating coolant and fuel were molten salts, and the fuel operated at around 1,200°F at close to normal atmospheric pressure, as opposed to requiring the pressurized containment structures used in today's popular electricity-producing pressurized water reactors.

This was a prototype for a breeder reactor based on the thorium fuel cycle. Thorium is naturally occurring and does not fission, but does produce fissionable ^{233}U in the reactor due to capture of neutrons, thus the term "breeder reactor." The last fuel cycle of the MSRE ended in March 1968, and the reactor was shut down in December 1969 when AEC decided that a thorium-based breeder reactor was not part of its future plans.[39] The MSRE was designated a nuclear historic landmark in 1994. Today, there is renewed interest in this technology and ORNL has hosted an annual workshop on Molten Salt Reactors, beginning in 2015.

DESALINATION

A big program at ORNL in the 1960s focused on desalination of seawater to help people in arid parts of the world procure drinking water. At the program's peak, there were around a hundred laboratory researchers working on this technology, starting early in the decade. Some worked on the physical chemistry of seawater and invented better evaporation tubes and filtration techniques to remove salt and other contaminants more efficiently than achieved earlier.[40] Others focused on using a nuclear reactor to efficiently provide the energy needed for the desalination process. The term *nuplex* (a portmanteau of "nuclear-powered agro-industrial complex") was born, referring to a proposed island-based reactor capable of generating the electricity needed to desalt seawater for irrigating crops on adjacent arid lands.

This concept gained political notice and support, as presidents Kennedy and Johnson supported the idea as a key part of their Middle Eastern policy. In fact, Lyndon Johnson met with Soviet Premier Nikita Khrushchev at the United Nations Conference on Peaceful Uses of Atomic Energy in Geneva in 1964 to discuss their mutual interest in the nuplex concept. However, the focus on and funding for this desalination program decreased precipitously in 1968, as AEC saw the rapidly increasing cost of building nuclear reactors and decided there must be more cost-effective ways to address the need for fresh water in impoverished—and arid—parts of the world. At the time, for instance, agricultural experts were developing crops that could be grown with much less water than had previously been needed. So, the desalination program at ORNL ended. Nevertheless, there is still interest in using nuclear energy to remove the salt from seawater.[41]

HIGH FLUX ISOTOPE REACTOR

A new world-class research reactor—HFIR—went critical at ORNL in August 1965 and achieved its full initial operating power of 100 MW. HFIR became a top research tool for decades to come and remains in operation at the writing of this book. It became the centerpiece of the National Heavy Element Production program, which sought to produce transuranic elements, those beyond uranium on the periodic table. None of these elements occurs naturally and can be produced artificially only through intense exposure of target materials to neutrons.

The HFIR target area was therefore designed to produce as many neutrons as possible for irradiation, and several design features enabled this. The fuel element of enriched uranium is about two and a half feet tall and one and a half feet in diameter, with irradiation target rods containing pellets fashioned from heavy elements americium, curium, or plutonium for producing heavy elements. Surrounding the fuel element is a 1,200-pound annulus of beryllium specially fabricated at Y-12 to reflect the neutrons that are generated by the fission reaction back into the reactor core.[42] After twelve to eighteen months of irradiations in HFIR, the irradiated material is removed and chemically purified at the nearby Radiochemical Engineering Development Center to isolate the desired heavy elements. This facility has processing cells with massive shielding (54-inch-thick walls of high-density concrete) and remote manipulators operated through viewing windows, each weighing ten tons due to the lead glass used for radiation shielding.

Although HFIR was intended to be used only for production of radioisotopes, Weinberg saw the potential of using the high flux for neutron scattering, and modifications were made to the original design to accommodate this new application. Holes were cut in the neutron reflector for the insertion of guide tubes to enable the neutrons' delivery to spectrometers for experiments to study the properties of various materials. Neutron-scattering experiments became an important part of the research programs at ORNL, UT, and many other institutions.

The four Union Carbide plants (ORNL, Y-12, ORGDP, and Paducah) built the initial neutron spectrometers, which featured computer controls for their safe and efficient operation. No longer would researchers have to attend the machine twenty-four hours a day and trade shifts with their colleagues during HFIR's twenty-three- to twenty-five-day experimental cycle. Each spectrometer had its own computer, first a PDP-5 and then a PDP-8, which not only controlled the machine, but also stored the collected data.[43] [44] This was an early example of computer-controlled experiments, and the PDP-8 is now a fond relic among physicists and computer geeks.

Reactor activities at ORNL included the technique of neutron activation analysis (NAA), which was employed in investigating one of the most-public murders in U.S. history. Indeed, a week after President Kennedy's assassination in 1963, the Federal Bureau of Investigation (FBI) asked that ORNL use NAA to study fragments of the bullets that struck the president and also paraffin casts taken of the hands and face of Lee Harvey Oswald, Kennedy's alleged assassin.

The FBI hoped ORNL researchers could match gunpowder particles on the paraffin casts with gunpowder from a rifle found at the crime scene, near the window of the now-infamous Texas School Book Depository. However, this particular analysis did not work well enough. The fact that Oswald had used a pistol to kill a Dallas police officer, J. D. Tippitt, soon after the assassination, and, that earlier tests had been conducted on the paraffin casts, complicated the research and made the analysis inconclusive.

The FBI also hoped that ORNL's analysis of the bullet fragments taken from the president's limousine could determine whether the bullets were fired from a single weapon, and this analysis proved more definitive. Lead bullets have traces of silver and antimony, and the laboratory's analysis of these traces indicated that the bullets did indeed come from the same rifle.[45] The results of this analysis at Oak Ridge were included in the summary government report on the assassination.[46 47]

In another high-profile case almost three decades later, NAA of hair and nail samples from the grave of President Zachary Taylor indicated he had not been poisoned by arsenic while in office, as one historian suspected, but had died of natural causes.[48]

TOWARD A MORE ACCESSIBLE NATIONAL LABORATORY

ORNL appreciated its dynamic relationships with the surrounding community. There was always curiosity on the part of the region's scientists and engineers about ORNL's often-secretive lines of research (all work there was classified by AEC rules until the early 1950s). Requests for attendance at laboratory-based seminars and use of the ORNL libraries were denied unless the interested individuals were part of AEC or its contractors. This changed in December 1962 when ORNL opened its unclassified seminars and library resources to the surrounding community.[49] These resources included the Central Library on the main ORNL campus as well as the Biology Library at Y-12.

Visitors hoping to attend scientific seminars were to contact ORNL's Office of University Relations for access, and industrial and commercial firms gained access through the Office of Industrial Cooperation established in 1962. The Public Information Office fielded Press and media inquiries. The ORINS library, an official depository library of all unclassified AEC reports, was already open to the public, as required of all depository libraries, but it did not contain the extensive resources of the combined ORNL libraries.

The decade of the 1960s was a time of changing research and production emphases at the national level and in Oak Ridge. Y-12 began its evolution into a facility in support of the country's nuclear weapons program. K-25 saw a decrease in its uranium-enrichment mission. And, ORNL saw a transition away from its strong reliance on nuclear reactors as its centerpiece toward a more balanced agenda serving the shifting national priorities.

In this decade, ORNL built new particle accelerators and opened its doors to the first university-based user community, allowing it to install its own measurement device for experiments at one of the accelerators. Construction started in October 1959 on ORIC, and it began operation in 1963, firing protons, alpha particles, and other light projectiles into various targets to enable research primarily into nuclear physics. The ORIC design was considered a major technological breakthrough in its day, and it marked the culmination of several years of research at ORNL into the design of an accelerator to operate as a fixed radiofrequency with strong and weak magnetic sectors to provide focusing of the beam of charged particles as they increase in energy and spiral outward in the cyclotron, making it isochronous.[50] A smaller version of the eventual ORIC was used with accelerated electrons to develop and test the principles of operation.

The Oak Ridge Isochronous Cyclotron

The Oak Ridge Isochronous Cyclotron (ORIC) is a circular device used to accelerate beams of nuclear particles to sufficient energies to initiate nuclear reactions and enable study of nuclei. The charged nucleus, e.g., a proton (a hydrogen atom with the electron removed), is bent into a circular orbit by a strong magnetic field confined by a 200-ton magnet (built by U.S. Steel) and generated by magnetic coils (built by Y-12). The beam is accelerated each time it crosses the gap in the "dees" by an electric field, which has to change polarity by the time the beam encounters the gap one-half an orbit later. In fact, this accelerating electric field must oscillate at radio frequencies so that the beam gets a boost in energy each time it hits the gap (twice per orbit). Dees get their name from their shape, which resembles the letter D.

As the beam gains energy, it spirals outward into a larger circular orbit each time it crosses the gap. The nature of the cyclotron is such that all spiraling particle beams at all radii get to the gap at the same time (isochronous) and uniformly get a boost in energy by the oscillating electrical field. This makes the cyclotron a highly efficient device to accelerate particles to sufficient energies to conduct experiments in nuclear physics.

Drafting the proposal for the $3.7-million ORIC construction was considerably easier in 1959 than it would have been in later decades. Alex Zucker was a research physicist in ORNL's Physics Division at that time and, in 2003, he recounted the process that he led to secure the ORIC project's award of funding.[51] He recalls that the only paperwork to get this project started was a one-page letter from ORNL director Alvin Weinberg to AEC chairman Glenn Seaborg requesting funds to build a cyclotron. AEC's favorable response was similarly brief.

ORIC enabled decades of research in nuclear physics and allied fields,

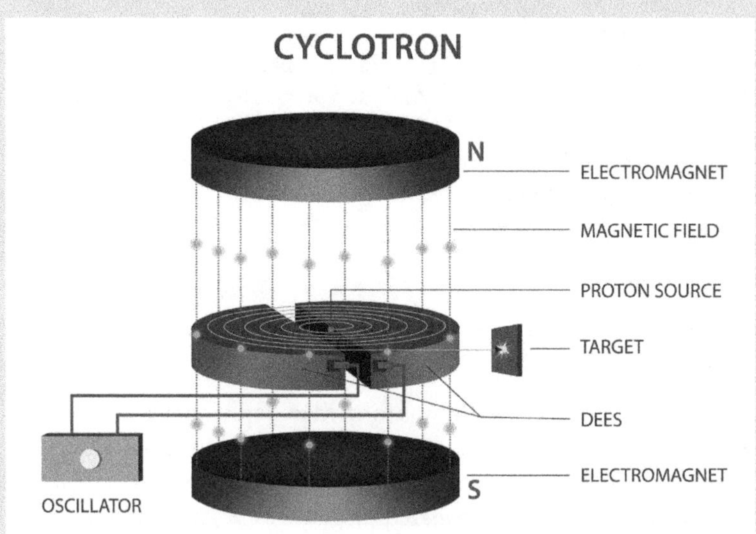

SKETCH OF THE CYCLOTRON PRINCIPLE. PURCHASED FROM *ADOBE STOCK*.

which was important for the nation and increasingly of value to universities, especially some in the Southeast. In the late 1960s, UT and Vanderbilt were part of the ORNL plan to win support for a bigger accelerator to enable a new area of research with beams of heavier projectiles (those above argon on the periodic table), which required higher energy than ORIC could produce.

With the support of universities throughout the Southeast, this accelerator concept began as a regional project. Initially, it was to be called CHEROKEE (after one of the Southeast's more noted Native American nations), but scientists could not find the words to form an appropriate acronym. So, it was named APACHE, the Accelerator for Physics and Chemistry of Heavy Elements.[52] Balking at its $25-million price tag, President Richard Nixon's Bureau of the Budget rejected the laboratory's regional APACHE concept in 1969.

This setback on building the next-generation accelerator at ORNL was offset in part by the positive benefits of getting strong high-level input from southeastern universities. This led to the successful idea of organizing a group of the region's universities to fund and operate a facility at ORNL utilizing beams of accelerated nuclear particles from ORIC. Led by physicists Joseph Hamilton of Vanderbilt University and William Bugg of UT, a consortium formed in 1968 among physicists from twelve universities to combine funding from their universities, state governments, and AEC to finance construction of the University Isotope Separator at Oak Ridge (UNISOR).

This separator was located on an ORIC beam line and started a new era at national laboratories in the United States, as it was the first university-financed and university-run facility at an AEC laboratory. UNISOR was dedicated in 1970 as the first user facility at a national laboratory and will be discussed in the next chapter. Fifty years later, ORNL has nine user facilities for the benefit of researchers from the United States and abroad.[53]

BIG BIOLOGY AND A NEW UT-OAK RIDGE GRADUATE SCHOOL

The Biology Division at ORNL was formed in September 1946 and evolved in the 1950s to become one of the top research units in the country. William and Liane Russell joined this division in 1947 and spent their careers developing the field of mammalian mutagenesis and, thereby, the science of genetic risk-assessment in humans. They built the Mouse House, a colony of a quarter million mice used for experiments on the effects of radiation (and later chemicals) on mammalian cells. Both Russells were later elected to the National Academy of Sciences, a profound honor for scientists in all fields.

Rachel Carson's *Silent Spring* heightened the country's concern over the role of chemical agents in biological and environmental degradation. Federal funding for the NIH was boosted significantly, and cooperative programs were developed between NIH and AEC laboratories. As discussed earlier in this chapter, in 1961, Norman Anderson of the Biology Division started adapting the centrifuge technology of the Oak Ridge Gaseous Diffusion Plant for use in separating impurities from the viruses causing polio and influenza A (H3N2), which caused a pandemic in 1968 and 1969. This enabled production of vaccines pure enough to administer to a wide set of patients. This work was performed under the AEC-NIH funded Molecular Anatomy Program in the Biology Division. Others in this program developed centrifugal analyzers to assay components of blood, urine, and other body fluids in minutes for help in medical diagnosis of disease, later a common technique used by medical laboratories. Work also proceeded on building high-resolution microscopes and other tools to investigate the complex biological events leading to cancer growth.[54]

In 1963, a grant to UT from the Ford Foundation (discussed later in this chapter) enabled a new level of cooperation between the university and the national laboratory and ultimately funded as many as fifty ORNL scientists to work on the UT campus for teaching and research on a one-day-per-week basis. This development occurred at about the same time that James Shannon, NIH director, became interested in building a graduate school near NIH headquarters in Bethesda, Maryland.

Shannon thought that establishment of a similar graduate program in biomedical sciences in Oak Ridge could expand NIH programs at ORNL.[55] An ORNL team led by Alvin Weinberg obtained AEC approval and that of President Johnson's science advisor, Donald Hornig, to consider such a program. Weinberg then approached UT President Andy Holt to learn if he was interested in pursuing this as a cooperative arrangement.

Weinberg told Holt that "our location in Appalachia and the strong contribution which a new biomedical program would make to President Johnson's Great Society should enlist the aid of our U.S. senators and representatives as well as the President." Holt was indeed interested and worked with Weinberg and others to develop a joint program.[56]

The UT Board of Trustees approved the Graduate School of Biomedical Sciences at its meeting on November 5, 1965, and this opened a new critical connection between UT and ORNL. This new entity became the twentieth school or college at UT and the first located in Oak Ridge. Federal funds would be sought to construct a building for the school, estimated at $1 million. As reported in the UT *Torchbearer*, Union Carbide Corporation, parent company of the operating contractor of the AEC Oak Ridge facilities, donated $100,000 to provide student center facilities for this new graduate program.[57]

Alvin Weinberg commented, "The establishment of the new UT-Oak Ridge Graduate School of Biomedical Sciences is a most important step in the development of graduate education in the Southeast. The school extends and reinforces the pattern of cooperation in graduate education between ORNL and one of its neighboring universities, a pattern that has already proved successful in many areas of science."[58] AEC Chairman Glenn Seaborg added that such a center would provide a new approach to the partial utilization of manpower at the national laboratories in educational pursuits and could result in the rapid establishment of a new area of educational excellence in a field hampered by lack of qualified people.

Launch of the Graduate School of Biomedical Sciences in fall 1967 was part of the expansive growth of the UT campus and the UT system. The years 1967 and 1968 were momentous years for UT and the state's higher education system. In 1967, the Tennessee Higher Education Commission was formed, and the Martin branch of UT was given autonomy from the

Knoxville campus by the legislature. In 1968, plans for a UT merger with the University of Chattanooga were implemented, and the UT System was created with primary campuses at Knoxville, Martin, Chattanooga, and Memphis, each governed by a chancellor.[59] Charles Weaver was appointed chancellor of the Knoxville campus, and Holt remained as president of the UT system until his retirement in 1970, when replaced by Ed Boling, who became the seventeenth UT president.

The benefits of the Graduate School of Biomedical Sciences to UT were significant. Led by Clinton Fuller, a scientist with ORNL's Biology Division, the new graduate program enabled UT to offer doctoral and master's degrees, along with postdoctoral research opportunities, in areas new to the university. The Oak Ridge facilities and research faculty were available nowhere else in the South and in few other places in the world, particularly in the field of mammalian genetics.

The faculty and students were based at ORNL, which enabled the students to spend about half of their time on research, beginning in their first year of graduate study, and allowed them quickly to become familiar with ongoing research and current progress in the field. Housing for the students in Oak Ridge was provided, in part, by UT's purchase of eight Manhattan Project-era apartment houses.[60] These E houses (their designation when built in the 1940s) at the intersection of Vermont and New York Avenues were torn down in the 1990s to make room for medical clinics.[61]

UT appointed an African American faculty member, Franklin Hamilton, as liaison with predominantly Black liberal arts colleges, about one hundred of which were located in the Southeast (of a total of 111).[62] Aided by the efforts of many individuals at both UT and ORNL, including Alexander Hollaender, former director of ORNL's Biology Division, the staff of the graduate school successfully initiated a program titled "Aid to Black College Students and Faculty Interested in Careers in the Biomedical Sciences." The National Institute of General Medical Sciences and the Carnegie Corporation of New York supported the program.

A number of undergraduate students and faculty members from Historically Black Colleges and Universities (HBCU) were supported for research training in the Biology Division, taking advantage of the Graduate School of Biomedical Sciences to prepare students for careers in that field. This summer program of research, courses, and seminars was quite successful. It started in 1971 and ran for five years. Of the sixty HBCU students from twenty-eight universities who participated in the program, twenty-three entered medical or dental school and eighteen entered graduate programs, including eight who joined the Graduate School of Biomedical Sciences at Oak Ridge and UT.

The UT-Oak Ridge biomedical sciences graduate school operated successfully for more than twenty-five years, starting in 1967 with its inaugural class of seven students. The directors were Clint Fuller, Tex Barnett, and Ray Popp, all based in the ORNL Biology Division. In 1993, there were twenty-six enrolled graduate students and 142 people were awarded PhD degrees since the formation of the school, in addition to twenty master's degrees. It was a big success by any academic measure, but a change in funding emphases led to its downfall.

DOE support of biological research was shifting in the mid to late 1990s, away from the mouse genetics emphasis of the ORNL Biology Division to functional genomics, including physical, chemical, and computational sciences, and technologies relevant to biology. This change in DOE-supported programs in the Biology Division made it more difficult for researchers to continue to compete for NIH funds that had been used as the backbone of the Graduate School of Biomedical Sciences. A number of these ORNL researchers retired or moved elsewhere as a result of this change in emphasis, and the school slowly lost its staffing and funding. Further challenging this graduate school's survival, the limited connections of the school's students and faculty with biology faculty at UT, Knoxville made this Oak Ridge based school an increasingly unpopular program at the university. Indeed, UT's biology departments felt that they had to compete with the Graduate School of Biomedical Sciences to recruit qualified students.

In the absence of substantial NIH grants, UT and ORNL supported the school for several additional years. However, the end of the program came in 1998, when biology leadership at the two institutions restructured it into the Graduate Program for Genome Science and Technology. This graduate program realignment resulted in greater involvement of researchers and more relevance to research initiatives and educational needs at both institutions.[63] Both of these biology-based graduate programs laid the groundwork for future UT doctoral offerings that leveraged ORNL people and facilities through the Bredesen Center starting in 2010.

THE NASCENT ENVIRONMENTAL MOVEMENT TAKES ROOT AT UT, ORNL

The early 1960s saw the beginning of the modern environmental movement in the United States, spurred in part by concerns about the effects of pesticides in the environment.[64] Sustainability, preference for organic foods, and the "back-to-the-land" movement continued to gain steam throughout the 1960s and greatly impacted the national research agenda.

Stanley Auerbach was hired at ORNL in August of 1954 and began research on applying systems theory—the notion that numerous interrelated smaller systems form a unified whole—to the field of ecology. This led to the hiring of Jerry Olsen, Bernard Patten, and George Van Dyne, who conducted research at ORNL and also taught a year-long course on systems ecology at UT from 1964 to 1967.[65] The National Science Foundation (NSF) appointed Auerbach as director of the ecosystems component of the International Biological Program for the eastern United States, and over eight years, NSF provided $1 million in annual funding to ORNL—marking the first major NSF-funded program based at a national laboratory. In general, it has been difficult for DOE national laboratories to attract NSF funding, since these grants are almost always awarded to universities.

The interest of UT graduate students in courses and research in ecology led to a desire for a formal program grounded in this relatively new field of science.[66] This resulted in a meeting at UT in 1967 involving Auerbach and David Reichle of ORNL, James Tanner of the UT zoology faculty, and Walter Herndon, UT vice chancellor for academic affairs. There was a common desire to establish an ecology program at UT, but the big questions involved the program's structure, administration, curriculum, and location. Alvin Nielsen (dean of the College of Liberal Arts) and Hilton Smith (dean of research) were opposed to setting up a new school or institute, and various department heads likewise opposed creating a new ecology department.

After much discussion and negotiation, UT leaders agreed on establishing an intercollegiate and interdisciplinary "program" in ecology, opening another critical connection between UT and ORNL. The ecology program involved faculty from any department or college wanting to teach in the program and direct graduate students in ecology. Faculty (twenty-eight) from seven departments joined the program, and most of the offered courses were existing ones in various departments plus a new year-long "Principles of Ecology" course. The program began in 1969 with Tanner as the director, succeeded by Frank McCormick, who served in this role from 1972 to 1981. McCormick maintained that this was the first U.S. program to offer a graduate degree in ecological science.

By 1985, fifty UT faculty were associated with the program plus thirty adjunct instructors, mostly from ORNL. The program started with fourteen first-year graduate students in 1969 and increased to forty students per year twenty years later. By 1980, around one hundred students received graduate degrees from the ecology program, including thirty-five doctorates. Today, ecology is part of the UT Department of Ecology and Evolutionary Biology, which is rated in the top 10 percent of ecology programs in North America, based on research impact. This has clearly been a successful UT program, which grew out of an early interest in ecology research at ORNL.

BRINGING ORNL RESEARCHERS ONTO THE UT FACULTY

By the mid-1960s, significant growth occurred in the UT physics department—growth in undergraduate and graduate enrollment, number of faculty, and research capability. U.S. competition with the Soviet Union (including the so-called "space race") created large programs providing federal support for graduate study in the sciences, including physics. The availability of adequate research funding for graduate work in physics, UT's proximity to ORNL, and the fact that graduate students were exempted from the draft during the early years of the Vietnam war resulted in enormous increases in graduate applications and enrollment. Typically, the UT Department of Physics had 130 to 180 graduate students at any one time during this period. For an eight-year interval in the 1960s, the department averaged 19.5 PhDs awarded per year.[67] An American Physical Society report ranked the UT physics department twentieth in the nation in PhD conferrals during this time.

The physics department, along with the chemistry and chemical engineering departments, retained close ties with ORNL for decades. Not only were many laboratory employees graduates of UT PhD programs or currently enrolled as students in those programs, many on-campus students conducted their research at ORNL using facilities under supervision by UT faculty or by approved ORNL personnel. Both UT and ORNL personnel taught graduate courses in Oak Ridge. In areas where the research interests of UT faculty corresponded with similar interests at ORNL, joint research projects were commonly created.

A mechanism sanctioned by the UT and ORNL administrations enabled UT faculty to be compensated for one day per week of work at ORNL, originally as consultants and later as part-time employees. This one-direction arrangement (facilitating engagement of UT faculty at ORNL) was important for the physics faculty, but it occurred to physics department head Alvin Nielsen that there would be great benefit to a bi-directional program for sharing faculty.

Nielsen approached the Ford Foundation with an innovative idea for that time: to bring outstanding ORNL personnel onto the UT faculty to get them directly involved in university education. He took the lead on developing a successful proposal. In 1963, UT President Andy Holt announced that UT had received a $750,000 grant from the Ford Foundation for graduate programs related to nuclear energy.[68] The program was designed to attract, on a part-time basis, some of the best researchers at ORNL to participate in teaching and research on the UT campus. This new critical connection marked a crucial step toward an eventual joint-faculty program with hundreds of shared researchers.

ORNL research scientists, who were eager to interact with students in a way not then possible at a national laboratory, enthusiastically received this new program. The program uniquely pioneered the concept of joint appointments between a university and a national laboratory. The idea was so revolutionary at the time that it required approval at the highest levels by AEC, which expressed concern about possible dilution of its laboratory's workforce.

From the university's viewpoint, the appointments were also unique in that they granted the ORNL appointees full privileges of regular faculty (except, of course, tenure). Typically, appointed ORNL scientists were limited by AEC to 20 percent of a fulltime workload (one day/week) at UT, while 80 percent of their time would continue to be invested at ORNL. The specific departments included in the Ford Foundation grant were physics, chemistry, and chemical engineering, with a few other departments added later.

The Ford Foundation funding initially covered the UT portion of the salary of the ORNL participants and was programmed to gradually decrease to zero in ten years, when those costs would be covered by university funds. The physics department profited immensely from the Ford Foundation program, with an average of six to eight such ORNL appointees, who contributed significantly to the outstanding growth of the department in academic excellence and research funding over the next forty years. The first appointments in physics under this program were a distinguished group from ORNL: Louis Roberts, Harvey Willard, and Ted Welton from the Physics Division, as well as Sam Hurst, Hal Schweinler, Loucas Christophorou, and Rufus Ritchie from the Health Physics Division. These divisions were the ones where most of the departmental collaborative research in physics was conducted.

ORNL chemists also joined the UT faculty, including Clare Collins, Milt Lietzke, Ralph Livingston, and Dave O'Kelley. Other ORNL staff participants included D.G. Thomas with UT's engineering mechanics department, A. S. Householder with mathematics and nuclear engineering, and M. S. Wechsler with chemical and metallurgical engineering.[69] Additional ORNL staff would follow in coming years. By 1965, the number of ORNL staff in these 20-percent UT faculty appointments grew to twenty-eight.[70]

Perhaps the most important lasting result of the program was the present-day acceptance by both institutions of the joint appointment as a means of increasing the effectiveness of both university education and laboratory research. Indeed, joint faculty hires have played a very important role in building the partnership between UT and ORNL, and the joint-appointment concept is now embraced throughout the nation.

The innovative Ford Foundation program was lauded throughout Tennessee and the nation. *Physics Today* described the program and named the new physics faculty from ORNL.[71] The director of the Southern Regional Educational Board reported on comments made by George Kistiakowsky of Harvard University in his testimony to a U.S. House of Representatives subcommittee on America's needs for improved scientific and educational centers of excellence. He called attention to "the vigorous efforts of the University of Tennessee and the neighboring Oak Ridge National Laboratory to create jointly such a center of intellectual activity."[72] He further urged the subcommittee to encourage other government research laboratories and universities to pool "their super-critical communities of scholars and educators," as UT and ORNL had done. *Physics Today* updated the ORNL participant numbers from nineteen in the first year to thirty in the second.[73]

Joint hiring of research experts has enabled UT and ORNL to build new strength in research areas important on the national scene. Over several decades joint hires helped grow strong programs in materials science, atomic and nuclear physics, and nuclear engineering. In the past two decades, a new focus on joint hires has been in the areas of nanoscience and additive manufacturing, which has led to well-regarded and well-supported joint research programs in these areas.

NSF GRANT BOLSTERS PHYSICS AT UT

A major boon for the UT physics department during the Nielsen era came with the award of an NSF Science Development Grant. Originally called "Center of Excellence Grants," these competitive national grants were motivated by the desire to assist in the improvement of the quality of science education in universities around the country. Initially, UT proposed a broad program involving ten departments.

Upon review, NSF responded that two departments, physics and chemical engineering, were better prepared to receive support and suggested that a more limited proposal be submitted. Nielsen and William Bugg, who succeeded Nielsen as head of the physics department in 1969, prepared the physics portion of the new proposal. It was accepted, and NSF awarded UT $1.45 million, which was split evenly between the two departments. According to interviews at the time, this was one of the largest awards ever received by UT. The NSF grant was designed to take both departments, already regarded as strong, up to the level of similar departments in the top twenty U.S. universities.[74]

Although perhaps the smallest university grant of this type funded by NSF, it was crucial to the future success of the UT physics department. Among the grant's salient provisions, it provided funds for the addition of new faculty whose support the university would later cover. Perhaps more important, the grant enabled research groups on campus to acquire sufficient state-of-the-art equipment to conduct high-level research, thus providing a springboard for increases in federal research funding, which occurred in the ensuing years.

As an example, the elementary particle physics group was able to acquire an automatic measuring machine (a spiral reader) for analyzing results of bubble chamber experiments at large accelerators. This addition of a forefront analytical tool allowed this research group immediately to compete with major universities and national laboratories in the field of elementary particle physics. This led to increased federal funding and recognition of UT as a leader in this field. The grant also funded the expansion of UT's atomic physics program (later one of the best in the nation) and gave a powerful boost to the strong research program in nuclear physics.

CRUCIAL STUDENT SUPPORT AT ORNL

Fellowships for graduate students and those in postdoctoral positions are the life-blood of any research institution, and, over the years, ORNL has been fortunate to host many outstanding candidates from various universities. Some came to ORNL on AEC fellowships in health physics or public health or with fellowships from ORAU (ORINS prior to January of 1966) or the NSF. The more fortunate graduate students managed to combine several of these awards to get their careers started, Lee Riedinger among them. AEC and ORAU fellowships enabled Riedinger to come to ORNL as a Vanderbilt graduate student to perform his dissertation research (1964–68) under the guidance of ORNL's Noah Johnson and Vanderbilt's Joe Hamilton. Later, an NSF postdoctoral fellowship enabled him to spend one year in Copenhagen to conduct research in nuclear physics, his chosen career path.[75]

In Oak Ridge during this decade, graduate students from various universities worked on dissertation research in different divisions. For example, in 1965, the Physics Division hosted twenty PhD students, including twelve from the UT.[76] The Health Physics Division hosted a number of graduate students funded by the AEC-U.S. Public Health Service Program and the AEC Health Physics Program, both created in 1949. The Health Physics Graduate Fellowship involved a summer program at ORNL and provided five weeks of field training in health physics, plus five weeks focused on research in an area related at least loosely to health physics. Because of this, Riedinger got started on nuclear physics research in the summer of 1965 at ORNL as a Vanderbilt graduate student. Other AEC facilities participating in these fellowship programs were Argonne, Brookhaven, Pacific Northwest, Lawrence Radiation Laboratory, the National Reactor Testing Station, and the Puerto Rico Nuclear Center.[77] [78]

ORNL's co-op program was initiated in 1952 as a way to give college students specialized training and experience for a semester as part of their education. ORNL's Employment Department administered this program, which was a valuable recruiting tool. Some of these co-op students later joined ORNL as employees after completion of their undergraduate degrees or following graduate work.[79] In 1962, significant enrollments came from UT (fifteen co-op students) and Virginia Polytechnic Institute (nine).

The Summer Student Traineeship Program, administered by ORINS, began in 1958 and involved ORNL as well as the UT-AEC Agricultural Research Laboratory, ORINS Medical Division, and Savannah River Laboratory. The ten-week program was designed for students in engineering, mathematics, or science, following completion of their junior undergraduate year. In the 1960s, the program was popular and highly selective. In 1967, for instance, the program accepted only 95 of the 446 applicants from 305 colleges and universities.[80] Prior to this class, more than 420 students had successfully carried out summer research projects in Oak Ridge and at Savannah River.

ORAU began a twelve-week NSF- and AEC-sponsored program in 1968 for undergraduate students focusing on quantitative ecology. Around ten appointments were made in the Radiation Ecology Section of ORNL's Health Physics Division for students interested in ecology or environmental protection. The program's primary activity was full-time study with senior scientists on projects focused on movement of isotopes or nutrients in the environment, radiation effects on plant and animal populations, or systems analysis of landscapes and fresh waters.[81] Programs of this type for undergraduates, sponsored by DOE and administered by ORAU, continue to this day.

There were also important programs to bring university faculty to ORNL generally for summer research. The Research Participation Program funded faculty studies through a special ORNL budget in its Education and

University Relations Department. It was conducted in cooperation with ORINS (called ORAU after 1966), which publicized the program and fielded the majority of applications. The program began in 1948, with about fifty faculty appointments made annually in many of ORNL's research divisions.

The majority of colleges and universities had only one faculty participant, while UT typically had several. Many participants continued their association with ORNL under consultant subcontracts or subsequent summer appointments outside of this program.[82] This was an excellent program to help faculty build their university-based research programs and for ORNL to have close associations with faculty and their students.

INTERNALLY FOCUSED EDUCATION ACTIVITIES

Continuing education for staff at the three AEC facilities in Oak Ridge was an important mission, and one ongoing effort was the Oak Ridge Resident Graduate Program, led by UT. It was heavily promoted at ORNL both in the form of courses offered and also with the selection of the staff researchers who served as instructors for some of the courses. As noted in *ORNL News*, curriculum and instructors were announced several weeks before each academic quarter began.

Courses during the last half of the 1960s were offered in physics, chemistry, mathematics, zoology, chemical engineering, metallurgical engineering, and microbiology. A total of twenty-one courses were offered in spring quarter 1969, the largest offering of the decade. The Resident Graduate Program was so popular that, in August 1967, it was extended to continue through June 30, 1972. In a policy change effective March 1, 1967, the Educational Assistance Program fully reimbursed employees after satisfactory completion of the courses,[83] an increase from the previous 50 percent.

An Oak Ridge education program directed toward staff eager to earn high school diplomas drew national media attention. A *New York Times* article in 1969 recognized that the Basic Education Improvement Program targeted the six hundred employees (of the five-thousand-person ORNL staff) who lacked high school degrees.[84] High school equivalency certificates could be gained by attending these classes at ORNL, and the laboratory paid employees for one hour of the two-hour class.

Vocational and skill training was developed in this decade through an inventive partnership program for Y-12 employees and for individuals even outside of Oak Ridge. From June 1966 to December 1968, a collaborative effort among ORAU (the project director), UT's Industrial Education Department, and the Y-12 Plant (operated by the Union Carbide Corporation, Nuclear Division) developed the Vocational-Technical Teacher Institute that included both in-service and pre-service elements. This institute awarded college credit for professional education courses taught by university faculty and for technical courses taught by the crafts persons, scientists, technicians, and engineers of Y-12. This program was included in the 1966 AEC Annual Report to Congress.[85]

Sixty teachers from ten southeastern states enrolled to receive instruction in machining, drafting, and electronics during the first in-service portion of the institute in the summer of 1966. Because of the overwhelmingly positive course evaluations, a second institute was held in the summer of 1967 for one hundred teachers in the same fields, with the addition of welding.

The first Pre-Service Vocational-Technical Teacher Institute, held during UT's 1966–67 academic year, drew nineteen attendees, while thirty-one attended the second session held the following academic year. This program provided one year of off-campus technical and professional study toward a bachelor of science degree in industrial education. These institutes were early examples of enhanced vocational training through the collaboration of government, industry, and universities. Others emulated this model, including the NASA Langley Research Center and Old Dominion University, beginning in 1968.[86]

The Training and Technology Project (TAT) was developed at Y-12 to provide vocational and technical training for both teachers and workers. Union Carbide Corporation's Nuclear Division and the UT Department of Industrial Education were responsible for the training. This was a large undertaking, with courses in six technical and craft areas. After a successful demonstration phase, the project was authorized for training about 300 individuals, and 42 percent of the first 190 enrollees were minorities. Y-12 was expected to hire most of the project graduates.[87] Surveys following three and seven years of TAT operation proved the success of the program, as it met employer demands for technical workers with newly acquired skills.[88] TAT was highlighted in *The New York Times*, which noted that the first five hundred graduates, trained at a cost of $1.5 million, returned an additional $2 million to the economy annually through their increased earnings.[89] Y-12 was not the only AEC facility to benefit. The National Accelerator Laboratory, under construction in Batavia, Illinois, sent twenty-three participants to TAT in 1969, nearly all from Chicago's inner city. The graduates of the program were subsequently employed in Batavia in the laboratory or in the private sector.[90]

The TAT program was so highly regarded that, even during a 1970 strike at Y-12, instructors were allowed to cross the picket line to keep the training program on schedule. The TAT program ran successfully until 1984.

OAK RIDGE AND LOCAL ECONOMIC DEVELOPMENT

By 2020, there was a strong emphasis on using federally funded facilities like those in Oak Ridge as engines to drive local and state economic development. The transfer of technology developed within a national laboratory to the private sector is now an important measure of a laboratory's success, but this was certainly not the case in the 1960s or earlier. The formation and quick success of the Oak Ridge Technical Enterprises Corporation (ORTEC) is an example of breaking new ground in 1960.

In the late 1950s, nuclear physicists at ORNL and elsewhere progressed rapidly in developing deeper understanding of nuclear structure and nuclear reactions. Hal Schmitt, a group leader in the Physics Division, was interested in measuring the fragments created when uranium and other nuclei fission due to excitation by an accelerated beam of nuclear particles. Detecting fission fragments with good energy resolution was difficult at that time, so Schmitt's group had to develop appropriately sensitive nuclear particle detectors.

The new idea was to design and build a silicon surface-barrier detector that could measure these fission fragments. One could not buy these devices off the shelf, and so the scientists had to design and build their own. Silicon surface-barrier detectors were developed in the Physics Division by John Walter and John Dabbs during the middle to late 1950s for use in their experiments, and they joined Schmitt's team to tailor the detectors for the fission fragment measurements. Meanwhile, James Blankenship, in the research group of ORNL's Instrumentation and Controls Division, studied techniques for more reliable fabrication of such detectors. Preliminary experiments with this new type of particle detector rendered promising results, and the ORNL developers came to realize that other laboratories might be interested in buying such a device for their own research programs.

Schmitt and colleagues asked the Instrumentation and Controls Division at ORNL to set up a laboratory to fabricate these detectors for their work and possibly that of other institutions. However, the director of that division refused to cooperate and so the physicists were forced to find other solutions, including possibly building the devices off-site. In May 1960, they met with Mansel Ramsey, ORNL associate director for administrative affairs, to ask about taking this technology development outside the lab. He was cautious but generally supportive, provided that the physicists followed ORNL procedures to avoid conflicts of interest and did not let this outside activity affect their laboratory work. They received approval and formed ORTEC two months later.

Money was needed to rent space, buy some equipment, and get this shoestring enterprise going, and the team managed to raise $14,500 from friends and colleagues, which turned out to be a good investment. As the nascent corporation began operations, no one foresaw the return of $150,000 for each $500 invested, when the small ORTEC company was bought by a large company (EG&G) in 1967. Indeed, within six years (1961–67), annual sales grew from $84,000 to $4.7 million. Unfortunately, the division director forbade colleagues in the Instrumentation and Controls Division at ORNL from investing in this start-up enterprise, so they missed out on the substantial investment windfall. ORTEC's success story was related in detail through interviews Ray Smith (one of the authors of this book) conducted with Hal Schmitt in 2011.[91]

Changes occurring in Oak Ridge culminated in April 1964 when Oak Ridge attorney Gene Joyce, a champion of community economic development, gave a talk summarizing some problems facing local community leaders.[92] He began his talk by encouraging long-term financial support by AEC of the community operations in Oak Ridge, stressing the importance of better interactions between the federal facilities and Anderson County government. He then focused on several broad operational issues relating to AEC and its contractors.

The consulting policies of Union Carbide enabled creation of several electronic and detector firms in and around the city, such as ORTEC. However, changes in 1963 barred scientists from consulting after hours or serving on boards of directors of companies related to their Union Carbide work activities or their areas of professional expertise. This restriction applied to salaried employees but not to hourly workers, perhaps due to bargaining restrictions. The policy was stricter than that applied at other AEC sites or to NASA contractors or other federal agencies. Joyce argued that these restrictions should change and remarked that there appeared little interest by Union Carbide in establishing a diversified economy in an AEC-dominated community.

Joyce called for scheduling more visits by congressional and administrative leaders to Oak Ridge, maintaining that past visits were the result of personal contacts and that a more formal and inclusive approach to attracting visitors would be needed on a long-term basis. If diversification of fabrication at Y-12 and research at ORNL were the goals, Joyce insisted, then there should be elected officials in Washington keeping track of the activities of relevant agencies to be sure that local interests were served. Joyce was a leader of these outreach efforts, and his work led to the local community becoming more active in lobbying legislators and stressing the capabilities of and opportunities provided by the federal facilities in Oak Ridge. These outreach efforts extended to the Tennessee government in Nashville, the corridors of the House and Senate in Washington, and the leadership of various federal agencies.

In 1973, amidst increasing funding stress on the federal programs in Oak Ridge,[93] Gene Joyce, Tom Hill (initial owner of the *Oak Ridger* newspaper), and Don Maxwell (Bank of Oak Ridge) started the Roane Anderson Economic Council as a forum for federal officials, their contractors, and community leaders to discuss and prioritize programs.[94] A good way to address problems of the city and of the federal facilities was to have open discussion among these various stakeholders in Oak Ridge and surrounding areas. It was perceived that this form of communication would provide an effective means for the broad community to shape local priorities and to influence and support federal initiatives.

Twenty years later, the council's organizational scope broadened to include five East Tennessee counties, and the name changed to reflect the expansion: the East Tennessee Economic Council (ETEC). This organization meets weekly on Fridays at 7:30 a.m. for presentations by local leaders, laboratory directors, UT administrators, and politicians. It still serves the East Tennessee community in the way that Gene Joyce envisioned in 1973, bringing together the important people from all walks of professional life to learn of important activities and research undertakings at the federal facilities, encourage community discussion, and shape a consensus view on where Oak Ridge and the wider East Tennessee region are headed.

Beginning in 1973, ETEC has honored individuals who have helped build the Oak Ridge community's economic and scientific base with the prestigious *Muddy Boot* award. The name reflects the Manhattan Project founders of Oak Ridge, who worked through adverse conditions (like muddy streets and sidewalks) to build the community. This award has been presented to elected officials, business leaders, DOE representatives, and others. From 1973 to 2022, Muddy Boot awards were given to 102 people, including many who contributed to the success of the UT-Oak Ridge partnership.[95] One non-scientist who received this award was Frank Munger, who covered the Oak Ridge scene (including the relationship with UT) for thirty-seven years as a reporter for the *Knoxville News Sentinel* and whose father was a dentist in Oak Ridge during the Manhattan Project. There are many references to Munger's articles in succeeding chapters of this book.

THE NEW DISCIPLINE OF INFORMATION TECHNOLOGY

In 1962, Alvin Weinberg, as a member of the President's Science Advisory Committee, chaired a group of experts to study the explosion of scientific and technical information. As stated in the group's report, the United States was then producing approximately 100,000 informal government reports per year in addition to 450,000 journal articles. Unlike journal articles, the reports had no bibliographical control or comprehensive collecting system. The Weinberg report called for investigation and implementation of robust modes of information retrieval that would depend on new information-gathering technology.[96] Thus was born a new field of information science and technology, which has become increasingly important in subsequent decades.

The Weinberg report also gave rise to a career pathway for Oak Ridger Bonnie Carroll. She read the report as a Columbia University graduate student and then joined ORNL on the library staff in 1971.[97] Later, she joined Weinberg's planning group and then was appointed to coordinate the laboratory's various information centers that had been established in response to the Weinberg report's recommendations. At age thirty, she left ORNL to work in government and the private sector. One job was in the DOE's Office of Scientific and Technical Information, which sharpened her expertise in information technology.

After working in the private sector for a few years, she started her own company at age forty, Information International Associates (IIa). The company grew to include more than 250 experts in information technology and security, people working in a half dozen different states or countries with a host of government customers including DOE and the U.S. Patent Office. After thirty years as chief executive officer (CEO) of IIa, Carroll sold her company in 2018 and stepped back to a less stressful life, although not finished with information technology. She took the lead on establishing the World Data System International Program Office, which is a worldwide consortium of people and institutions with expertise in data systems. She helped bring this new initiative to the Oak Ridge Innovation Institute, which is a joint venture between ORNL and UT, and, in fact, served as interim executive director until a permanent hire was made. So ended the official career of Bonnie Carroll, although her work as one of Oak Ridge's foremost entrepreneurs will surely continue.

TENNESSEE TRIES FOR A NATIONAL ACCELERATOR

A national competition occurred in 1965 for a large accelerator to enable the next step in high-energy physics and to understand the constitution of matter at its most elemental level. At that time, the largest accelerator in the United States was at Brookhaven National Laboratory, boasting a maximum energy of 33 GeV (billions of electron volts). Physicists wanting to study the properties of elementary particles with greater accelerated energy and intensity needed a larger machine, and AEC started a process to site and fund a 200-GeV accelerator. States proposed two hundred potential sites and an AEC panel narrowed the list to eighty-five, including Memphis and Oak Ridge.

The proposed Oak Ridge site was near the K-25 Gaseous Diffusion Plant. The AEC team visited Memphis on November 29 and 30 of 1965 and Oak Ridge on December 1 and 2 to assess minimum requirements for a successful site.[98] However, in May of 1966, these two Tennessee sites were removed from the list of potential locations. Although these sites seemed to meet most of the advertised requirements, some local observers blamed the loss on Tennessee's ranking as forty-eighth among states in spending on education.[99] Illinois won the competition and the accelerator (National Accelerator Laboratory; AEC added Fermi to the lab name in 1969) was located in Batavia, near Chicago. The state of Tennessee competed more seriously for the next-generation high-energy accelerator—the Superconducting Super Collider—in 1987. Texas won that competition, but the Tennessee proposal was said to be ranked second.[100] The anticipated cost of that laboratory ballooned from $4.4 billion to $12 billion and the U.S. Congress terminated construction in October of 1993 after $2 billion had been spent.

CHANGING TIMES IN OAK RIDGE IN SCIENCE AND TECHNOLOGY

On the one hand, the federal facilities in Oak Ridge were large and well supported in the 1960s. Ranking first among national laboratories in the 1966 AEC annual report in terms of completed plant and equipment, ORNL was listed as receiving an AEC investment of $320 million, with FY 1966 operating costs of $85 million. Investments at the Oak Ridge Gaseous Diffusion Plant were $823 million, and those at Y-12 totaled $380 million, providing the state of Tennessee with the highest AEC investment of any of the fifty states. AEC had total operating costs in Tennessee of $220 million, and UT was awarded about $1.5 million in AEC research contracts.[101]

On the other hand, research priorities shifted in the United States, away from nuclear energy and toward environmental science. In a commentary in *Nuclear Applications* in July 1967, Alvin Weinberg summarized the national discussion about redeploying the national laboratories to address challenges that have little direct connection to nuclear energy.[102] Weinberg observed that Congressman Chet Holifield, chairman of the Joint Committee on Atomic Energy, urged AEC to devise ways to reduce air pollution. Weinberg noted ORNL's efforts in desalination, civil defense, and chemical carcinogenesis. And, by the close of the 1960s, 20 percent of the laboratory budget was devoted to issues of nuclear safety, and 14 percent of ORNL work was non-nuclear in scope and funded by agencies besides AEC (other national laboratories drew less than 1 percent of their budget from outside AEC at that time).

The national press noted this broadening of activity. As reported in *ORNL News*, Jonathan Spivak wrote an article for *The Wall Street Journal* about ORNL venturing well beyond the purposes prescribed by AEC for its laboratories and becoming a "national" federal facility capable of conducting a broad range of meaningful R&D.[103] Crediting Alvin Weinberg with the vision of diversification—or redeployment—Spivak noted that Weinberg applied ORNL's capacities to resolve social or scientific problems that had resisted solution. Examples provided were ORNL's efforts to develop purer vaccines, produce potable water from the sea, and understand biological processes occurring in cells that passed hereditary diseases on to future generations. ORNL did all this while not forsaking its nuclear roots. In fact, the Biology Division was the largest division in terms of staff at ORNL by the end of the decade.

To help plan for a changing future, Weinberg brought David Rose, a physics professor from MIT, to ORNL during two summers (1969 and 1970) for seminars and study groups on future directions for the laboratory. Rose joined the ORNL staff in 1969 to lead the new Long-Range Planning Office. Rose led seminars on environmental issues and the role of science in public policy. Three mid-career laboratory scientists worked closely with Rose on these studies: John Gibbons (nuclear physicist and later science advisor to President Clinton), Claire Nader (a social scientist and sister of Ralph Nader), and James Liverman (Biology Division and first interim director of the Oak Ridge Graduate School of Biomedical Sciences). These summer studies led to the idea of realigning national laboratories to address environmental problems in a holistic and interdisciplinary manner. Weinberg and Rose carried this idea to Senators Howard Baker (R-Tennessee) and Edmund Muskie (D-Maine) who liked the concept and sponsored a Senate resolution advocating for the establishment of a National Environmental Laboratory at Oak Ridge.[104] [105]

This idea of an AEC national environmental laboratory did not go far at that time. Representative Chet Holifield (D-California), chairman of the Joint Committee on Atomic Energy, blasted the Baker-Muskie resolution and was rumored to have said, "Let Muskie get his own laboratories!" A rider to the 1970 AEC authorization act effectively put a stop to the Baker-Muskie initiative.[106]

Despite the setback, the growth in support for environmental awareness and protection would help shape the national agenda in the 1970s. The Rose summer studies led to the formation of the Environmental Sciences and Energy Divisions at ORNL in that decade. The National Environmental Protection Act of 1970 passed and the Environmental Protection Agency (EPA) formed in that same year. The first Earth Day took place on April 22, 1970, and helped prompt passage of the Clean Water, Clean Air, and Endangered Species acts (Muskie and Baker led the Senate work on both the Clean Water and Clean Air acts). The U.S. environmental movement was fully afoot by the early 1970s, and ORNL would be fully onboard.

Budget problems occurred late in the 1960s, in part due to the costly war in Vietnam. ORNL operated on a flat budget from 1965 to 1970, the equivalent to a 25-percent loss due to inflation. And, the holiday season of 1969 was reminiscent of a similarly difficult time in 1947. The Bureau of the Budget ordered across-the-board cuts of ORNL staff from 5,300 to under 5,000, the thermal breeder program was cut by two-thirds, and plans for a new accelerator (APACHE) were scrapped. At the time, Associate Laboratory Director Bill Fulkerson said that "the lab lost its innocence in 1969," as new national priorities focused on the environment replaced centerpiece programs of the past (mostly related to nuclear energy). ORNL entered the 1970s underfunded and understaffed, but, nevertheless, the laboratory needed to meet new challenges in the coming decade, including two energy crises and the rapid ramp up of environmental programs.

SUMMARY AND LOOK FORWARD

The 1960s ended as a decade of transition in Oak Ridge. There was considerable attention paid to joint UT-ORNL programs relating to university-based education. New policies at both UT and the three Oak Ridge AEC institutions, particularly ORNL, made increased collaborations possible. Changes in funding patterns, the closure of old facilities, and construction of new reactors and accelerators were all part of the evolution of the laboratory's scientific research agenda. Societal and political challenges came more to the forefront of national discussions as the Great Society programs reshaped the post-Sputnik era.

Y-12 became engaged in a $170-million expansion to support the U.S. anti-ballistic missile system, the Oak Ridge Gaseous Diffusion Plant responded to increased demand for uranium enrichment services for the electric power industry, and ORNL moved away from a dominant emphasis on nuclear energy toward a balanced research enterprise, with an increasing focus on environmental science. This change in focus led to the formation of a strong ecology program at UT, which grew substantially in size and importance in future decades. The Oak Ridge Graduate School of Biomedical Sciences began in this decade as a new doctoral program at UT, leveraging the expertise in ORNL's Biology Division. This school ran for nearly thirty years, and, twenty years after its closing, its design and structure were resurrected in creating the Bredesen Center with its two joint UT-ORNL doctoral programs.

The ecology program at UT and the Graduate School of Biomedical Sciences were two new critical connections built in the decade, in addition to the Ford Foundation grant that brought a cadre of Oak Ridge researchers to part-time faculty roles at UT.

The 1960s saw the construction of a forefront cyclotron for nuclear physics research, and twelve universities in the southeast (led by UT and Vanderbilt) exploited this device to create the first university-controlled and -funded user facility at any national laboratory. Increased future emphases on user facilities at ORNL (and at other national laboratories) would facilitate numerous possibilities for faculty and student research at these specialized facilities.

The decade of the 1960s saw significant change in all three federal facilities in Oak Ridge as UT grew greatly in size, student enrollment, and faculty. This decade set the stage for the 1970s, when ORNL became a more balanced national laboratory, which resulted in expansion of its partnership with the university. Gone were the days of ORNL as a primarily reactor lab, as two new research divisions would open in the 1970s—the Environmental Sciences and Energy divisions—based on new national emphases. This led to formation of a UT research center, the Energy, Environment, and Resources Center, started by an Oak Ridger, John Gibbons. Nuclear physics flourished as a research discipline at UT based on expansion of experimental facilities in Oak Ridge.

5

THE DECADE OF THE 1970S

The Nation Celebrates Its First Earth Day, DOE Is Established, and ORNL and UT Further Bolster Programs Aimed at Energy Efficiency and Environmental Protection

IN THE 1960S, UT doubled in size, while ORNL experienced a declining role for its reactor programs coupled with the emergence of new national emphasis on environmental protection. The decade of the 1970s opened with the formal recognition of that new emphasis with the celebration of the first Earth Day on April 22. By then, both ORNL and UT had established new programs focused on issues and challenges in the field of environmental sciences. With a growing national focus on energy in this new decade, ORNL became a multiprogram laboratory with a diverse research portfolio, and the university expanded research options for its students.

Several larger-than-life personalities in leadership positions at the two institutions left the scene in this decade. Alvin Weinberg had been a highly respected ORNL director since 1955 but was forced out of his job at the end of 1972. Herman Postma, a noted ORNL fusion physicist and a future leader in strengthening the partnership with UT, succeeded him. At the university, Andy Holt, who had presided over the doubling of student enrollment and campus size, retired and was replaced by Ed Boling, who started his eighteen-year tenure as president in 1970. The UT system was formed in 1968, after which each of the four campuses—Knoxville, Chattanooga, Martin, and Memphis—was led by a chancellor reporting to Boling. Jack Reese, who served as UT, Knoxville chancellor for fifteen years (1973–89) developed a close partnership with Postma.

Nuclear physics grew as a research discipline at ORNL and at UT through the 1970s. UT teamed with Vanderbilt and ten other universities in the Southeast to create the first user facility at the national laboratory: UNISOR. This was a new critical connection that opened the door to a number of shared UT and ORNL facilities over future decades. UNISOR played a critical role in ORNL capturing a next-generation accelerator to enable expanded research capabilities, leading to growth in the nuclear physics group at UT and establishing one of the strongest areas of joint research between the two institutions for the following thirty years. And, UNISOR started the strong trend toward more user facilities at ORNL and other national laboratories.

John (Jack) Gibbons played a large role at both institutions in this decade. He worked as a nuclear physicist at ORNL for fifteen years before migrating at Alvin Weinberg's request to work on energy and environment, leading to his appointment in 1973 as the first director of the U.S. Federal Office of Energy Conservation. In 1975, he returned to Tennessee to create and lead UT's Energy, Environment, and Resources Center. Gibbons' pioneering research in these closely interrelated fields—energy, environment, and resources—eventually led to his appointment as President Bill Clinton's science advisor in the 1990s.

The national interest in nuclear energy decreased in the early 1970s, in part due to the increased focus on the health of the environment and early concerns about nuclear reactor and nuclear waste safety. This change led to decreased projections of the number of reactors that would be built for providing electrical power and resulted in suggestions by President Richard Nixon and others that the federal government should sell its three gaseous diffusion plants (in Oak Ridge, Tennessee; Paducah, Kentucky; and Portsmouth, Ohio) to private industry because of their enormous demand for electricity.[1] [2] [3] Research on new, more efficient strategies for enriching uranium advanced in this decade—including enrichment through centrifuges and the use of lasers. However, these new enrichment strategies did not bear fruit in this or the next decade, ultimately leading to the closure of the Oak

Ridge Gaseous Diffusion Plant (K-25) in 1983. Through the decade, Oak Ridge still maintained a significant nuclear presence, which led to a 1972 airplane hijacking and threatened crash into the High Flux Isotope Reactor.

Hijackers of Southern Airlines Flight Target Oak Ridge

The hijacking of Southern Airlines Flight 49 by three men began on November 10, 1972, and ended thirty hours later, garnering headlines and triggering news bulletins around the world. The nuclear facilities in Oak Ridge became a focal point in the riveting saga. The flight of the twin-engine DC-9, with thirty-four people onboard, originated in Birmingham, Alabama, and the drama ended in Havana, Cuba. The crew, plane, and passengers would eventually be returned to the United States.

One of the hijackers, Melvin Cale, was from Knoxville and had escaped from the Nashville Community Work Release Center, where he was serving a five-year sentence for grand larceny.[4] The other two hijackers lived in Detroit, Cale's half-brother Louis Moore, 27, and Henry Jackson, 25, who Detroit police alleged had perpetrated a number of sexual assaults. After Moore and Jackson made bail on the sexual assault charges, they headed to Tennessee to rendezvous with Cale, who had long been fascinated with plane hijackings.[5] The three men, heavily armed, drove in a stolen car to the Birmingham, Alabama Municipal Airport. They were able to board Southern Airways Flight 49 with their weapons, which included handguns and grenades, without airport security's detection.[6]

Once onboard and en route to Detroit, the hijackers took over and demanded a ransom of $10 million, to be paid by the city of Detroit, as restitution for Moore's and Jackson's alleged harassment at the hands of the city's police force. When Detroit refused to pay the ransom and efforts on the part of the airline to raise the funds faltered—and following stops in Cleveland and Toronto—the hijackers had reached their limit and threatened to fly the plane to Oak Ridge and crash it into HFIR. Having grown up in Knoxville, Cale must have been aware of the reactor.

On the previous day, a civil defense preparedness test was conducted in the Oak Ridge area,[7] and all but a couple hundred of the Oak Ridge facilities' thirteen thousand on-duty employees were evacuated. The reactors also were shut down.[8] As nightfall approached, a decision was made to turn off all lights at the Oak Ridge facilities as a precautionary measure in response to the hijackers' threat to attack HFIR. In some cases, power to the lights were severed with cutting tools.[9] A quick analysis indicated the worst-case damage to HFIR in a plane crash would not disperse radioactive materials beyond the Oak Ridge Reservation, and, besides, HFIR already had been shut down.[10]

With the commandeered jetliner circling overhead, word reached the hijackers that Southern Airlines had managed to raise the ransom, which would be waiting for them in Chattanooga. The plane subsequently left the airspace over Oak Ridge. After receiving the ransom—$2 million, not the demanded $10 million—the plane departed Tennessee bound for Cuba.

Following the incident, a representative of Southern Airways flew to Cuba but officials refused to release the $2 million ransom the airline had handed over to the hijackers. On August 15, 1975, two and a half years after the hijacking, President Fidel Castro finally authorized the release of the ransom money back to the airlines. The hijackers were sentenced to eight years in prison in Cuba and returned to the United States in 1980, where they pled guilty in a 1981 trial. Louis Moore and Melvin Cale were sentenced to twenty years and Henry Jackson to twenty-five.[11]

The hijacking of Southern Airways Flight 49 and the threat to the facilities in Oak Ridge made it clear that airplanes could be used as weapons, prompting the Federal Aviation Administration to mandate screenings for all U.S. airplane passengers, effective January 5, 1973.

Through the tumultuous decade of the 1960s, student anti-war protests that affected other universities were largely—but not completely—absent at UT. This changed on May 28, 1970, when President Nixon joined evangelist Billy Graham in a large revival-type gathering in UT's Neyland Stadium, resulting in the arrest of several protesters.[12] This demonstration was yet another sign that the university was changing in profound ways, with increasing avenues for faculty research, greater opportunity for UT students and faculty to pursue science at the facilities in Oak Ridge, and, at least for a time, a politically agitated student body willing to openly express its views.

UT Knoxville chancellor Charles Weaver invited evangelist Billy Graham to campus for a nine-day (May 22–31, 1970) crusade in Neyland Stadium. Students and faculty had raised issues over the apparent violation of church-state separation, but Weaver assured them that this was not officially a UT-sanctioned event. Graham's "East Tennessee Crusade" drew tens of thousands of people daily to hear the noted evangelist preach. The crescendo of this event was a visit by President Richard Nixon on May 28.[13] Anti-war student protesters chose to attend this rally *en masse*, part of a crowd of one-hundred thousand people, most of whom supported Graham and Nixon. Activist faculty Charles Reynolds (religious studies) and Kenneth Newton (psychology) led more than four hundred students and about a dozen other faculty into the stadium, and many of them carried signs reading, "Thou Shall Not Kill." Law-enforcement crackdown on the protesters was swift, and forty-seven people were arrested during the Graham crusade, with one case—that of Reynolds—almost reaching the Supreme Court.

Other faculty tried to disrupt the Nixon appearance from outside the stadium, and among them were physics professors Paul Huray and Gil Nussbaum. The UT physics building sits adjacent to Neyland Stadium, and, in 1970, one could peer into the stadium from the building's roof. (The north end zone was later enclosed with double-decked stands, blocking the rooftop view.) Huray and Nussbaum tried to rig up a set of large speakers to mount on the building's roof to broadcast anti-war messages into the stadium during Nixon's speech. Although both excellent experimentalists, neither was competent enough in audio electronics and their plan failed—the speakers did not work. William Bugg was head of the Department of Physics at the time, but he knew nothing of their foiled plan and was relieved that the FBI never came to visit his department.[14]

THE RISE OF ENVIRONMENTAL PROGRAMS

By the end of the 1960s, ORNL's Biology Division was the largest at the laboratory, with a staff of 450.[15] Late in 1969, funding for the molten-salt reactor program was cut by two-thirds, and a staff reduction across the laboratory was ordered due to budget shortages. At about the same time, the proposal for a next-generation accelerator for nuclear physics was axed, and ORNL entered the decade of the 1970s, significantly altered in terms of research mission and staffing. Indeed, ORNL's workforce, which had peaked at 5,500 in 1968, dwindled to 3,800 by early 1973, when AEC formally ended the molten-salt reactor program.[16]

The increased focus on the environment that had started to evolve in the 1960s only expanded in the 1970s, with implications for the laboratory, the university, and the nation. MIT physicist David Rose led 1969 and 1970 summer seminars at ORNL to explore possible future directions for the laboratory.[17] As noted in chapter 4, Rose became an ORNL employee while on leave of absence from MIT. These explorations led to ORNL evolving into an important environmental sciences laboratory through the 1970s. In 1970, the National Environmental Policy Act (NEPA) passed, and the Environmental Protection Agency was established. In March of that year, Weinberg expanded Stanley Auerbach's Ecology Section into an Ecological Sciences Division with focus on studies of terrestrial, aquatic, and forest ecology.[18]

Groups specializing in radiological assessment and geosciences were added, and, in 1972, this research organization was renamed the Environmental Sciences Division. In 1969, UT launched a graduate ecology program affiliated with the Environmental Sciences Division. This program evolved into the Department of Ecology and Environmental Biology with close ties to ORNL.[19]

NEPA required all federal agencies to conduct studies and issue environmental impact statements before initiating projects that posed significant effects on "the quality of the human environment."[20] The law sought to anticipate the potential environmental consequences of these projects before the first shovelful of soil was turned, and few organizations were better equipped to conduct these studies than ORNL. Additionally, NEPA generated a considerable amount of funded work for the new Environmental Sciences Division. Among numerous other environmental impact studies undertaken by ORNL staff, Bill Fulkerson, Dub Shults, and Bob Van Hook led the examination of the impacts of existing fossil-fuel power plants on the "human environment."[21] Each of these scientists would later become ORNL leaders (division directors or associate laboratory directors) and work to promote the UT-ORNL partnership.

THE EARLY PROMISE OF CHEAP NUCLEAR POWER FAILS TO MATERIALIZE

In the 1950s, nuclear reactors were widely touted as the cheap, efficient, and safe answer to the world's increasing need for energy. Countries poor in fossil fuels or absent of vast water supplies to provide hydroelectric generation of electricity could have reactors that would cheaply yield abundant energy—emphasis on "cheaply."

This euphoria over nuclear power started to change in the 1960s, perhaps in connection with the rise in the national emphasis on environmental issues. Public concern over the safety of reactors and their short- and long-term impacts on the environment increased.[22] Though there was not yet an accident to ignite public concern—that would come on March 28, 1979, with the partial meltdown of a reactor at the Three Mile Island nuclear facility in Pennsylvania—by the early 1970s, in some public quarters, concern increased about nuclear power's two most-obvious drawbacks: the risk of an accidental release of radiation and the safe disposal of spent reactor fuel rods.

Scientists and engineers have devised technical solutions to these issues, but they come at considerable expense, which has dramatically driven up the actual cost of nuclear energy, and even the most-conscientiously designed protective solutions come with no absolute guarantees. Indeed, no risk assessor can ensure that an accident related to a nuclear reactor will not occur.[23] Likewise, no one can offer assurance that a hydroelectric dam will never structurally succumb to the pressure of the impounded water it holds back. But, in our legalistic society, one needs to prove that the probability of such accidents is extremely and acceptably low. The precise meaning of these two adverbs is debatable, but, as technicians approach the ideals they represent, the associated costs of the generated electricity dramatically increase.

Even before the 1960s, research programs, including those at ORNL, focused on the safety of nuclear reactors. Through the 1960s, ORNL was primarily a reactor R&D facility, long dedicated to designing and building new kinds of reactors for various purposes (civilian power, naval ships, nuclear aircraft, desalination). ORNL director Alvin Weinberg, one of the most-noted reactor physicists/engineers in the world, was the key leader and proponent of this research focus, which included an emphasis on safety. During this period, an increasing portion of the laboratory's effort was dedicated to studying the possible failure of reactors' various components: the fuel rods, the cooling systems, and the containment vessels. Indeed, by the end of the 1960s, 20 percent of ORNL's reactor budget was devoted to nuclear safety, engaging one hundred scientists and engineers.[24]

This growing emphasis on the safety systems for nuclear reactors increasingly preoccupied Alvin Weinberg. In fact, he opened his 1970 State of the Lab speech with the following observation:

> We in the nuclear community have been comfortable in the belief that our work—providing a great new source of energy—is an unmitigated and obvious good. It therefore comes as a perplexing shock to realize that the nuclear community is confronted with what seems to be a crisis of public confidence. Opposition to nuclear energy, which was first expressed publicly seven years ago by David Lilienthal, has mushroomed. . . . Where we insist nuclear energy is clean, our critics claim it is dirty. Where we insist nuclear energy is safe, our critics claim it is unsafe. Where we insist it is needed for our ultimate survival, our critics say it is unnecessary. . . . We at ORNL have a particular responsibility in this matter . . .[25]

Lilienthal was chair of TVA in 1941 and the first chair of AEC in 1946. Lilienthal's questions about nuclear safety indicated that a few well-positioned American policymakers had doubts about this power source. In a 1971 talk that Weinberg gave at a meeting of the American Nuclear Society, he first discussed the "Faustian bargain" that the builders of nuclear reactors had made with society. In this instance, society's reward for striking the Faustian deal? Nuclear engineers and executives are the guardians of an inexhaustible source of energy, one cheaper and less polluting than energy from the burning of fossil fuels. However, "the price that we demand of society for this magical source is both a vigilance from and longevity of our social institutions that we are quite unaccustomed to."[26] In other words, nuclear engineers provide the public with an unlimited source of energy, and, in return, the public must place its trust in the engineers and their technology.

In the years of the Cold War following World War II, many Americans may have trusted government enough to accept this deal, but, by the early 1970s—after the cultural upheaval of the 1960s, the Vietnam War, and Watergate—Americans' distrust of government began a descent that continues to the present. So, the transactional arrangement that Weinberg described does not work in today's world. Discussion of this bargain made Weinberg increasingly unpopular in the nuclear industry; he was viewed almost as a turncoat against nuclear energy, which, of course, was far from the truth. But, this view started to erode the confidence that Representative Chet Holifield and the AEC nuclear reactor office had placed in Weinberg and would contribute to his downfall a year or so later.[27]

Another source of increasing conflict between Weinberg and his AEC superiors concerned the type of breeder reactor design that the country should pursue.[28] In the heyday of the emergence of nuclear power in the 1950s, the assumption was that reactors would become the dominant source to satisfy the public's ever-increasing demand for energy. When that occurred, experts could foresee an eventual shortage of economically recoverable uranium from the Earth. In short, the worry was that we would run out of accessible uranium that could be enriched into ^{235}U to drive the nuclear reactors.

Thus came the idea to build a breeder reactor, a device that creates more fissionable material than it uses. Enriched ^{235}U initiates the chain reaction in the reactor, and the capture of a fission neutron by the plentiful ^{238}U (which accounts for more than 99 percent of natural uranium) in the fuel rod produces ^{239}Pu, which does not exist in nature but is an even better fissionable material than ^{235}U.[29] In breeder reactors, the fuel rods are removed and reprocessed to extract and effectively recycle the un-fissioned ^{235}U and the created ^{239}Pu to generate fuel for the next set of fuel rods. The reprocessing of used nuclear fuel rods also removes much of the long-lived radioactive isotopes from the rods and thus makes long-term disposal of nuclear waste easier and safer—a strategy that France uses to this day.

All water-cooled reactors breed ^{239}Pu in the mostly ^{238}U fuel rods, but the amount bred is small. To increase the plutonium breeding ratio, water is not used as a coolant because it absorbs some of the emitted fission neutrons and makes fewer neutrons available to be captured by ^{238}U to produce ^{239}Pu. Instead, a liquid metal (such as sodium) is used to take the heat out of the reactor core and transfer it to the electricity-producing mechanism. Water also serves as a moderator in a normal reactor to reduce the energy of the emitted neutron and make a chain reaction in ^{235}U more likely. In the absence of water, more highly enriched ^{235}U (15 to 20 percent) is used in a breeder reactor than that used in a normal reactor (3 to 5 percent) to sustain the chain reaction. With no neutron-moderating material, the average energy of the emitted fission neutrons remains higher (faster), which improves the chances for capture of a neutron by ^{238}U to produce ^{239}Pu. Fast-neutron fission of ^{238}U also contributes a little to the chain reaction dominated by ^{235}U. This device that produces fission is thus called a liquid-metal fast breeder reactor.

A debate ensued concerning the specific type of breeder reactor to design and build as the foundation for the future of U.S. reliance on nuclear energy. AEC favored the liquid-metal fast breeder reactor to produce fissionable ^{239}Pu as a substitute for the declining amount of uranium to be mined. This breeder reactor was designed primarily at Argonne National Laboratory, and, in 1972, the chosen site for a breeder reactor plant was in Oak Ridge along the Clinch River not far from the K-25 site, thus the name the Clinch River Breeder Reactor (CRBR).[30]

Weinberg of course supported the CRBR project, and, indeed, ongoing research at ORNL addressed various breeder reactor design needs. However, he strongly favored a parallel path of developing a thermal breeder reactor based on the molten-salt fuel and coolant systems that ORNL had worked on in connections with the earlier nuclear aircraft project.[31] In this scheme, the reactor fuel would be thorium (^{232}Th), which is abundant in the Earth's crust. Thorium does not fission, but, in a reactor core, it would capture a neutron, leading to creation of the highly fissionable isotope ^{233}U, which is not found in nature. In short, this reactor is capable of breeding fissionable ^{233}U from un-fissionable thorium. Weinberg described this reactor scheme as "burning rocks" for energy. The specific molten salt reactor that uses a thorium molten salt as the coolant and fuel runs at higher temperature and lower pressure than the pressurized water-cooled reactor at the heart of most nuclear power plants. The reactor's lower-pressure operation reduces mechanical stress on system components, simplifying reactor design and improving safety.

Weinberg pushed hard for ORNL's molten-salt thermal breeder reactor as an alternative to the fast breeder that AEC preferred.[32] The MSRE at ORNL went critical in June 1965, and the fuel was switched from ^{235}U (plus thorium) to the newly bred ^{233}U (plus thorium) in October 1968. These experiments ran well until December 1969, when project funding ended. Weinberg advocated for a thermal breeder demonstration project, using the same building that had housed the experimental gas-cooled reactor project until terminated by AEC in 1966. However, AEC's Milton Shaw, director of the commission's Reactor Development and Testing Division, focused on the fast breeder development and demonstration project on the Clinch River.

The ORNL molten-salt project limped along until January 1973, when AEC abruptly ordered that work on it end.[33] In later years, Weinberg explained Shaw's focus on the fast breeder concept to the exclusion of the thermal breeder scheme in two ways.[34] First, the fast-breeder concept was the first to come to the attention of AEC, and it was based on the existing uranium-reactor technology. So, the AEC thinking about a breeder got focused quickly on the *fast* breeder reactor. The second reason was that a molten-salt reactor scheme was a new concept and policymakers knew less about it and, thus, trusted it less, even though a molten-salt reactor had been built and operated at ORNL. This new technology greatly differed from the existing light-water reactor scheme. It is not uncommon for a new, better technology not to be initially accepted since it differs from the current state of the art. Weinberg fervently believed in the molten-salt reactor as a power source and as a breeder, ideas that AEC rejected in the early 1970s. At the writing of this book, renewed interest occurs on the part of several laboratories, companies, and countries in the molten salt reactor. ORNL has hosted Molten Salt Reactor workshops since 2015.

CLINCH RIVER BREEDER REACTOR DOOMED BY COST PROJECTIONS

In 1972, Shaw chose Oak Ridge as the site of the demonstration of the liquid metal fast breeder reactor. The resultant Clinch River Breeder Reactor had been authorized in 1970, and initial appropriations were provided in 1972. After termination of Weinberg's favorite thermal breeder reactor program, research at ORNL focused on issues related to the fast breeder technology. This started to have a significant impact on the region, as, for example, the formation of the Technology for Energy Corporation (TEC) in 1975 in a new complex along the recently constructed Pellissippi Parkway, which linked Oak Ridge and West Knoxville by a four-lane divided highway. The first part of this parkway opened in 1971 and was finally completed as a shorter route to the Knoxville airport in 1987. Governor Lamar Alexander (1979–87) pushed through the funding for the completion of the parkway and saw this as a corridor that would give rise to high technology developments perhaps akin to Silicon Valley in California or the famed Route 128 corridor in Boston.

Norbert (Bert) Ackermann Jr., then an ORNL employee, and two UT professors formed TEC in 1975 to provide instrumentation in support of nuclear reactor operation, focused in large part on subcontracts from the big CRBR project. Ackermann's father had been the captain of the 1940 national champion UT football team. Bert, like his father, came to UT to play football and, in the early 1960s, was the Vols' starting center.[35] He graduated with an engineering degree in 1965 and then a PhD in nuclear engineering in 1971. TEC was a local company that did well for some years until CRBR's termination in 1983.

Early on, tough questions were raised about the need for a breeder reactor. For example, Thomas Cochran[36] wrote a book in 1974 taking a critical look at the economic and environmental arguments that had been made in favor of an early introduction of the liquid metal fast breeder reactor as a central component of the U.S. electrical energy system.[37] Jimmy Carter became the U.S. president in January 1977 and opposed the breeder reactor project because of concerns about nuclear proliferation. He served one term, and Congress continued funding the CRBR project over his objections.

Carter's successor, President Ronald Reagan, supported the project, but opposition was mounting in various circles. The cost of this first-of-a-kind project escalated. AEC's initial estimate in 1971 was $400 million, but, a year later, the number grew to $700 million. By 1981, $1 billion of public money was spent on the project, and the estimated cost to completion grew to $3 billion, with another $1 billion needed for an associated spent nuclear-fuel reprocessing facility.[38]

Ground was broken at the site in Oak Ridge along the Clinch River, but opposition increased for three primary reasons: the escalating cost of the project; the fear of nuclear proliferation should terrorists somehow steal the stream of reprocessed fissionable uranium and plutonium; and the new calculation that the supply of uranium ore on Earth would last far longer than anticipated in the 1960s, largely because nuclear energy was firmly in a retrenchment mode by 1981.[39] Still, Tennessean Howard Baker, who became Senate Majority Leader as Reagan became president, kept the project alive. A *Time Magazine* article quoted a U.S. House CRBR opponent as observing that the strongest support for the project came from Baker; "After that, you don't need any other lobbyists."[40] However, even Baker could not turn the tide, and the Senate terminated the CRBR project in October 26, 1983. This dealt a big financial blow to Oak Ridge and the East Tennessee region.

DISAGREEMENTS OVER REACTOR DESIGN, SAFETY CONCERNS LEAD TO WEINBERG'S DISMISSAL

Alvin Weinberg's public concern about the safety of nuclear reactors and his insistence on a molten-salt breeder program led to increasing tension between Weinberg and Milton Shaw, director of AEC's Reactor Development and Testing Division.

Shaw was a strong and controversial AEC administrator and a protégé of Admiral Hyman Rickover, known as the father of the U.S. nuclear Navy. Shaw, actually a Knoxville native, was born in 1921; his father, William Shaw, was a professor of agricultural chemistry at UT. Shaw studied mechanical engineering at UT and, upon graduation, joined the Navy. After World War II, he continued his career with the Navy and became part of Rickover's team sent to the Oak Ridge School of Reactor Technology in 1950–51. In 1964, Shaw left the Navy and joined AEC. He remained personally close to Rickover and shared his mentor's gruff and autocratic management style.[41]

Shaw particularly criticized Weinberg's emphasis on continuing research into reactor safety issues. In his view, if a reactor were built properly and according to specifications, there should be no safety concern, and, thus, Shaw saw no need for continuing R&D on reactor safety. He did not agree with the new emphasis on calculating the probability that a reactor would fail, seeking instead to minimize the probability. And, further, Shaw had decreed that the liquid metal fast breeder reactor was the strategy for the future. Since then, many have questioned Shaw's emphases and leadership at this period in the early 1970s.[42]

Weinberg's tenure as ORNL director started to unravel in 1972 when he and ORNL Deputy Director Floyd Culler were called to meet with Congressman Chet Holifield, chairman of the House Atomic Energy Subcommittee and the Joint Committee on Atomic Energy. Holifield strongly supported Milton Shaw and, in the meeting, apparently criticized Weinberg's de-facto "rebellion" against Shaw and the Oak Ridge leader's continuing focus on reactor safety. Further, Holifield suggested that it might be time for Weinberg to leave the ORNL directorship.[43]

Weinberg was irate over this exchange and pledged to continue his role as director, but, later in 1972, John Swartout, vice president for research at Union

Carbide Corporation, told Weinberg that he needed to leave the director's office. On January 1, 1973, Weinberg began a leave of absence, while Culler assumed the role of acting director. This effectively marked the difficult end of Weinberg's long and brilliant career at ORNL.

Events in fall 1973 significantly shaped Weinberg's future. On Yom Kippur, which fell on October 6, Egypt and Jordan attacked Israel, and the United States quickly provided aid to Israel, which was able to regain the initiative and defend itself. Arab anger over this intervention was immediate, and, on October 16, the Organization of Petroleum Exporting Countries (OPEC) voted to raise the price of oil by 70 percent per barrel. Three days later, members of OPEC voted to cut off all oil supplies to the United States. Gasoline prices rose quickly and severely in the United States, as the cost of petroleum rose from $3 per barrel to nearly $12 by March of 1974. Long lines at gas stations were common, car-pooling to work was mandated, and a national 55-mile-per-hour speed limit was implemented. The age of cheap oil was over, forever.[44] [45]

The Nixon administration focused on U.S. energy issues, even as the Watergate hearings began in May 1973. In June of that year, President Nixon established the Energy Policy Office, and one of its key missions was to increase domestic energy supply by developing the newly discovered vast oil reserves at Prudhoe Bay on Alaska's Arctic coast.[46] A pipeline would have to be built to move the petroleum to Valdez, largely free of the obstructing sea ice found further north, where tankers could load and carry the oil away.

The oil crisis initiated late in 1973 triggered a flurry of activity in the United States and inspired the serious exploration of lighter-weight vehicles and alternative sources of energy. These efforts continue today. One of many outcomes of the OPEC oil embargo was an emphasis on forming energy think tanks that would allow scholars and policymakers to examine energy issues and devise potential solutions. Alvin Weinberg began considering creating such an institute in late 1973. After visiting various national laboratories, searching for a place to locate the planned organization, Weinberg talked to William Pollard about establishing his new energy institute within Oak Ridge Associated Universities. An agreement was reached, and the Institute for Energy Analysis (IEA) began operating in January of 1974.[47]

As part of the frantic response to the OPEC oil embargo, the Nixon administration created the Federal Energy Office, a more robust organization than the Energy Policy Office established earlier in 1973. Weinberg was nominated to direct the Office of Energy Research and Development, a subunit of the larger energy office.[48] He was rapidly confirmed for that role and began work on January 10, 1974. While Weinberg addressed energy issues from his post in Washington, his friend and former ORNL colleague H. G. MacPherson set up and served as director of IEA in Weinberg's stead. MacPherson had attended the Clinton Training School in 1946–47 while employed by Union Carbide, and he joined ORNL in 1956. Weinberg had appointed MacPherson to direct the Molten Salt Reactor Experiment and then made him ORNL deputy director from 1965 to 1970. MacPherson left ORNL in 1970 to assume a professorship in nuclear engineering at UT, a position he held until 1976. His tenure at the university was interrupted only by his year at the helm of IEA.

After a one-year hiatus in Washington, Weinberg served as the IEA director from 1975 until his retirement in 1984. The institute served as a center for the study of diverse issues ranging from arms control to the atmospheric buildup of carbon dioxide and the associated global climate change. Weinberg had a rich and productive career in his institute after the disappointment of having to leave ORNL.

URANIUM ENRICHMENT FACES MYRIAD CHALLENGES

Oak Ridge and the surrounding area had challenging times in the 1970s. There were intense discussions about enrichment at the gaseous diffusion plants in Oak Ridge, Paducah, and Portsmouth because of projections for the need to double the supply of enriched uranium in the next decades. AEC announced in October 1970 that sales of enriched uranium to utilities in the free world generated more than $100 million for the U.S. Treasury.[49] This followed discussions by the Nixon Administration and AEC beginning in 1969 about the potential sale of gaseous diffusion plants to the private sector; Tennesseans uniformly panned this notion.[50]

The 1973 OPEC oil embargo led to numerous actions—and reactions—in the United States. The cornerstone of President Nixon's Project Independence plan was the quick and rather cheap construction of more nuclear power plants. At that time, AEC predicted that, by 2000, nuclear plants would meet a whopping 60 percent of the country's electricity needs. The actual percentage turned out to be a third of that lofty projection. However, with the energy concerns of late 1973, nuclear power represented an obvious partial solution, which led to deep thinking about the extent of the country's uranium enrichment capability.[51] The enrichment needed for use in nuclear power plants was typically 3 to 5 percent in ^{235}U compared with greater than 90 percent for use in nuclear weapons. The three existing gaseous diffusion plants (all federally owned) certainly could not handle the demand of an expanded nuclear power industry, so discussions explored how to increase this capability and whether the government or the private sector should run the expansion.

Alternatives to gaseous-diffusion enrichment utilizing centrifuges and lasers (i.e., atomic vapor laser isotope separation [AVLIS]) emerged. Enrichment by centrifuge relies on the action of centrifugal force to separate isotopes of gases, in this case uranium hexafluoride (UF_6), whereby the molecules with the desired uranium isotope ^{235}U (about 0.7 percent of the mixture) are forced to the inside of a rotating container, while those with the heavier ^{238}U isotope are forced to the outside. This process yields higher concentrations of ^{235}U while using significantly less energy compared with gaseous diffusion. Since energy is a significant cost of any enrichment process, centrifuge separation showed promise for producing enriched uranium at a significantly lower cost.[52]

In February 1976, Exxon Nuclear announced plans for a $1.5-billion centrifuge enrichment plant on the Tennessee River between Spring City and Dayton.[53] Boeing secured a contract for centrifuge development and began construction in Oak Ridge in 1980. Although construction was suspended in September of that year for budgetary reasons,[54] construction later resumed, and Boeing was supporting about 450 local jobs when centrifuge construction was finally curtailed in 1985. Boeing shifted work at the facility to more defense-related projects and doubled its employment in 1986.[55]

In 1975, President Ford asked Congress to allow production and sale of enriched uranium by private industry.[56] [57] Options began to emerge for acquiring enriched uranium—at a cost much cheaper than that produced through the U.S. gaseous-diffusion process—from overseas suppliers, as well as potentially from U.S. industry, as domestic prices of electricity continued to increase.[58] As a result, energy-intensive enrichment processes became increasingly less attractive. The Oak Ridge Gaseous Diffusion Plant ceased operation in August 1985, and similar operations at Portsmouth, Ohio stopped in 2001 and at Paducah, Kentucky in 2013. According to DOE's Office of Nuclear Energy, foreign countries produce 90 percent of the uranium fuel used today in U.S. reactors.[59] Though laser enrichment of uranium never gained traction, DOE maintains a long-term interest in uranium enrichment by centrifuge, focused on the Portsmouth, Ohio site of a former gaseous-diffusion plant. Oak Ridge experts on centrifuge technology moved to ORNL when the K-25 plant closed and took the lead on a cooperative agreement with Centrus Energy Corporation, which currently works to construct the nation's first production facility for high-assay low-enriched uranium[60] in

Ohio. Some of the reactor designs being prepared for future deployment will need fuel with a ^{235}U enrichment above 5 percent but below 20 percent. This is still far below the enrichment required to make weapons or to power U.S. submarines and aircraft carriers.

THE ELUSIVE PROMISE OF FUSION ENERGY

In 1974, Herman Postma was appointed ORNL director, and he served in this role until 1988. He was important for the further development of the laboratory's partnership with UT. Postma, a fusion physicist who joined ORNL in 1959, became a leader of the laboratory's increasing emphasis on experimental and theoretical investigations into nuclear fusion energy. Early optimism for fusion power reactors was high. In fact, Homi Bhabha, the leader of India's nuclear energy project, opened the 1955 Geneva Conference on the Peaceful Uses of Nuclear Energy by confidently predicting that controlled nuclear fusion would be demonstrated within twenty years.[61] Likewise, in 1965, a professor confidently assured graduate student Lee Riedinger that fusion reactors would generate electricity in thirty-five years. While advances have continued over the past half century, the prediction remains that viable commercial fusion power—which represents a daunting technological challenge—is yet thirty-five years away.

Fusion vs. Fission

Nuclear fusion and fission both produce energy in the process of the nuclear reaction. The *fission* of a heavy nucleus like uranium results in two nuclear fragments (fission products), each with roughly half the mass of the uranium nucleus, plus two or three neutrons, plus energy. The emitted neutrons are important, since they can cause adjacent uranium nuclei to fission and support a chain reaction. A problem to overcome in a fission reactor is that the fission products are very radioactive and so the spent fuel elements need removal and sequestering for centuries.

Nuclear *fusion* works by the forced joining of two light nuclei (e.g., hydrogen) producing energy. The forcing of fusion of hydrogen nuclei takes place in stars due to high temperature and pressure. A fusion reactor simulates this stellar process by confining isotopes of hydrogen in a toroidal ring (a tokamak) and injecting energy into the plasma by various means to cause fusion of hydrogen nuclei. The advantage of fusion over fission is that the fuel (hydrogen) is abundant on Earth (in the form of water), and the fusion products do not pose a radioactive hazard (owing to short half-lives). One difficulty of building a fusion reactor is the huge engineering challenge to design and construct a large torus (like a doughnut) with sufficient size and efficiency to produce more energy than is needed to confine and heat the plasma of circulating hydrogen nuclei. Hopefully these challenges will be overcome and commercial fusion reactors will be built by mid-century.[62]

Through the 1960s, laboratories, including ORNL, experimented with how to form, heat, and confine a plasma of hydrogen nuclei with sufficient density, temperature, and confinement time to allow fusion to occur between the hydrogen nuclei. This process takes place efficiently in the cores of stars—including our Sun—and is the source of the energy that causes stars to burn and radiate heat. Replicating that environment on Earth is difficult, and many laboratories were focused on this challenge in the 1960s. A breakthrough in magnetic confinement strategies occurred in 1968 when Soviet scientists announced the invention of the tokamak, a device that uses a powerful magnetic field to confine the hot plasma in the shape of a torus.

That year, six ORNL fusion researchers traveled to a conference in Novosibirsk in the Soviet Union and also visited the Kurchatov Institute in Moscow. Scientists in the West were skeptical of the rumored success of the tokamak confinement device but were impressed with the results they saw in the Kurchatov laboratory. The leader of this facility, L.A. Artsimovich, came to MIT to give a series of lectures in early 1969, and Postma, who had become director of ORNL's Thermonuclear Division in 1968, sent three of his staff to hear the lectures and talk with Artsimovich.[63] In spite of the Cold War, Soviet scientists gave ORNL considerable advice on how to build a tokamak and improve the plasma confinement parameters. Postma then made an appeal to AEC for funds to build a tokamak at ORNL. Quick approval for funding came from AEC, with an order to make the device operational by 1971. The Oak Ridge Tokamak (ORMAK) was a huge undertaking in part because of the large-scale precision machining it required. The Thermonuclear Division was located at Y-12, and the ORMAK team was able to make use of the plant's exquisite machine shop.

A considerable number of technical problems had to be overcome, and the team of physicists, engineers, and machinists worked around the clock for over a year to resolve them. ORMAK started running in 1971, meeting AEC's deadline, and promising results arrived by 1973. This was the first tokamak to achieve a plasma temperature of 20 million °F, measured in one

INSIDE OF ORMAK, AN EARLY TOKAMAK, WAS GOLD PLATED BECAUSE GOLD IS INERT TO ADSORPTION OF COMMON GASES. COURTESY OF OAK RIDGE NATIONAL LABORATORY.

way by directing a laser beam into the plasma and observing the spread in wavelength of the laser light scattered by electrons in the plasma (this gives a direct measurement of the electron velocity distribution, and hence the temperature). After the success of these fusion experiments, the device was enlarged and renamed ORMAK II in 1973. Alvin Weinberg was optimistic about the early results from ORMAK I and II and anticipated making improvements for the eventual ORMAK III version, but Postma worried about the damaging effects of bombardment of fast neutrons on the walls of the torus. Further, there was still too much leakage of plasma from the tokamak's confinement vessel, and so the next version of ORMAK was not built.[64]

These experiments and those at other laboratories made steady progress over the next forty years in improving the fundamental parameters of the confined plasma (e.g., density of the hydrogen nuclei, temperature, and length of the confinement time), but the world has not come close to a scaled-up fusion power plant. The forefront fusion device in 2020 is being built in southern France by a scientific collaboration among six nations (China, India, Japan, Russia, South Korea, and the United States), plus the European Union. It is called ITER (Latin for "the way"), and ORNL leads the U.S. contributions to this ambitious project. First plasma is expected in 2025, beginning a series of experiments on how to demonstrate controlled fusion in this $60-billion project.

ITER is still two developmental steps away from an on-line fusion reactor to generate electricity for the power grid. A next step after ITER may be to build a smaller tokamak in the first attempt to extract heat and generate small amounts of electricity, followed by a scaled-up device that would more closely approximate the large power-generating fusion reactor that would eventually be part of the national power grid. Perhaps this can all be accomplished by mid-century.

For a period in the 1970s, the fusion program was the largest DOE-funded program at ORNL. UT contributed strength to this program, including electrical engineering faculty member Igor Alexeff and physics faculty Ed Harris, Bob Scott, and Owen Eldridge. However, when these Oak Ridge programs waned in the 1980s, the UT faculty recruited for fusion research retired or left for other institutions. In 1974, Postma was promoted from his job as director of the Thermonuclear Division to ORNL director, beginning a highly acclaimed fourteen-year tenure.

Early Fusion Research

One of seven UT faculty who came to Oak Ridge to join the Manhattan Project was E. D. (Ed) Shipley. In 1936, he joined the UT faculty and helped form and then head the Department of Electrical Engineering.[65] After WWII ended, Shipley stayed in Oak Ridge and built a long and successful career, especially in the area of nuclear fusion. Initially, he was the director of research and development at the Y-12 Plant and shifted to ORNL when responsibility for three research divisions was transferred from Y-12 to the laboratory in 1950.[66] The ORNL Thermonuclear Experimental Division was formed in 1957 with Shipley as the first director. This division was located at Y-12, due to the availability of high-voltage electrical power. This is where experimental nuclear fusion research started in Oak Ridge, and the first fusion device constructed in 1957 was called the Direct Current Experiment (DCX).

Two versions of this device were built and proudly exhibited at the 1958 Second United Nations International Conference on the Peaceful Uses of Atomic Energy in Geneva, Switzerland. Soviet fusion scientists were especially interested in seeing and even operating the DCX at this conference. In 1961, ORNL reorganized its fusion program to create the Thermonuclear Division, led by Art Snell with Shipley serving as deputy director. Shipley later became a member of ORNL's senior research staff, retiring in 1972. Herman Postma, who succeeded Snell as division director in 1967, was a dynamic leader of the fusion program at ORNL and then the director of the laboratory for fourteen years. Postma and others built the fusion program partially on early contributions by Shipley, the former UT engineering professor.

ORNL HELPS THE NATION CONFRONT INCREASING ENERGY WOES

The OPEC oil embargo in October 1973 shook the psyche and the economy of the United States deeply. The average U.S. cost of a gallon of gasoline, 36 cents in 1972, climbed to 57 cents by 1975 (almost $3/gallon in today's dollars). A second shock occurred after the start of the war between Iraq and Iran in 1979, and, by 1981, a gallon of gas cost $1.38 (equivalent to around $4 today).[67] The first shock, in particular, led to numerous federal actions targeting energy use, distribution, efficiency, and a greater national focus on new energy strategies.

ORNL's Energy Division was formed in 1974, in part a response to the nation's mounting concern over its energy future following the OPEC oil embargo of 1973. Another driver in the division's creation was strictly domestic in nature and predated the embargo by three years. The passage of the National Environmental Policy Act in 1970, first referenced earlier in this chapter, led to the requirement to conduct studies and release environmental impact statements before federal construction projects could commence.

This legislation rather suddenly affected reactor construction projects, leading to public conflict between the nuclear power industry and antinuclear groups like the Union of Concerned Scientists. This conflict culminated in a year-long series of hearings held by AEC in 1972 in Bethesda, Maryland. The

Hearings on Acceptance Criteria for Emergency Core Cooling Systems for Light Water-Cooled Nuclear Power Reactors were sparked by an application for a reactor to be built at Calvert Cliffs, Maryland.

These hearings focused on the robustness of the water-cooling systems for large reactors (over 400 MW) being designed and built for the civilian nuclear energy industry. ORNL experts served as key witnesses during this public spectacle, as did other experts, including those representing groups opposed to nuclear power. AEC had issued interim criteria for emergency core-cooling systems, and the purpose of the hearings was to get input from various groups and stakeholders on these criteria.[68] One question involved the highest temperature that nuclear fuel rods could withstand without experiencing failure. Reactor vendors were confident that 2,300°F was safe, but others argued for 2,100°F, necessitating more robust core cooling systems.[69] The twenty thousand pages of testimony from the hearings resulted in the establishment of more stringent safety criteria and eventually led to the congressionally mandated abolition of the AEC in 1974. The hearings also created a research organization, the Energy Research and Development Administration (ERDA), and a watchdog unit focusing only on commercial nuclear reactor safety and licensing, the Nuclear Regulatory Commission.

The new safety criteria for reactor construction prompted the need to revisit and revise environmental impact statements for ninety-two recent AEC applications. AEC requested the rapid assistance of several national laboratories, and, in response, ORNL formed a group of seventy-five people, led by Ed Struxness and Tom Row, to perform these statement revisions. These efforts represented a key emphasis of the new Energy Division at the time of its creation. Sam Beall—who had a long history at ORNL in reactor development—was tapped to form and lead this division. He served in this role for a year, before retiring; Bill Fulkerson replaced him.

In 1971, Congress authorized AEC to investigate all energy sources, not just nuclear. This led to a host of new programs at ORNL, especially in its Energy Division. After the energy crisis of late 1973, the federal government was interested in funding R&D programs that promised quick payoffs in terms of benefits to the nation. In response, ORNL initiated programs in energy conservation, efficient use of fossil fuels, and transportation. Subcontracts to UT faculty and students brought added workers to these projects. One outcome from these programs was a substantial improvement in the energy efficiency of home appliances, including heat pumps, refrigerators, water heaters, and ovens. This research led to enhanced federal energy-efficiency standards for appliances, which, in turn, led to a decrease in household energy consumption for these appliances.

NEW CENTERS INCREASE UT'S FOCUS ON ENERGY AND ENVIRONMENT

Historically, universities have existed as institutions defined by a set of distinct and largely independent academic disciplines, whereas national laboratories are, almost by definition, more interdisciplinary by virtue of their mandate to serve the country's diverse R&D needs. However, over the past fifty years, universities have increased their focus on interdisciplinary collaboration through creation of research centers that arch across departments and colleges and draw talent and inspiration from all corners of the institutions. Such was the case at UT with the establishment of three interdisciplinary centers: the Water Resources Research Center in 1964, the Transportation Center in 1970, and the Environment Center (later renamed the Energy, Environment, and Resources Center) in 1973.[70]

People responding to new challenges and opportunities usually drive change, and such was the case with UT chemistry professor Friedrich Schmidt-Bleek, who, in 1969, began to explore opportunities for interdisciplinary environmental research at the university. Schmidt-Bleek identified more than a hundred faculty who wanted to participate in this new effort and perhaps join the staff of a new center. In 1970, he represented the university in a tripartite meeting with ORNL and TVA to discuss a joint environmental agenda. Representing the laboratory were physicist John (Jack) Gibbons, environmental scientist Roger Carlsmith, and social scientist Claire Nader. The meeting was fortuitous in that it prompted Gibbons' initial engagement with UT.

Gibbons joined ORNL as a nuclear physicist in 1954 and worked on research for fifteen years, focusing on the role of neutron capture in the nucleosynthesis of heavy elements in stars. In 1969, Alvin Weinberg appointed Gibbons director of ORNL's new environmental program. In that role, Gibbons initiated and directed research on energy efficiency in buildings, transportation, and electricity generation, while also exploring the environmental impacts of energy production and supply and resource use.

Schmidt-Bleek led the drafting of a 1971 grant proposal to the NSF that showcased the research emphases and expertise of sixty-five UT faculty members. The proposal addressed the Appalachian region's heavy reliance on coal production and the environmental, economic, and social difficulties that had resulted from it. This initial proposal was too broad for NSF financing, but a slimmed-down version was eventually funded in 1972. The UT Environment Center was established early in that year to house the Appalachian Resources Project (directed by Schmidt-Bleek) plus other faculty activities related to this project. University leadership searched for a dynamic leader to hire as director of the Environment Center and found the perfect person in Gibbons, who opened the center for business in March 1973.[71]

The Environment Center under Gibbons' leadership soon began to focus continually more on energy issues, following the OPEC oil embargo in October of 1973. The embargo and ensuing national energy crisis prompted the university and the state to look to Gibbons for help on formulating policies to address energy use and conservation. The federal government also sought his guidance and brought him to Washington, D.C. Gibbons was named the first director of the Federal Office of Energy Conservation by President Richard Nixon in September 1973, just a month before the oil embargo began. In response, Gibbons launched the first national campaign to reduce reliance on fossil fuels and promote energy independence and security. While Gibbons was engaged in Washington, Schmidt-Bleek served as the interim director of the Environment Center.

JOHN GIBBONS. COURTESY OF OAK RIDGE NATIONAL LABORATORY.

In 1974, Gibbons returned to UT to resume his role as director of the center, now largely focused on energy-related issues. As Gibbons resumed his duties at UT, Schmidt-Bleek returned to Germany to direct research on hazardous waste management for the West German Environmental Agency in Berlin. He later became a European leader on environmental issues, founding and directing the Factor 10 Institute, which conducted research on social and economic policy, and serving as honorary president of the World Resources Forum.[72] From 1992 to 1997, Schmidt-Bleek headed the Wuppertal Institute for Climate, Environment, and Energy.

One example among many collaborative efforts between ORNL and UT's Environment Center was the Tennessee Energy Conservation in Housing project, which began in 1975.[73] ORNL's Energy Division developed an innovative home heating and cooling technology called the annual cycle energy system, known more familiarly as a heat pump. The system relied on a pump that extracted heat from a large, insulated tank of water in the winter and changed the water into ice for cooling during the summer.

Gibbons arranged to use university land along Alcoa Highway, between Knoxville and Maryville, to build three demonstration houses to compare heating and cooling strategies. One incorporated the new ORNL heat pump, a second relied on solar power, and the third employed a conventional heating and cooling system. The houses were funded by the Energy Research and Development Administration and managed by UT, ORNL, and TVA. Two additional houses were added in 1979 and 1981. Following President Ronald Reagan's election in 1980, federal funding for research on renewable energy and energy efficiency—deemed by the new administration as the domain of the private sector—decreased drastically. This ended research on these special houses, and control of the structures was ceded to UT's agricultural college, which managed the land on which they were built.

In 1979, Gibbons returned to Washington to head the Office of Technology Assessment. From 1972 to 1995, the office served as an independent, nonpartisan agency that provided analysis for Congress of the benefits, costs, and risks of various efforts to address the scientific and technological challenges facing society. In his last official action as director of the Environment Center, Gibbons changed its name to the Energy, Environment, and Resources Center (EERC) to reflect the center's multidisciplinary—and multisector (government, universities, industry)—research agenda. Gibbons later served as President Bill Clinton's science advisor from 1993 to 1998.

Edward Lumsdaine, a UT engineering faculty member and an expert on solar energy, succeeded Gibbons as EERC director in 1979. In 1983,

Lumsdaine left UT to become dean of the College of Engineering at the University of Michigan at Dearborn. Lumsdaine's name lives on in East Tennessee; his son Arnold joined ORNL in 2009 and is well known as a mechanical engineer in the fusion community. He serves as a joint faculty member at the UT/ORNL Bredesen Center.

In 1983, UT hired Bill Colglazier as a professor of physics and he served as the next EERC director for the following eight years. Colglazier oversaw a significant expansion of EERC's scope and funding and led it to heightened national prominence. He left in 1991 for the National Academy of Sciences, where he served in a variety of roles, including executive officer. In 2011, Colglazier became the science and technology adviser to the U.S. Secretary of State. Jack Barkenbus, who had served for ten years as senior scientist and deputy director of Alvin Weinberg's Institute for Energy Analysis in Oak Ridge, succeeded Colglazier as EERC director. In 2005, the University of Tennessee combined several research centers, including EERC, to form the new Institute for Secure and Sustainable Environment. Gibbons and EERC thus opened a new critical connection between UT and ORNL in research pertaining to energy and environment.

Programs of the Energy, Environment, and Resources Center

The University of Tennessee formed the Environment Center in 1973 to focus on interdisciplinary issues relating to the environment, e.g., those of the Appalachian region. With the energy crisis that started to occur later that year, director John Gibbons expanded programs addressing energy and conservation issues. An example was the construction of demonstration houses along Alcoa Highway in Knoxville to test and compare three strategies for heating and cooling residences. The name of the center was changed to the Energy, Environment, and Resources Center (EERC) in 1979, reflecting this increasing emphasis on energy. EERC leadership and staff played a prominent role in formation of the 1982 International Energy Exposition that became Knoxville's World's Fair which ran for six months that year. Three international energy symposia, organized in part by EERC staff, took place as part of the World's Fair. A National Solar Energy Conference was organized by EERC and held in Knoxville in August of 1982.[74]

Federal support of research programs focused on energy and environment waned in the 1980s during the Reagan administration, and EERC found itself in difficult times. New director Bill Colglazier started a successful broadening of EERC programs late in 1983. One of his emphases was waste management analyses and policies, relating to waste recycling in Tennessee, the federal Superfund program to remediate toxic waste sites, and the Resource Conservation and Recovery Act of 1984 to improve management of future waste sites. In a high-profile case, Colglazier led the analysis of the impact of the proposed Monitored Retrievable Storage facility, an Oak Ridge site that would provide temporary storage of high-level nuclear waste (until a permanent storage facility could be built, which has not yet happened). The analysis of the pros and cons of this facility led to Tennessee Governor Lamar Alexander opposing its construction in his state.

The Tennessee Centers of Excellence program was formed in 1984 and funded the new Waste Management and Research and Education Institute as part of EERC.[75] At that time, this institute was the broadest U.S. research center focused on all types of hazardous waste: solid, chemical, and nuclear. The policy studies of this EERC institute have directly impacted Tennessee regulations related to waste, e.g., recycling, deep-well disposal, solid waste management, and radioactive disposal. The Tennessee Solid Waste Management Act of 1991 was based in part on an eighteen-month study led by EERC. The work of the waste management institute continues today.

Energy policy research and the publication of findings have been closely intertwined between UT and ORNL from the 1970s onward. In 1989, the Joint Institute for Energy and Environment (JIEE) was created as an initiative of UT, ORNL, and TVA. Among the institute's many successes, JIEE received a $2.5 million NSF grant in 1995 to create the National Center for Environmental Decision-making Research, led by UT's Milton Russell, a former EPA assistant administrator for policy, planning, and administration, and a shared faculty member with ORNL. This same emphasis on energy policy would feature prominently in the agenda of UT's Howard H. Baker Jr. Center (created in 2008) and in the curriculum of the Bredesen Center's UT/ORNL joint PhD program in energy science and engineering (created in 2011).

A MOUSE COLONY, FROZEN EMBRYOS, AND THE STORIES WRITTEN IN GENES

The cover of the October 27, 1972, issue of *Science* features a photo of black mouse pups born to their white surrogate mother. The black pups resulted from research collaboration among ORNL Biology Division researchers Peter

Mazur and Stanley Leibo and British collaborator David Whittingham, who spent a year at ORNL after a stay at the Marshall Laboratory in Cambridge, England, and before joining the new Mammalian Development Unit at University College London. They successfully implanted the thawed embryos of black mice into white female mice, and, on June 17, 1972, Mazur and his colleagues celebrated the births of the first mammals from embryos that had been frozen to -321°F in their cryogenic research laboratory. This was the first successful freezing and thawing of mouse embryos without incurring cell damage, and the technique involved identifying a protective solution that allowed the cells to dehydrate as the temperature was lowered less than 1 degree per minute.

This research led to methods to preserve embryos from superior cattle specimens and implant them into the uteruses of other animals, spurring a revolution in animal husbandry. In a 2012 interview with Frank Munger, columnist for the *Knoxville News Sentinel*, Mazur noted that the same procedure described in this article has also been used to preserve embryos

SPECIALLY BRED OAK RIDGE MICE FEATURED ON THE COVER OF 1972 *SCIENCE* JOURNAL. COURTESY OF OAK RIDGE NATIONAL LABORATORY.

of at least twenty-five other mammalian species. In the case of humans, from 2005 to 2009, over thirty-seven thousand pregnancies were produced in the United States by transfer of cryopreserved human embryos.[76] Mazur, a pioneer in cryobiology, joined Oak Ridge in 1959 and was named an ORNL Corporate Fellow in 1985. He retired from ORNL in 1999 and became a research professor in UT's Department of Biochemistry and Cellular and Molecular Biology for nineteen years. Leibo left ORNL in 1981 and worked in industry and academia to improve the success of cryopreservation of many species.[77]

The first U.S. president to visit ORNL was Jimmy Carter on May 22, 1978.[78] This turned out to be a fruitful and well received visit, in spite of Carter's firm opposition to the Clinch River Breeder Reactor program. A roundtable discussion gave a number of laboratory researchers the chance to explain their programs to the president. Briefing topics included analysis of the president's proposed National Energy Plan, alloy properties at high temperature, design of proliferation-proof nuclear fuel elements, effects of air pollution on vegetation, the future of magnetic fusion, and genetic effects of coal products on mice. Liane Russell, who came to ORNL with her husband William (Bill) in 1948, covered this last topic well. Together, they developed and led for decades a program to study the genetic effects of radiation and other environmental hazards on mammals, using a large colony of mice for their groundbreaking experimental research.

Liane Russell

Liane B. Russell made fundamental scientific contributions to basic genetics, teratogenesis, and mutagenesis. Among her more renowned achievements are discovery of the chromosomal basis for sex-determination in mammals by establishing that the Y chromosome determines maleness, her demonstration that only one of two X chromosomes in a cell is active, and her discovery that mutations induced in reproductive cells at different stages differ qualitatively as well as quantitatively. Her findings, and their implications for humans, have been the benchmark for studies into mammalian mutagenesis and for genetic risk assessment worldwide.

In 1943, Liane Russell saw a mouse's fertilized egg under a microscope during a summer program at Jackson Laboratory in Bar Harbor, Maine, led by her future husband, Bill. That experience determined her future career path.[79]

Following World War II and his marriage to Liane, Bill Russell looked for an institution that would employ him and his wife, and Clinton National Laboratory was the only institution that would employ them both. They left Jackson

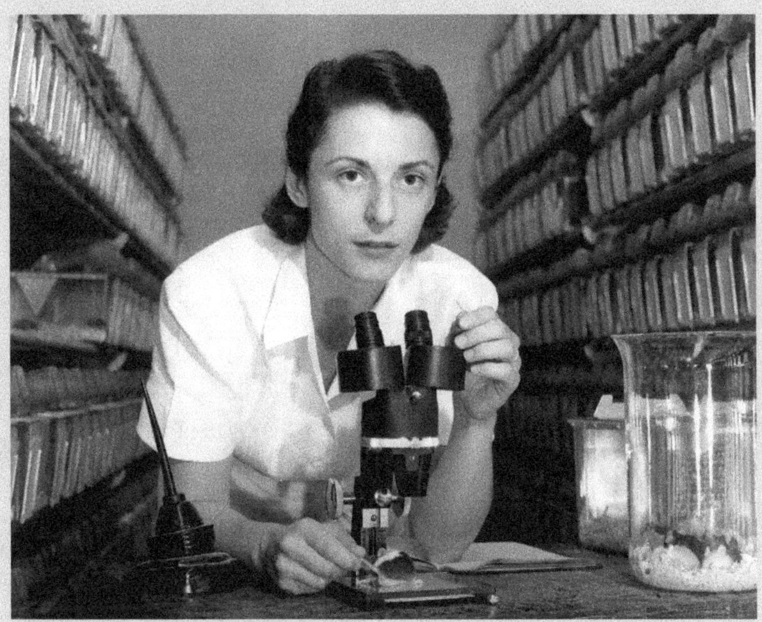

LIANE RUSSELL. COURTESY OF OAK RIDGE NATIONAL LABORATORY.

Laboratory, bound for Oak Ridge, just after a fire had destroyed their research facilities and asphyxiated the mouse colony they hoped to use as a basis for their future research in Tennessee. They soon moved into the ground floor of a Y-12 building and began to build a mouse colony on their own. Eventually, the Russells' mouse colony and research equipment would occupy three entire floors of the large structure.[80]

Radiation effects on embryos and fetuses had not been extensively studied at this time, and so Liane Russell began her research on mice to better understand how in-utero exposure to radiation influenced birth defects and related developmental problems. She was the first to identify periods during embryonic growth that were critical to normal development of body organs and skeletal elements, which led to warnings against exposure to diagnostic radiation during the early stage of pregnancy. Liane continued to study developmental effects of radiations and chemicals on reproductive cells affecting future generations.

Liane had a second career as a volunteer activist for the protection of wild and natural lands and rivers in Tennessee and the nation. Backed by the organization Tennessee Citizens for Wilderness Planning (TCWP), which she had helped found in 1966, Russell's efforts have led to, among other conservational successes, the creation of the 125,000-acre Big South Fork National River and Recreation Area in 1974 and the Obed National Wild and Scenic River in 1976.[81] In 1973, she and TCWP planned and developed the North Ridge Trail, an eight-mile walking path in the city of Oak Ridge.[82]

Among Liane's many honors, she was named an ORNL Corporate Fellow in 1983 and a Senior Corporate Fellow in 1987. She was elected to the National Academy of Sciences in 1986 and was awarded DOE's Enrico Fermi Award in 1993. She died in 2019 at age ninety-five.

ORNL, UT SCIENTISTS USE NEW TOOLS TO GAIN GREATER UNDERSTANDING OF NUCLEAR PHYSICS

Oak Ridge was born out of the need to harness nuclear fission to create the world's first atomic bomb in a frantic race against the scientists of Nazi Germany, which was pursuing a parallel path. Following the war, nuclear physics grew at ORNL and became increasingly focused on the fundamental understanding of the atomic nucleus. Indeed, research on the structure of nuclei has had a rich history in the laboratory's Physics Division, much of it focused on experimental work with particle accelerators. At the Van de Graaff Laboratory starting in the 1950s (see the sidebar on accelerators in chapter 3 of this book), Paul Stelson and Francis McGowan performed measurements using ionized beams of light nuclei (protons, deuterons, alpha particles) to yield early knowledge of how football-shaped nuclei can rotate when bombarded with a projectile.

As discussed in chapter 4, the construction of ORIC introduced a new series of measurements, because of the higher energy of the beams of nuclei produced by this accelerator and also because of the heavier projectiles that could be usefully accelerated (as heavy as the element argon, atomic number 18). This facility also accelerated the recruitment of nuclear physicists by UT and Vanderbilt University and helped establish nuclear physics as one of the most active fields of collaboration with ORNL (and with each other) for the next twenty years.

Joseph Hamilton joined the Vanderbilt physics faculty in 1959 and in the early 1960s began collaborations with Russell Robinson at the Van de Graaff Laboratory at ORNL on measuring the properties of nuclei. Lee Riedinger, then a 19-year-old graduate student, came to Vanderbilt in 1964, joined

Hamilton's group, and took the initial steps that would lead to a long career in nuclear physics. Riedinger was the first of Hamilton's graduate students sent to ORNL, and he spent two years at the laboratory working on research with Noah Johnson and his group for a doctorate, which he completed in fall of 1968. Hamilton has had a remarkable career, sending many students to Oak Ridge for experiments, publishing more than a thousand papers, helping to discover element 117 (tennessine), and initiating a host of joint activities between universities and the national laboratory. Hamilton's collaboration with the laboratory's Physics Division lasted for sixty years, and, in 2002, ORNL honored him as the first Visiting Distinguished Laboratory Fellow. Hamilton retired at age 90 in 2022 after sixty-four years on the Vanderbilt faculty, teaching and mentoring graduate students throughout his long career.

The early research successes of ORIC led to a desire for an even larger device that could produce more energy and accelerate nuclei heavier than argon. ORNL proposed APACHE, but funding for the project was denied by AEC in 1969.[83] This led to a proposal for a new accelerator that ultimately would serve a national group of users and involve numerous universities. A steering committee to draft the proposal for the next heavy-ion accelerator

JOSEPH HAMILTON. COURTESY OF VANDERBILT UNIVERSITY.

formed, including William Bugg of UT and Joseph Hamilton of Vanderbilt. The two scientists not only contributed their ideas for the next accelerator project, but they also led an initiative to form a consortium of twelve universities[84] dedicated to funding, building, and operating an isotope separator on-line to ORIC, UNISOR.

These twelve universities plus Oak Ridge Associated Universities and ORNL contributed funding to build UNISOR, officially beginning in July of 1971 and dedicated in June of 1972.[85] ORNL expanded Building 6000 to create space for this new beam line and isotope separator. While UNISOR was in operation, beams of heavy-ion projectiles (up to argon) from ORIC would strike a target, producing a host of short half-life radioactive nuclei to study. These reaction products were rapidly ionized, extracted by an electric field, and sent through the magnetic separator to allow the dispersal of the many reaction products by the mass of the produced nuclei. This separated stream of radioactive nuclei was then focused on an aluminum foil that would collect the mass-separated species for one or two half-lives before moving them to a detector station where the desired radioactive nuclei could be counted. This process enabled discovery of several new isotopes and new information on many unstable nuclei.

UNISOR was unique in a couple of ways. It was the first isotope separator located at a heavy-ion accelerator, and it was the first university-owned and -operated user facility at a national laboratory. UNISOR's creation eventually led to a change in focus at national laboratories, many of which are now organized around large user facilities open to scientists from across the country (and around the world), based on review of their proposed experiments.

UNISOR operated successfully for thirty years, and, over that time, countless graduate students and faculty from the twelve partnering universities spent summers or sabbatical years conducting research at the facility, leading to the publication of numerous scientific papers. And, the family of collaborators that formed around UNISOR would later conduct more advanced nuclear-physics projects at ORNL, utilizing new accelerator facilities as they came online.

The nuclear physics groups at most of the member universities expanded due to the UNISOR collaboration, especially those at Vanderbilt and UT. Carrol Bingham was hired onto the UT physics faculty in 1967 and subsequently took the lead on the university's UNISOR developments. In 1971, Riedinger joined the UT faculty to help expand the university's emphasis on nuclear physics, and he frequently relied on UNISOR in conducting his research. The robust university collaboration inspired and sustained by

UNISOR became a key selling point in ORNL's proposal for a larger accelerator facility, the National Heavy Ion Laboratory, which was approved for funding in 1974 and began operation in 1980. Upon opening, it was named the Holifield Heavy Ion Research Facility.[86]

The UNISOR project helped to launch the new era of user facilities at national laboratories, and the activity of university faculty and students in nuclear physics at ORNL increased substantially through the 1970s. The presence of this new group of researchers, many far from their homes and often engaged in week-long experiments, suggested the need for an on-site facility where they could get some sleep. Vanderbilt University's Hamilton was one of the prime movers in this initiative, which resulted in the construction of a small visitor facility near the building that housed the ORIC cyclotron and the soon-to-be-built Holifield Heavy Ion Research Facility.

Ground was broken on July 10, 1979, in a ceremony attended by leaders of ORNL (Alex Zucker, associate director, and Herman Postma, director); UT, Knoxville (Chancellor Jack Reese); Vanderbilt (Chancellor Alexander Heard); Union Carbide (Roger Hibbs, president of Nuclear Division); and DOE (James Leiss, associate director for high energy and nuclear physics, Office of Energy Research). Funds to construct the building came from Union Carbide, UT, and Vanderbilt. The earth-sheltered facility, designed by ORNL staff architect Hanna Shapira with technical input from Paul Barnes (also of ORNL), provided temporary living facilities for visiting researchers conducting experiments. This small facility included six bedrooms, a kitchen and lounge area, and five offices that made life during a week-long experiment comfortable for visitors. The building also boasted a number of energy-conservation features, including a roof covered by soil and grass atop a partially submerged structure. It came to be the first building of the Joint Institute for Heavy Ion Research, which would be established in 1982.

PIONEERING ORNL RESEARCH INTO HEAVY ELEMENTS

The opening of the HFIR in 1966 led to a new area of research at ORNL and expanded service to the nation.[87] Situated next to HFIR is the Radiochemical Engineering Development Center (REDC), which is a multipurpose processing and research facility that includes laboratories, glove boxes, and heavily shielded hot cells. HFIR and REDC together greatly expand the availability of such heavy elements as berkelium (Bk, element 97), californium (Cf, element 98), einsteinium (Es, element 99), and fermium (Fm, element 100) for industrial and research applications.[88] These transuranic elements, with atomic numbers above that of uranium (element 92)—the heaviest element found in nature—must be human created. The transuranium elements are produced in the HFIR through the multiple capture of neutrons and separated and purified at the Transuranium Processing Facility for distribution to scientists in the United States and abroad. The Transuranium Research Laboratory (TRL) was a research center where studies were performed on these heavy elements. A major TRL research emphasis sought to explore the chemistry of these newly created heavy elements.

Chemistry of Transuranic Elements

Neutron capture in ORNL's HFIR, the highest flux reactor in the nation, made possible the production of small amounts of elements beyond uranium. Studies of the physics and chemistry of these heavy elements were facilitated by laborious separation of small amounts of these elements from the target material irradiated in HFIR, a challenge in view of the extremely radioactive nature of this material. TRL at ORNL was opened in 1967 and built to allow studies of the chemistry of the heavy elements, berkelium (element 97) through fermium (element 100). Two UT faculty, Joe Peterson from chemistry and Paul Huray from physics, played key roles in experiments at TRL.[89]

Experiments on these elements were difficult for at least two reasons. The amount of the produced and separated heavy element is small, requiring special techniques for experimenting on them—effectively, single-atom chemistry. Also, these elements are radioactive with short half-lives, requiring careful handling and shielding. The special facilities of TRL made possible large advances in knowledge of elements at the upper reaches of the periodic chart, the so-called actinide elements. This lab operated for almost thirty years.

Measuring fundamental chemical and physical properties of the newly created actinide elements gave scientists knowledge about the ordering of the atomic electron orbits for these heavy elements. One way to study the filling of electrons in atomic shells is to measure the atomic magnetic moment, which reflects the strength of the magnetic field that is produced in the atom by its circulating electrons. Working with his UT grad student Stanley Nave and TRL colleague Dick Haire, Huray built a superconducting device that could detect the extremely small magnetic flux that is generated when a microgram-sized sample is placed in a magnetic field. Their measurements showed the specific atomic shell occupied by the valence electrons in the actinide elements, an important result.

Joe Peterson did his PhD work in the 1960s at Berkeley, a university and a

JOE PETERSON, DICK HAHN, AND PAUL HURAY.
COURTESY OF OAK RIDGE NATIONAL LABORATORY.

national laboratory both deeply immersed in the study of heavy elements using accelerators. ORNL capabilities in this field attracted him to the UT chemistry faculty and he spent much of his career conducting heavy element chemistry experiments at TRL, which ORNL's Dick Hahn directed for ten years starting in 1974. Hahn then moved to Brookhaven National Laboratory to work on a more esoteric aspect of physics, the properties of neutrinos coming from the Sun. He led the U.S. effort in the international collaboration that mounted a solar-neutrino experiment underground in the Gran Sasso massif in the Apennine Mountains of Italy.[90] These experiments helped to solve the "missing solar neutrino" problem that had plagued physics for decades by showing that there are actually not one but three types of neutrinos in nature.

In the decade of the 1970s, important heavy-element experiments were performed at the ORIC accelerator facility. The discovery of elements 102 through 105 was proposed from difficult experiments at large accelerators in various countries, but it was always a challenge to be sure which newly found alpha-particle emissions are associated with which elements. In response, ORNL physicist Curt Bemis led a series of measurements at ORIC using beams to produce a few atoms of each of these heavy elements and then observe the time coincidence between the emitted alpha particles (that had been published from other experiments) and the x-rays emitted in that very short-lived radioactive decay process.

The energy of the x-ray is a definitive indication of the element involved. Bemis' complex experiments used considerable accelerator running time, which did not make him popular with other groups wanting to perform their own measurements using ORIC. The experiments were successful, as, for example, the 4.5-second alpha decay of an isotope of element 104 (rutherfordium-257, ^{257}Rf) did indeed lead to the characteristic x-rays of the daughter nucleus of element 102 (nobelium), ^{253}No.[91] [92] The ORNL experiments were the penultimate steps that led to the official naming of these heavy elements by the usual international panel, the International Union of Pure and Applied Chemistry.

HEAVY ELEMENTS, WEIGHTY TOPICS: SCIENCE VS. RELIGION

Religion and science generally do not mix well, especially when one is used to try to explain or justify the other. One example of this tension revolves around experiments performed by Robert Gentry while he was a visitor to ORNL's Chemistry Division for more than a decade. His research dealt with the observation of halos that had formed in minerals such as mica and biotite. The heavy elements undergo radioactive decay in part by the emission of alpha particles, and, when this occurs in radioactive inclusions found in mica, a halo forms in the mineral as the alpha particles slow down and then deposit much of their energy in the mica when their activity ceases. This process causes a change in the mineral's crystal structure and produces a visible circular pattern around the point of alpha particle emission.

It is well known that the heavier the element the higher the energy of the emitted alpha particles, and the radius of the halo is proportional to the energy of the emitted alpha particle.[93] In some cases, the halos were larger than expected, and this finding had been published by various scientists. Gentry spent a number of years investigating these "giant" halos in his collaboration with ORNL. The "giant" term relates to the fact that the radius of the observable halo is larger than expected for known actinide elements, including uranium and thorium, that exist in nature and decay by alpha emission, all having long half-lives (spanning millions or billions of years).

Gentry measured many mica and biotite samples, systematized the data, and published several papers in the prestigious journal *Science.*[94] Besides the giant halos, there seemed to be a conundrum in explaining halos that apparently relate to the comparatively rapid alpha decay of two polonium isotopes (whose half-lives are measured in days), not known to be part of the chain of alpha emissions of uranium, thorium, or the long-lived actinide elements. In the last sentence of his 1974 *Science* article, Gentry questions whether these polonium halos can be explained by presently accepted cosmological and geological concepts relating to the origin and development of Earth over its 4.5 billion years.

In the nonrefereed and nonscientific literature, Gentry took a bold and controversial step in his conclusions relative to the origin and development of Earth. He was a faculty member at Columbia Union College, a small school associated with the Seventh Day Adventist denomination, a protestant faith that observes the Sabbath on Saturday and believes in the imminent return of Jesus Christ. Gentry had a master's degree in physics from the University of Florida and later enrolled at Georgia Tech for PhD work, following his conversion to Seventh Day Adventism. However, Gentry left the doctoral program when his department would not support a dissertation addressing the age of Earth. He started working on halos on his own as a Columbia Union College faculty member and published some early papers on giant halos. ORNL invited him to work at the laboratory over summers, based on his more conventional and scientifically valid research on heavy elements.[95]

Ultimately, Gentry used his detailed research on halos to argue that Earth was not billions of years old, as scientists contend, but, instead, had been created far more recently. In short, Gentry maintained that if polonium isotopes, with their short half-lives, were creating these halos in some of the oldest rocks on the planet (from the Precambrian period), in the absence of the extremely long-lived elements of uranium and thorium, these rocks—and, thus, Earth—must be younger than previously believed, dating to thousands, not billions, of years ago. In this regard, Gentry's hypothesis fell more in line with what some Christians call the "Biblical age of the Earth," based largely on their interpretations of the creation stories in the Bible.

Gentry was not able to get his creationist views published in the reviewed scientific literature, so he self-published the book *Creation's Tiny Mystery* in 1986. Considerable controversy ensued, and many scientists examined his work and came to far different conclusions on the observed halos in mica and biotite, all compatible with Earth's well-known age of 4.5 billion years. ORNL did not renew Gentry's visitor's appointment in 1981. Gentry spent the next thirty years working on his controversial ideas and died in January 2020 in California. A memorial service was held for him at the Seventh Day Adventist church in Knoxville in March 2020. So ended a saga relating to the use of science to try to prove a questionable religious tenet.

Oak Ridge Women's Basketball Enters the Modern Age

Never before were the fortunes of Oak Ridge and UT athletics more closely intertwined than during late summer 1976. On August 13, the Tennessee Secondary School Athletic Association (TSSAA) was sued in Federal District Court in Knoxville by Oak Ridge attorneys Dorothy Stulberg and Ann Mostoller on behalf of Victoria Ann Cape, a rising junior on the Oak Ridge High School basketball team, over its rules limiting high school women's basketball games to half-court play. At the time, Tennessee was one of five states that required six players on the court for each women's team, with three forwards on offense and three guards on defense, and no players were allowed to cross the half-court line during the games. Cape claimed the current rules deprived her of full enjoyment of the game and hampered her future prospects as a college player.[96]

At the trial on October 25, 1976, TSSAA insisted that it only had the best interests of the young women at heart and was providing greater opportunity for "clumsy, awkward girls" to participate. Pat Head (later Pat Head Summitt), the head coach of the UT Lady Vols, testified and declared that "Tennessee schools forced 'a mental and physical handicap' on girls" and that the women were not being adequately pushed physically: "Don't tell me we're not strong enough to run full court."[97] [98]

On November 24, 1976, Judge Robert Taylor issued his ruling, finding in favor of Victoria Ann Cape, but he did not order an injunction because he fully expected TSSAA to abide by his ruling. By a voted majority, TSSAA member schools insisted that no changes were necessary.[99] The Capes asked for an injunction, which Judge Taylor granted on December 27.[100] TSSAA challenged this in a U.S. Court of Appeals, where they were successful in overturning Judge Taylor's decision on October 3, 1977.[101]

Even though Tennessee high schools, independent of TSSAA, approved the proposed rule changes, the TSSAA Board of Control refused to make the change and the issue festered into the next year.[102] A complaint was filed with the U.S. Department of Health, Education, and Welfare (HEW), and, in January 1978, HEW admonished the Oak Ridge school system to treat female and male athletes equally or lose $750,000 in federal funds.[103]

The TSSAA remained entrenched and voted against any changes to the rules until an interview with Vol's coach Pat Head was published on March 4, 1979. Head insisted that she would no longer recruit from within the state of Tennessee unless the high school women's game went full court.[104] She added that the Lady Vols would recruit the two players needed for the next year's team from Virginia and Georgia, bypassing Tennessee.

Head's pronouncement shook the state legislature and soon members began drafting resolutions calling on TSSAA to change course.[105] This finally led to a breakthrough, and the nine-member TSSAA Board of Control voted to have the women play full-court basketball in the next season, three years after the Cape lawsuit. It took the strong remarks of UT basketball icon, Pat Head Summitt, to drag Tennessee into the modern era of women's high school basketball. This story is told in more detail in a book published in 2018.[106]

SUMMARY AND LOOK FORWARD

Emphasis on issues of energy and the environment shifted to front and center in the decade of the 1970s, in the United States, in Oak Ridge, and at UT. April 22, 1970, marked the first celebration of Earth Day, and the OPEC oil embargo of 1973 triggered the nation's first major energy crisis. These events changed the direction of the country in substantial ways. Locally, at ORNL, Alvin Weinberg, who was forced out of his lab directorship role at the end of 1972, established the Environmental Sciences Division and the Energy Division. At UT, the Ecology Program gained popularity, and the Energy, Environment, and Resources Center formed, both by virtue of the university's critical connection to Oak Ridge. The landscape of nuclear reactor research changed at ORNL, and, as the molten salt program was terminated, the fast breeder reactor came to define the nation's future direction in nuclear power. Meanwhile, an increased emphasis on reactor safety research evolved.

ORNL became a more balanced laboratory in the 1970s and, in various ways, this helped build the partnership with UT. The laboratory's new emphasis on energy and environment certainly helped the development of allied programs at the university. Fundamental research at ORNL in nuclear physics grew in this decade as new accelerators received funding, which enabled universities in the southeast to have access to research tools similar to those in other parts of the country. This prompted the development of a strong nuclear physics group in the UT Department of Physics and also at Vanderbilt University. New university-national laboratory synergies grew as UT and Vanderbilt took the lead (with ten other universities) in forming the UNISOR user facility at the ORNL cyclotron, which initiated thirty years of evolving new programs and new joint facilities.

Research in fusion flourished during this decade at Oak Ridge and at other national facilities, and this led to the hiring of new faculty at UT. However, by the end of the decade, everyone came to realize that there would not be a quick transition to fusion power reactors but rather a long slow progression in the technology, perhaps producing fusion power on the grid by the middle of the twenty-first century.

A crucial part of the federal landscape changed in this decade. The Atomic Energy Commission was abolished in 1974 and the Energy Research and Development Agency was started in 1975. This agency was replaced when President Carter established the U.S. Department of Energy in 1977 as the twelfth cabinet-level department.

A productive relationship with a Tennessee governor had yet to develop, as Governors Dunn and Blanton had little or no interest in or impact on the UT-Oak Ridge partnership. However, that would soon change with the election of Lamar Alexander in November 1978.

6

THE DECADE OF THE 1980S

The Partnership Comes of Age

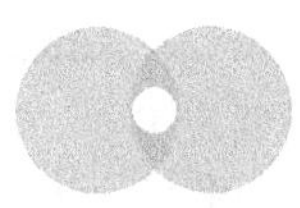

THE 1980S SAW a substantial increase in the interaction between UT and ORNL, and new opportunities and new programs led to heightened expectations for what this partnership could become. In many ways, the 1980s represented UT's coming of age as a competitive research university and a more respected partner of ORNL. The groundwork for this elevated status had been laid in the four previous decades, but many opportunities and activities converged to allow the university to take a giant leap in the 1980s.

The advances in the partnership through this decade came along with, or were made possible by, big changes in the nature of ORNL's structure, staffing, and mission through the 1970s. Emphasis on programs related to energy and environment took hold through the 1970s, driven by national and international trends and events. The partial meltdown of the core of one of the nuclear reactors at Three Mile Island in March of 1979 shook public confidence in nuclear energy, greatly increased the scrutiny of nuclear plants, and led to implementation of ever-stricter safety standards controlling the plants' operation.

Two big changes in federal policy relative to national laboratories occurred around the beginning of this decade. In 1980, Congress passed the Stevenson-Wydler Technology Innovation Act, which U.S. President Jimmy Carter signed into law on October 21, 1980. This was the first major U.S. technology transfer law, and it required federal laboratories to actively participate in and budget for technology transfer activities to nonfederal entities. The other change came in 1979 when DOE announced that its national laboratories should be open to outside users for cooperative and even proprietary research programs, ushering in the era of user facilities at national laboratories across the country. Both of these new policies served to open up national laboratories to companies and universities both for research and for technology transfer.

ORNL and UT entered this user era earlier than others through the UNISOR collaboration at ORNL led by UT, Vanderbilt, and ten other regional universities, starting in 1971. In 1980, ORNL started three new user facilities, all well aligned with research interests of UT and other universities in the Southeast (and across the country). One such facility was the National Environmental Research Park, developed to encourage environmental research on the DOE reservation in Oak Ridge. Another was the Holifield Heavy Ion Research Facility, developed to spur new research in fundamental nuclear physics (an area of expertise among UT physics faculty). The third, the National Center for Small-Angle Scattering Research (funded by DOE and the NSF), sought to encourage use of ORNL's HFIR for neutron scattering on large molecules.

These three new user facilities, all operational at ORNL by 1980, reflected scientific strength among its facilities and its research staff in three important areas: environmental science, nuclear physics, and materials science. ORNL was no longer a "reactor laboratory" but rather a broad multipurpose facility with special strength in these three areas. By 1988, the number of user facilities at ORNL grew from three to twelve, and the number of guest researchers tripled. At the same time, the budget to support nuclear reactor research and development decreased to one-fourth of its previous levels, down to only 3 percent of the laboratory's budget in 1986. In 1987, DOE shut down all ORNL reactors because of management deficiencies and allowed only two (HFIR and Tower Shielding Reactor TSR-2) to reopen a few years later.

Through the 1980s, this triad of scientific focuses—environmental science, nuclear physics, and materials science—also formed the foundation of UT's strength and the basis for new collaborations with ORNL. The first three UT-ORNL joint hires of top research talent were in these areas. In fact, the two institutions built at least some of this strength together by virtue of the developing partnership. And, the institutions' achievement in building these joint areas of strength in the 1980s reinforces so well the ingredients essential to the partnership's successful evolution and outcome.

Discussed in this chapter are some of the crucial people driving this change, among them UT Chancellor Jack Reese and ORNL Director Herman Postma—respectively, an English professor and a fusion physicist. Opportunities for change and development of the partnership came in this decade, and a cadre of people responded to the resultant opportunities. These included university leadership and especially some Department of Physics faculty who explored the possibility of UT competing for the management of the Oak Ridge facilities in the early 1980s. The successes of this decade came with significant contributions from political leadership, including Tennessee Governor (and, by decade's end, UT President) Lamar Alexander and U.S. Senate Majority Leader Howard Baker. A crucial outcome of the work of the 1980s was the inception of a new UT-ORNL Distinguished Scientist Program, which recruited talented joint hires capable of spawning new research initiatives.

These four high-level leaders—Alexander, Baker, Reese, and Postma—were crucial for key developments of the partnership through this decade. These were four very different men, two from the political arena and two from the academic realm, and the impact they had on this partnership cannot be overestimated. At the end of this decade, a higher-ranking individual—President George H. W. Bush—visited Tennessee to learn about the partnership's advances.

STATE, UNIVERSITY LEADERSHIP PLAYS A CRITICAL ROLE

Lamar Alexander, the son of a school teacher and a high school principal, grew up in Maryville, a small, cozy town not far from Knoxville. He learned the piano at an early age and over the years has impressed many with his musical skills. Alexander went to Vanderbilt for his undergraduate degree and to New York University for law school.

After a year of clerking, Alexander, a Republican, began his political career by working as a legislative assistant to Senator Baker in 1967. A few years later, he returned to Nashville, built a law practice, and ran for governor in 1974,

LAMAR ALEXANDER, U.S. SENATOR FROM TENNESSEE, 2003–2021.

losing to Democrat Ray Blanton. Alexander returned to work on Baker's staff in 1977 when Baker was elected Senate minority leader. This led to Alexander's second run for the governorship in 1978, and this time he bested Blanton. Alexander took office three days earlier than usual in 1979. At the time, the Blanton administration was under investigation over a cash-for-clemency scandal.

On January 15, 1979, Governor Blanton announced pardons of fifty-two inmates, including twenty-four murderers.[1] It was reported that Governor-elect Alexander "flirted with and then rejected" an early oath-taking because it would be "totally inappropriate for me to assume power wholly on my own initiative."[2] According to an opinion by the state attorney general's office, this oath-taking could be conducted with Alexander appearing before a magistrate. Alexander was sworn in on the evening of January 17, 1979, following a federal prosecutor's warning that Blanton, given more time, might free more dangerous convicts. U.S. Attorney Hal Hardin spoke with state officials including Lt. Governor John Wilder, House Speaker Ned McWherter, as well as Alexander. Alexander later noted, on behalf of himself, Wilder, and McWherter, that the U.S. attorney "had substantial reason to believe that

Governor Blanton is about to release one or more persons, prisoners who are targets of the United States investigation into alleged payoffs for pardons and commutations of sentences. That information causes each of us to believe that it is in the best interests of the people of Tennessee that the governor-elect assume the office immediately."[3] On June 9, 1981, Blanton was found guilty on eleven counts of liquor license fraud.[4] Blanton served twenty-two months in a federal penitentiary; nine of the original eleven charges were later overturned.[5]

Early in his second term, Governor Alexander took the lead on a bill to raise the state sales tax specifically to support education. Tennessee had long ranked near the bottom of states in spending per student. The higher-education part of the bill called for the establishment of Centers of Excellence at state universities. Universities' proposals for creation of these centers underwent rigorous review, with the limited number of awards based on merit.

In 1973, Jack Reese was named chancellor of UT, Knoxville and served in that capacity for sixteen years, until 1989. Before that, Reese served as vice chancellor for academic affairs and associate dean for graduate studies in the College of Liberal Arts. Reese began his academic career at UT in 1961, when he joined the faculty of the English department. Reese, a native of Hendersonville, North Carolina, received his bachelor's degree from Berea College and a doctorate from the University of Kentucky. He was highly respected

JACK REESE. COURTESY OF BETSEY B. CREEKMORE SPECIAL COLLECTIONS AND UNIVERSITY ARCHIVES, UNIVERSITY OF TENNESSEE, KNOXVILLE.

as a chancellor who always kept the best interest of students, faculty, and administrators at the top of his agenda.[6] He led the university through times of student unrest, took steps to improve campus race relations, and facilitated increases in UT's number of African American students, faculty, and administrators. Other universities experienced similar racial issues, but Reese was particularly effective in dealing with community relations as well as the university's relationship with members of the state legislature in Nashville.[7]

During his tenure as chancellor, Reese led the effort to establish new, more stringent entrance requirements to raise admission standards and enhance the student body's academic pedigree. He was among the first to realize the potential for closer (and more formal) ties to ORNL.[8]

On May 3, 1982, Union Carbide Corporation announced that it would withdraw from its contract with DOE for operation of the Nuclear Division facilities.[9] Union Carbide's management contract started in 1943 for the Oak Ridge Gaseous Diffusion Plant (K-25) and was expanded to include Y-12 in 1947, ORNL in 1948, and Paducah, Kentucky's Gaseous Diffusion Plant in 1950. With Union Carbide's announced departure, DOE started a process to find a new contractor to operate these four facilities.

After this announcement, UT officials immediately started discussions about what the university should do in response. In late May, Vice Chancellor for Research Evans Roth arranged a visit to the University of Chicago to learn the details of its management of Argonne National Laboratory.[10] Roth learned of the complexities of a consortium of universities operating Argonne, leading the University of Chicago to return to the model of one university managing this national facility.[11]

In June 1982, Chancellor Reese established a university task force to make recommendations on how UT should respond to the impending process to select a new managing contractor for the four DOE facilities. Reese, despite the considerable existing demands of the chancellorship, chaired this task force. In managing his now greatly expanded workload, Reese leaned heavily on his right-hand man, Don Eastman, his executive assistant and executive director of university communications.

Eastman would serve Reese and UT for a decade, before working at two other universities and becoming president of Florida's Eckerd College in 2001. Reese's task force included an administrative support group, a science and technology support group, a national laboratories study team, and an academic programs impact study group. These five groups, which included many key administrators and faculty, assisted the chancellor in deciding how UT should respond to the management opportunity in Oak Ridge. The existence of this task force was first publicly disclosed on June 23, 1982.[12]

PAUL HURAY. COURTESY OF BETSEY B. CREEKMORE SPECIAL COLLECTIONS AND UNIVERSITY ARCHIVES, UNIVERSITY OF TENNESSEE, KNOXVILLE.

Reese appointed his friend and tennis buddy, physics professor Paul Huray, to lead the Oak Ridge task force. Huray grew up in Oak Ridge, attended Oak Ridge High School, and was a member of the 1956 football team that won the school's first state championship. A high school friend later recalled that Paul was a special student, different from all the others, and one day skipped a whole day of school so that he could stay at home and build his own oscilloscope.[13] In college, Huray became an engineering physics major, earned a PhD from UT, and joined the UT physics faculty in 1968.

Huray was an excellent teacher and researcher, working in superconductivity at UT and ORNL. Like many other faculty members, Huray eschewed more formal attire and arrived for work each day clad in jeans and a sweatshirt. However, Huray's wardrobe preferences changed abruptly in 1980 when he was appointed associate dean for research in the UT College of Liberal Arts (renamed the College of Arts and Sciences in 1994). His practice of wearing a suit and tie began promptly the day after his appointment was announced. This appointment was an important step for Huray, who was well positioned to lead Reese's task force.

UT EXPLORES ITS OPTIONS IN PURSUIT OF THE MANAGEMENT CONTRACT

Chancellor Reese's task force was called the National Laboratories Study Team, a small but coherent group that was chaired by Huray and included two university finance administrators, one research assistant, and three faculty:

William (Bill) Snyder of the College of Engineering (Snyder would become chancellor ten years later) and Ivan Sellin and Lee Riedinger from the Department of Physics. Team members visited other national laboratories in the fall of 1982 to study how the management contracts operated and identify any potential pitfalls. Nobel Laureate Glenn Seaborg (1951, chemistry), a preeminent science leader at Lawrence Berkeley National Laboratory and former AEC chair, visited Knoxville in August to offer advice. He encouraged UT to bid on the contract alone, citing the University of California, which then managed three national laboratories—Berkeley, Livermore, and Los Alamos—as a model.

On August 16, 1982, DOE requested expressions of interest in the management contract. On August 19, word arrived that DOE was considering a change in the single-contractor management system then in place in Oak Ridge and might, instead, select up to four separate prime contractors for the four government facilities. On the October 15 deadline, UT submitted its expression of interest, as did fifty-seven other organizations, including Oak Ridge Associated Universities and a host of other universities.[14] [15] DOE left open the possibility of an institution or consortium bidding on one or up to four available management contracts.

The UT task force spent considerable time through the fall of 1982 talking to companies that would be potential partners on a bid to manage one or all four available DOE facilities. Task force members sought a company with considerable management capabilities, since the university did not have sufficient experience to oversee a large and complex science laboratory—including a nuclear reactor—on its own (much less the production facilities at K-25 and Y-12). Rumors circulated that managers in DOE's Oak Ridge Operations office did not view UT as an experienced enough institution to be part of a bidding team. Meanwhile, others felt that the university should definitely form or join a team to mount a bid.

Reese's task force determined that, if DOE offered separate contracts for the four facilities, UT would focus primarily on the ORNL contract and partner with an industrial company in submitting its bid. Teaming with other organizations is a commonplace activity in government procurements, as strengths of one partner can overcome the weaknesses of other partners. For example, UT had little experience in managing large industrial facilities and needed as a bidding partner a company that had been a seasoned manager in construction and operation of technically complex facilities. Discussions were held with a number of such companies, including Eagle-Picher, Ebasco (teamed with Bendix), Burns and Roe, Sverdrup, Foster Wheeler, and United Nuclear Corporation—all of which initially sought an exclusive partnership with UT. All of these corporations were respected managers of large industrial laboratories and production facilities.

Discussions were held with other companies that had less interest in such a relationship: Westinghouse, Lockheed, EG&G, and Cabot. It seemed that Battelle Memorial Institute (more widely known simply as Battelle) was interested in bidding but not in teaming with UT, a story that would differ considerably in 1998. This nonprofit applied-science and technology-development company headquartered in Columbus, Ohio, was the managing contractor of Pacific Northwest National Laboratory since 1965, when the facility was established near DOE's Hanford site.

Amid this uncertainty over whom to partner with and whether the four facility contracts would be bid separately, Chancellor Reese asked the UT task force to develop a list of the assets, capabilities, and collaborative opportunities that the university should seek either from the teaming partner or, should UT opt not to enter the bidding process, from the next managing contractor at ORNL. In other words, precisely what did UT hope to gain through its relationship with ORNL, whether UT was a managing partner or not? In fact, Governor Alexander was mulling over this same question. Eager to strengthen the UT-ORNL partnership in whatever form it might take, he sought ideas on how to expand the relationship in ways that would benefit both institutions—but as governor, he was particularly keen to see UT benefit from the partnership. A group of UT physics professors would soon answer the call, and, ultimately, Alexander would enlist his former boss and mentor, Senator Howard Baker, in seeing the effort through and thereby enhancing the university's partnership with the national laboratory.

A NEW IDEA, CONCEIVED AROUND A KITCHEN TABLE, BEARS FRUIT

On a Sunday morning in October 1982, Huray hosted fellow physics faculty Ivan Sellin and Lee Riedinger at his home, where they clustered around a kitchen table. The three scientists explored various new ideas rooted in the UT-ORNL partnership and finally came up with a concept that would bolster the university's science faculty and also infuse ORNL programs with fresh talent. Their proposed Distinguished Scientist Program[16] would hire renowned researchers into fifty-fifty joint appointments between UT and ORNL. The recruited scientists would be based at the university, receive tenure, and receive the then outlandish salary of $100,000 for twelve months' work. (For comparison, Riedinger's salary at that time was $30,000, and those of the ORNL director and the UT president were under $100,000.)

In addition to their salaries, the Distinguished Scientists would receive $28,000 in annual benefits plus financial support ($72,000) for their research. The university and the laboratory would split these costs.[17] The idea was to use this substantial package to attract the most accomplished and well-known scientists and engineers in fields of interest to both institutions. No one was clear on where these funds would come from at the university or the national laboratory—and some feared that these high salaries would prompt a storm of protest among current researchers at ORNL and faculty at UT—but all involved realized that this unique new program could have a profound impact on the partnering institutions. In return for the generous salaries and benefits packages awarded to the Distinguished Scientists, UT and ORNL would expect the participants to build leading research programs at both places and serve as magnets in attracting new funding and top people to work in their research groups.

The proposed Distinguished Scientist Program became the cornerstone of the benefits that UT hoped to gain from the next contractor of ORNL, regardless of whether the university was involved in the winning management team. It was not expected that DOE would announce the bid configuration (whether there would be one or up to four managing contracts for the four facilities) until mid-January of 1983, and the official request for proposals was not likely to be issued before April. Under the leadership of Chancellor Reese and his executive assistant Don Eastman, the task force developed a list of new joint programs that would be part of any management proposal. The idea then evolved that Reese should take this list to ORNL Director Herman Postma to get some of them started before the new contractor assumed management responsibilities.

Herman Postma and Jack Reese were vastly different men, in terms of background, temperament, and fields of expertise, but, despite their differences, they worked well together. Postma was born in North Carolina into a farming family of Dutch ancestry.[18] Neither of his parents finished high school, but, early on, their son demonstrated exceptional skills in physics and math, placing first in a statewide high-school competition in physics and second in the math category. He pursued his undergraduate degree in physics at Duke University and worked as a researcher at ORNL during two summer breaks. Postma went on to earn a PhD from Harvard before accepting a research position at ORNL in 1959. He excelled in plasma physics and hoped eventually to harness fusion to provide for society's power needs.[19]

HERMAN POSTMA. COURTESY OF OAK RIDGE NATIONAL LABORATORY.

He rose quickly through the ranks and, in 1968, became director of ORNL's Thermonuclear Division. In August 1968, Postma attended a conference in Russia where the world's most powerful tokamak—an experimental device that uses magnetic fields to confine a plasma—was revealed, and Postma returned to ORNL to contribute to the design and construction of a similar device in Oak Ridge. While working on the project, Postma took Russian classes to enable close collaboration with Soviet scientists, and he became proficient and nearly fluent in the language.

Continuing his rapid rise through the system, Postma became ORNL director in 1974 at age forty-one and stayed in that position for fourteen years. Although a long-time national laboratory researcher and leader, Postma valued the work of universities—including the research taking place at UT—and sought opportunities to collaborate with university-based science and engineering programs. Over the years, Postma, the fast-rising fusion physicist, and Jack Reese, the quiet English professor turned chancellor, developed a productive and collegial relationship.

On December 7, 1982, Reese sent Postma a long letter listing his desired outcomes from an expanded partnership between the two institutions.[20] Postma responded to Reese on January 19, 1983, and appointed various ORNL leaders to work with UT faculty in pursuit of expanded university-laboratory programs. Just under a month later, on February 11, Reese and Postma signed a memorandum of understanding (MOU) to establish the Distinguished Scientist Program and define the manner of hiring and funding these scientists, which established an important new critical connection between the two institutions. This MOU has endured as the program's baseline agreement in all the decades since—a reality that has caused considerable heartburn among attorneys at the two institutions, since no lawyers were involved in drafting or signing the MOU. Relying on this trust-based—rather than legally binding—approach allowed the two leaders to agree upon, draft, and sign the memorandum quickly.

On January 19, 1983, DOE announced its plan to award one contract for management of all four facilities (three in Oak Ridge and one in Paducah), which dampened UT's interest in participating in a management consortium whose responsibilities included Y-12, a nuclear weapons facility. On April 18, twenty-seven corporations submitted a second round of statements of interest and intent to respond to DOE's recent request for proposals. UT was not mentioned as a managing partner in any of them, which ended the substantial initiative at the university to compete for the management of the Oak Ridge facilities. For the time being, UT was out of the bidding process. However, sixteen years later, the university—better positioned and with far greater depth in research and administration—would directly vie for a new contract for management and operation of ORNL.

A NEW STATE CENTER OF EXCELLENCE: UT-ORNL SCIENCE ALLIANCE

Though UT withdrew from the bidding process, much good came from this unsuccessful attempt to engage in the management of the Oak Ridge facilities, including creation of the Distinguished Scientist Program. A few days after Huray, Sellin, and Riedinger had met at Huray's home and conceived the program, Huray passed the details to Chancellor Reese and then to Governor Alexander, as he knew that the governor intended to meet soon with Senate Majority Leader Howard Baker. Since Baker had a meeting scheduled with Energy Secretary Don Hodel the next week, Alexander hoped Baker could use this opportunity to get the new concept planted at DOE. The scheme worked, and Baker presented the idea to Hodel with Alexander present. The DOE secretary saw the wisdom of this request from the majority leader, and agreement was reached. This resulted in communication between Hodel and DOE Oak Ridge Operations Director Bob Hart in support of the program's launch. On February 7, 1983, Hart, Jack Reese, and Herman Postma met and agreed to move forward to recruit the first of what would become a cadre of twenty to twenty-five Distinguished Scientists, at a total cost per scientist of $200,000 per year—including salary, benefits, and pledged funds for research.[21]

The first meeting of the UT-ORNL group to discuss how to hire the Distinguished Scientists took place at UT on June 16, 1983. Meeting participants discussed fields of science important to both institutions and even floated names of potential program recruits. However, one big mystery remained: how would the two institutions pay for these high-priced hires? The answer to this question came in the fall of 1983. Governor Alexander (a Republican), then early in his second term, worked with the legislature and especially Tennessee Speaker of the House Ned McWherter (a Democrat) to pass legislation to raise the state sales tax specifically to bolster education. The higher-education part of this bill established centers of excellence, which would be available to state universities with unique capabilities and ideas to propose for support from a limited pool of resources.

Huray took the lead on writing the proposal to state government for creation of the Science Alliance, a state center of excellence that would promote greater cooperation between UT and ORNL. The Science Alliance would become a new critical connection and the first to seek advancement of the UT-ORNL partnership over a wide range of fields and programs. Two key parts of this proposal were the hiring of Distinguished Scientists and the creation of joint institutes in specific fields of common strength. The Science Alliance was awarded the first of these center-of-excellence grants, and, by the fall of 1983, money was suddenly about to appear to finance UT's half of the Distinguished Scientist Program. This led to a more intensive meeting on September 13, 1983, to begin to identify prospective recruits.

Serendipity, by its nature, is impossible to quantify or to orchestrate, but its arrival in the midst of a planning process can yield unexpected, yet entirely beneficial, results. In this case, the serendipity centers on Howard H. Baker Jr. Baker was born in Huntsville, Tennessee, (about fifty miles northwest of Knoxville and thirty miles north of Oak Ridge). His father served as a Republican member of the U.S. House of Representatives from 1951 to 1964, representing a traditionally Republican district. Baker graduated from Tulane University and served in the Navy from 1943 to 1946.[22]

Baker's choice of a graduate program at UT reflects a degree of serendipity—and a bit of impatience. In the fall of 1946, he intended to enroll in the College of Engineering, but, on that particular day, the line to sign up for engineering courses was too long and so Baker walked down the street to a shorter queue that had formed in front of the law school. Thus, Howard Baker began his pursuit of law and graduated from the UT College of Law in 1949. He practiced law for a time but then followed in his father's footsteps and ran for elected office. In 1964, he lost his first bid for office—a seat in the U.S. Senate—but handily won a seat two years later, in 1966, and became the first Republican senator from Tennessee since Reconstruction.[23]

Baker served three terms in the Senate with great distinction. He was the Senate minority leader (1977–81) and then Senate majority leader (1981–85). Ira Shapiro describes the Senate's bipartisan achievements in those years in his book *The Last Great Senate: Courage and Statesmanship in Times of Crisis.*[24] Baker decided to run for president in 1988 and, consequently, decided that he would not stand for re-election to the Senate in fall of 1984. This led to a curious—and serendipitous—chain of events relating to the theme of this book.

Though a trained lawyer, Baker had a deep interest in and understanding of scientific and technical issues, in part due to his East Tennessee roots and his fondness for Oak Ridge. Unlike most elected officials, he liked having a trained scientist on his staff. One such Baker staff member was Fred Bernthal, a nuclear physics faculty member at Michigan State University before coming to Capitol Hill in 1978 on a Congressional fellowship from the American Physical Society. When the accident occurred at the Three Mile Island nuclear reactor in April 1979, Bernthal earned respect and influence as a Senate staffer because of his understanding of nuclear physics. When Baker decided privately in the spring of 1983 to forego the next campaign for his Senate seat, he nominated Bernthal to a seat on the Nuclear Regulatory Commission and asked Bernthal to find another scientist to come and serve as his science advisor for the senator's final year in office.

Bernthal and Lee Riedinger were friends and contemporaries in the field of nuclear physics, both receiving PhDs in late 1968 (Bernthal from the University of California, Berkeley, and Riedinger from Vanderbilt University). The surprise—and, yes, a bit more serendipity—came in June 1983, when Riedinger was spending the summer with his family in Liverpool, England, working with British colleagues in nuclear physics. Bernthal called Riedinger to suggest that Riedinger come to Washington for a year and serve as Baker's science advisor. Riedinger thought of a dozen good reasons to pass on the offer—his team of graduate students, demands of family, pending experiments—but promised to call and ask the opinion of his physics department head at UT, William Bugg. Bugg said he would check with Chancellor Reese and call back to Liverpool the next day to advise Riedinger on a preferred course of action. Bugg's reply was only partially tongue in cheek, as he explained that the chancellor would revoke Riedinger's academic tenure if he failed to seize the opportunity. As it turned out, this would be only the first instance in which Riedinger feared his tenure might be in jeopardy!

Riedinger, the son of a dry cleaner who had not attended high school due to the Great Depression, was the oldest of six children in a family with few resources. However, a scholarship to a local college and an AEC fellowship to graduate school at Vanderbilt enabled him to earn a PhD in nuclear physics by age twenty-three. Riedinger joined the UT physics faculty in 1971. In 1983, with the support of Bugg and Reese, he accepted the one-year assignment to Baker's staff and, in the process, learned much about the workings of government and developed deep respect for Howard Baker, the man and the legislator.

THE FRUITLESS PURSUIT OF A "REPLACEMENT" PROJECT AFTER CRBR IS TERMINATED

After the 1983 cancellation of the CRBR project in Oak Ridge, the hope was that Senator Baker could deliver a "replacement" project to Oak Ridge. Leaders of Oak Ridge, ORNL, and Union Carbide journeyed to Washington to present their ideas on the next big science project that could perhaps be sited in Oak Ridge. Riedinger was the first person assigned to talk to the Oak Ridge visitors as a prelude to meeting with the majority leader. It was to Riedinger's advantage to be able to meet and get to know this set of Tennessee leaders, as that initial contact would benefit him in his later efforts to build the UT-ORNL partnership.

In the process, the meaning of the term "lame duck" became crystal clear. While a powerful Senate majority leader might normally have the ability to secure funding for a big-ticket item to replace something lost, it did not work in this case. The problem was that Baker announced in early 1984 that he would not stand for re-election to his Senate seat later that year, because he intended to run for president in 1988. While a sitting senator can mount a long and expensive presidential campaign, Baker felt that he could not neglect his leadership position in the Senate sufficiently to mount a successful campaign for the Oval Office. After his announcement not to seek re-election to the Senate, Baker became a lame duck and no longer had the power (or inclination) to find funding for a replacement project for Oak Ridge.

Although he intended to, Howard Baker did not compete for the presidential nomination in 1988. In 1987, President Reagan asked Baker to come to the White House to be his chief of staff after a series of staffing problems.

Ever the loyal soldier, Baker accepted this job and, in the process, had to support the successful run of sitting Vice President George H. W. Bush for the presidency in 1988. Baker served as U.S. Ambassador to Japan from 2001 to 2005 before retiring to Huntsville so that he could spend considerable time working at UT's Howard H. Baker Jr. Center for Public Policy until his death in 2014.

DISTINGUISHED SCIENTISTS: THE PUSH TO HIRE THE BEST AND BRIGHTEST

By 1984, the state's Centers of Excellence program fully funded the Science Alliance and UT and ORNL set out to recruit their first Distinguished Scientists. The search process produced two exceptional candidates: George Bertsch, a nuclear physicist from Michigan State University, and Gerald Mahan, a solid-state physicist from Indiana University. However, convincing the two well-entrenched scientists to leave their home universities to accept newly created positions in a nascent joint research program presented a considerable challenge. Bertsch said that he would come to UT and ORNL on one condition: that the *New York Times* would cover the move. But how to achieve that? Since Riedinger was working on Baker's staff at the time, UT devised a plan: if Majority Leader Baker, with his national stature and sterling reputation, would preside over the announcement of these two hires into this new UT-ORNL program, the "Gray Lady" just might take notice. Riedinger discussed the idea with Baker's chief of staff and press secretary, and they both firmly said "no."

Discouraged but not daunted, Riedinger decided to execute an end run. One day while he was in Baker's office, briefing the senator on an unrelated matter, Riedinger directly asked the senator if he would preside over the ceremony announcing the new Distinguished Scientist Program hires. Baker quickly assented, and the wheels were set in motion. Baker wrote to Bertsch to explain his role in the origin of the Distinguished Scientist Program, encouraged him to accept the appointment, and suggested a signing ceremony be held in his office. The reception took place on June 11, 1984, but not in the senator's office. Baker announced the hiring of the first two UT-ORNL Distinguished Scientists in the Mansfield Room of the U.S. Capitol, which was filled with dignitaries, including Governor Alexander, ORNL Director Postma, UT President Edward Boling, Chancellor Jack Reese, Energy Secretary Hodel, U.S. Representative Al Gore, and a host of others. Though the story was not covered in *The New York Times*, it did garner substantial notice in journals

SEN. HOWARD BAKER AND LEE RIEDINGER IN THE MAJORITY LEADER'S OFFICE ON CAPITOL HILL, OCTOBER 1984. IN POSSESSION OF THE LEAD AUTHOR.

devoted to physics, including coverage in *Physics Today*,[25] and Bertsch followed through on his commitment to join the program, as did Mahan.[26] This was a huge step forward for the evolving UT-ORNL partnership.

Materials Science

Materials science is a foundational strength at ORNL and flourished at both ORNL and UT in the 1980s. At the laboratory, this was due in part to an excellent team of experimentalists and the unique tools they built. In 1980, Bill Appleton (later ORNL deputy director) created the Surface Modification and Characterization Research Center, which, along with other avenues of research, explored how to implant accelerated ions of certain elements into surfaces of materials to improve their properties.[27] For example, ions implanted below the surfaces of semiconductor materials are crucial for making effective computer chips. Artificial hip joints last much longer if implanted with nitrogen ions, averting the need to replace failed joints while avoiding patient discomfort and saving significant amounts of money.

Steve Pennycook perfected the scanning transmission electron microscope, which enabled researchers to visualize the structure of a material on the scale of the atom. The laboratory's work was highlighted in a presentation to President Reagan during a visit in 1985. ORNL scientists at the High Temperature Materials Laboratory, a user facility that began operating in 1987, produced improved alloys capable of withstanding corrosive high-temperature environments.[28] ORNL scientist Doug Lowndes pioneered the use of lasers to make thin films of "high-temperature" superconductors[29] for potential application in improving electricity transmission.

UT and ORNL also sought to build strength in the theoretical aspects of materials science through the hiring of Gerald Mahan as one of the first two Distinguished Scientists in 1984. Luring Mahan to Tennessee from Indiana University was a real coup for the two institutions, aided by the work of U.S. Senate Majority Leader Howard Baker. Mahan served in this special role for twenty years and was elected to the National Academy of Sciences by virtue of his extraordinary achievements in ion transport and optical properties of materials, especially relating to solid-state devices.[30] He finished his career at Pennsylvania State University.

GERALD MAHAN. COURTESY OF BETSEY B. CREEKMORE SPECIAL COLLECTIONS AND UNIVERSITY ARCHIVES, UNIVERSITY OF TENNESSEE, KNOXVILLE.

A NEW ACCELERATOR FOR ORNL AND NEW PARTNERSHIP OPPORTUNITIES

Nuclear physics was long a main area of research for ORNL, as well as for UT through faculty and graduate students using Oak Ridge accelerators for their research. ORIC was the centerpiece facility for this line of research after its construction in the early 1960s. But, the need for a wider variety of nuclear projectiles and higher energies of these beams called for a larger accelerator to enable an expanded program of research in nuclear physics.

TOWER HOUSING THE VERTICAL ACCELERATOR OF THE HOLIFIELD FACILITY AT ORNL. COURTESY OF OAK RIDGE NATIONAL LABORATORY.

The Holifield Heavy Ion Research Facility opened in 1982.[31] This national user facility featured a vertical 25-MV particle accelerator. The accelerator enabled a twenty-year run of experiments on the fundamental properties of nuclei as studied when heavy-ion projectiles accelerated and then smashed into a thin target of a chosen element. At the time of its creation, the Holifield accelerator featured the highest terminal voltage of any such device in the world and enabled nuclear reactions initiated by projectiles as heavy as tin (an element with the atomic number 50).

The Holifield facility gave nuclear physicists a new tool to study the shape, dynamics, and stability of the tiny atomic nucleus, which contains most of every atom's mass, with a large "microscope" consisting of a beam of projectiles to probe and excite the nucleus. At the opposite end of the length scale in nature, a similarly revolutionary new scientific tool, the space-based Hubble telescope, was launched into Earth orbit in 1990. Like Holifield, Hubble enables astrophysicists to study shape, dynamics, and stability, not of nuclei, but of large distant objects—galaxies, groups of stars, and even planets orbiting stars. These two scientific tools, situated at the extremes of scale, actually unite in one important aspect: Holifield facilitated research into nuclear astrophysics, allowing scientists to probe the pathways by which neutrons are rapidly absorbed to make successively heavier nuclei and elements when a supernova explodes.

The Holifield facility benefited the research programs of many universities, especially UT and Vanderbilt as a result of their proximity to Oak Ridge and the history of collaborations in this field of research.

Rotating Objects in the Universe

Holifield was used to study the rotation of nuclei. Rotation of objects is common in the universe, from the smallest to the largest size. A common question is how the rotational motion affects the shape of these bodies. Everyone is familiar with the daily rotation of the Earth, which translates to a point on the equator moving at around 1,000 miles per hour (mph). This amount of rotation does not appreciably elongate the Earth because it is solid. The Sun is around one hundred times larger than Earth and rotates once every twenty-seven days, translating to an equatorial speed of around 4,000 mph. However, the Sun does not stretch as it rotates so rapidly because it is gaseous and dynamic and undergoes differential rotation where the period of rotation is smaller at the equator than near the poles. Neutron stars are very small (around 10 km in radius, six hundred times smaller than Earth) and, on average, they rotate once per second—extremely fast. This means a velocity at the equator of the dense neutron star is around 140,000 mph. But, with the tools and technologies available today, it is impossible to observe what happens to the shape of a neutron star as it rotates so rapidly.

However, the nucleus of an atom rotates even faster than a neutron star, though it is much smaller in size. The radius of a nucleus is around 10^{-16} km and rotates at more than 10^{18} times per second, which translates to a huge rotational velocity of around 800,000 mph. Studying the possible stretching of a nucleus while rotating can tell us if the nucleus is solid (like the Earth) or more gaseous (like the Sun).

The Holifield accelerator facility was used in 1982 for such a measurement, producing short-lived ytterbium-160 (^{160}Yb) by colliding a beam of titanium-48 (^{48}Ti) on a target of cadmium-116 (^{116}Cd), in an experiment led by Lee Riedinger and his UT research group. The Yb nucleus was produced in an instantaneous state of high energy and high rotational velocity, and it rapidly cooled by emitting a sequence of gamma rays, which were detected in a spherical array of counters. This experiment (and others before this one) proved that the nucleus is not solid like the rotating Earth nor gaseous like the Sun, but stretches and gradually elongates its slightly prolate (football) shape as the rotational velocity increases. This Holifield experiment produced ^{160}Yb to a higher state of rotational spin than had not yet been observed at other laboratories.

This experiment and others of this type led to various accelerator laboratories building even better spherical arrays of gamma-ray detectors. The result in 1986 was the first observation (at a laboratory in England) of nuclei starting out as a normally deformed rotor (less elongated than an American football) but then flipping over, due to fast rotation, into a super-deformed mode where the nucleus has roughly a football shape.[32] Other experiments showed that some nuclei even fission (split into two pieces) due to fast rotation. So, rapid rotation greatly affects the nucleus, which reveals information on the nature of its interior. Perhaps in some future astronomical measurement of neutron stars, we will learn if their rapid rotation leads to such profound changes as seen in atomic nuclei.

STATE APPROPRIATION BUILDS THE FIRST JOINT INSTITUTE

The Science Alliance developed two new programs of cooperation between UT and ORNL—the joint hiring of distinguished scientists and the formation of a joint institute. The purpose of a joint institute is to enable specialized research projects and inspire new collaboration between users and laboratory staff in scientific areas defined by the tools and technologies available in an adjacent user facility. A joint institute is formed around a strong experimental facility at ORNL in part to help users from universities build more extensive programs at that centerpiece facility. The first two distinguished scientist hires were in two areas of mutual strength at the two institutions: nuclear physics and materials science. The first joint institute was built around programs of nuclear physics at UT, ORNL, and Vanderbilt University: the Joint Institute

DIGNITARIES AT THE CEREMONY TO OPEN THE JOINT INSTITUTE FOR HEAVY ION RESEARCH: (FROM LEFT) JOE LAGRONE (HEAD OF DOE IN OAK RIDGE), JOE WYATT (VANDERBILT CHANCELLOR), GLENN SEABORG (NOBEL LAUREATE AND KEYNOTE SPEAKER), JOSEPH HAMILTON (VANDERBILT PROFESSOR), HERMAN POSTMA (ORNL DIRECTOR), MARILYN LLOYD (U.S. CONGRESS), JACK REESE (UT CHANCELLOR), ED BOLING (UT PRESIDENT), AND JIM LEISS (DOE). OCTOBER 1984. COURTESY OF OAK RIDGE NATIONAL LABORATORY.

for Heavy Ion Research (JIHIR). The centerpiece ORNL facility for this joint institute was the Holifield Heavy Ion Research Facility.

While the Science Alliance would supply much of the operating budget of this joint institute, funds for construction of a building were needed. The state of Tennessee supported this effort with a special appropriation of $350,000 to fund construction of a 6,000-square-foot building to house the institute and, in the process, forge a new critical connection. However, significant challenges remained, primarily in constructing a state-owned building on DOE (federal) property to be funded and operated primarily by the two universities, UT and Vanderbilt. This had never been done before at a DOE facility and a host of legal and regulatory issues had to be resolved. In the years since, it has become far more common for states or the private sector to construct their own facilities on DOE sites, in part because of the important groundwork laid in creating this first joint institute.

The official opening of the first joint institute occurred on October 15, 1984, and numerous luminaries attended the ceremony. Nobel Laureate Glenn Seaborg served as the keynote speaker. Leadership of UT and ORNL was well represented and Chancellor Joe Wyatt of Vanderbilt University made remarks.

Once the facility was operational, Russell Robinson, Joseph Hamilton, and Lee Riedinger assumed the lead in managing it. This became an important space for faculty, graduate students, and visitors at ORNL to conduct experiments and pursue collaboration in nuclear physics connected with the Holifield Facility.

Users from around the world came to ORNL to conduct their experiments at Holifield. The joint institute provided university researchers and other visiting scientists with space to plan new lines of experiments, convene workshops and symposia, and take advantage of office and sleeping facilities available during experiments. It was indeed common for university professors to sleep soundly in the joint institute's bedrooms, until graduate students dared to come and awaken them in the middle of the night to help solve a problem or answer a critical question.

The joint institute also had a nuclear theory component to supplement the fine experimental programs clustered around Holifield. Indeed, one of the first two Distinguished Scientists hired in 1984 was George Bertsch. Through Bertsch's appointment, and through the strategic hire of other scientists with expertise in both experimentation and theory, UT and ORNL added considerable strength to their research enterprises.

The nuclear physics programs at JIHIR's two partnering universities—UT and Vanderbilt—had been strong for decades and the joint institute facilitated

FOUNDERS OF THE FIRST JOINT INSTITUTE LOCATED AT ORNL: (FROM LEFT) LEE RIEDINGER (UT), RUSSELL ROBINSON (ORNL), AND JOSEPH HAMILTON (VANDERBILT), OCTOBER 1984. COURTESY OF OAK RIDGE NATIONAL LABORATORY.

numerous experiments and collaborations between the universities' researchers and Holifield scientists. Many other universities likewise benefited from Holifield and the opportunities for joint institute collaborations. In fact, by the end of the 1980s, about a quarter of all PhDs granted in the United States in the area of low-energy nuclear physics involved experiments conducted at Holifield. The user program at Holifield was terminated in 2012 and the American Physical Society designated the facility as a Historic Physics Site in 2016. DOE closed Holifield as it embarked on building a new accelerator facility (Facility for Rare Isotope Beams) at Michigan State University.

JIHIR's success created the model for additional UT-ORNL joint institutes fifteen years later, when UT-Battelle LLC would assume responsibility for managing and operating ORNL. As with Holifield, they were structured around existing ORNL facilities in computational science, biological science, and neutron science. Though physically situated on the ORNL campus, these four joint institutes occupy space in buildings constructed by the state of Tennessee. In the years since, partnerships between other universities and national laboratories have adopted this model.

A NEW OAK RIDGE CONTRACTOR AND MORE DISTINGUISHED SCIENTISTS

The remainder of the 1980s saw other key events, including DOE's award of the contract for management of its facilities in Oak Ridge and Paducah, Kentucky. The finalists in the competition were Westinghouse, Martin Marietta, and Rockwell Scientific, as determined by a DOE process based on the companies' proposals and their defined capabilities and expertise in managing other large technical complexes. UT leaders (especially Paul Huray, who headed the task force) spent considerable time with representatives of these three companies, to ensure that the relationship with the university would be expanded no matter which organization won the competition.

Leaders of these three companies visited Senator Howard Baker in 1984 to tout their expertise, each assuming that the majority leader would play a significant role in the DOE decision. Future Tennessee U.S. Senator Fred Thompson, who worked for Westinghouse at that time, focused some of his attention on Baker's science advisor, Lee Riedinger. Thompson served as minority counsel to assist the Republican senators (especially Baker) during the Watergate hearings of 1973. He treated Riedinger to a number of costly lunches near Capitol Hill to explain why Westinghouse should win the DOE award. In the end, Baker played no part in the decision, and DOE awarded the management contract to Martin Marietta,[33] which assumed its role in April 1984, almost two years after Union Carbide announced its plan to withdraw. Thompson was elected to the U.S. Senate in 1994 and served until 2003, competed for the Republican presidential nomination in 2008 (unsuccessfully) and appeared in a number of movies and TV series.

During the Martin Marietta tenure at ORNL, more distinguished researchers were hired jointly in various fields of science and engineering, as the partnership grew substantially. President Ronald Reagan visited UT on September 24, 1985, to engage in a panel discussion titled "Teaming Up for Economic Growth."[34] Jack Reese, Herman Postma, Ralph Gonzalez (UT engineering professor who had escaped from Cuba as a child), Warren Neel (dean of the UT College of Business), Dick Ray (a leader of ALCOA), and

Randy Henke (biology professor and entrepreneur) made panel remarks. Riedinger gave the first presentation at the panel discussion about the Science Alliance.

David White, the third UT-ORNL Distinguished Scientist hired, moved to East Tennessee in the spring of 1986. White had built a stellar research career as a faculty member at Florida State University before accepting the joint appointment. He had received more lucrative offers from the University of Maryland and Georgia Tech, but he chose the UT-ORNL position because of the vast array of resident expertise and research opportunities at the two partnering institutions and the invitation to bring his entire research team—twenty graduate students, postdoctoral research associates, and research faculty—to UT and ORNL with him. White and his team's research focused primarily on microorganisms, or microscopic organisms, that exist as unicellular, multicellular, or cell clusters.

David White's Many Hats

Many ways exist to describe David White throughout his seventy-seven years of active life: physician, scientist, entrepreneur, innovator, and teacher. White was a unique individual with training in both science and medicine. He held a medical degree from Tufts and a doctorate in biochemistry from Rockefeller University. He spent a decade on the medical school faculty at the University of Kentucky and then eleven years at the Florida State University medical school. He was licensed to practice medicine in New York, Kentucky, Florida, and Tennessee.

White was hired in 1986 as the third UT-ORNL Distinguished Scientist because of his expertise as a researcher but also because his field explored environmental biotechnology and, thus, would contribute to ORNL's increasing emphasis on environmental science. At UT, White forged a new direction for the Department of Microbiology.

White spent a career studying microbes, which are widespread in nature and benefit all living things, powering all the geochemical cycles by which life on Earth is maintained. His research led to discovery of signature patterns—biomarkers—inside the cell structures of these microbes. These biomarkers allow scientists to distinguish among the types of microbes that are present and help interpret microbial activity, notably including changes in microbial communities in response to contaminants present in soil and water. Because of White's considerable array of sophisticated scientific tools and devices, he earned a reputation as quite an "equipment" man, specializing in mass spectrographs.

Science was always at the center of David White's life. He authored more than five hundred peer-reviewed publications, had his papers cited more than six thousand times, received significant external funding for his research, won multiple teaching awards, and was recognized internationally for excellence in his field. He died in 2006 when he tried to drive across a busy highway on his way to ORNL, was struck by oncoming traffic, and died a few days later.

David White's name lives on through awards endowed by his family, including the David and Sandra White Professor Chair, bestowed on the UT microbiology department, and the David C. White Research and Mentoring Award, presented annually by the American Society for Microbiology.

While David White's hire infused the UT-ORNL research enterprise with considerable talent in microbiology and biomarker analyses, the size of White's team and its need for abundant laboratory space to house the researchers and their considerable complement of tools and equipment created a problem. ORNL provided White with ample space in several of its buildings and research directorates, but, at the time, the university campus was woefully short on accommodations for its current research faculty, let alone for the space-hungry new hires. When the issues began to seem all but intractable, Homer Fisher, then UT vice chancellor and later UT senior vice president, devised a solution.

Fisher earned a bachelor's degree in economics and a master's of business administration at Auburn University before serving thirty-eight years in higher education administration—sixteen years at Florida State University and twenty-two years at UT after coming to Knoxville in 1977.[35] Though not a scientist, Fisher understood the unique needs of science faculty and their potential for delivering new discoveries, large grant awards, and international prestige, provided they had sufficient space and equipment to conduct their research. Often, high-level administrators at large state universities are risk adverse and not able, or willing, to seek innovative solutions to address critical needs. Fisher decidedly was not such a man.

Fisher, previously vice president for administration at Florida State University, served there with White on the faculty. And, with White and his team's arrival in Knoxville, Fisher sought a novel approach to finding space and funding construction of the facilities that White and others on campus would need to advance their ground-breaking—and high-prestige—research. Rather than rely on the more-traditional appeal to the state for funding for new on-campus facilities, Fisher took a different approach. He signed a contract with a private developer to build laboratory and office space on an off-campus site on Dutchtown Road, roughly halfway between campus and the national laboratory.

Construction on the facility was completed by 1988 and White and his team moved in. The contract committed the university to pay rent on the new space, which ultimately would require the need for some retroactive support from the state—a somewhat sticky situation, since Fisher's off-campus research facilities had not been reviewed or approved by the state building commission. In securing the state's backing, Fisher leaned heavily on UT Executive Vice President Joe Johnson—known for his finesse and powers of persuasion—who secured the funding.

With the space issue resolved, Fisher turned his attention to another challenge: how to finance the extensive list of expensive equipment and tools White and his team required to conduct their research? In this instance, Fisher turned to another fellow UT administrator—Vice President Emerson Fly—who succeeded in obtaining an interest-free multi-million-dollar line of credit. University administrators surmised, correctly, that the loan would be paid off over time through increased overhead costs (so-called facilities and administration fees) recovered on new grants and contracts resulting from the investment in new space and equipment. This is now a common practice, but, at the time, it was a new—and sometimes criticized—approach for UT.

Nevertheless, it worked. White and his team were able to set up a considerable array of mass spectrometers and other pieces of equipment to pursue White's passion: research on biomarkers. In an adjacent laboratory at the Dutchtown Road facility, UT faculty colleague Gary Sayler, also a microbiologist, built a successful and well-funded research effort. Fisher's strategic gambit represented an impressive example of risk-taking and entrepreneurship that, previously, had not been hallmarks of the university, and, as such, it marked a critical step in UT's coming of age, particularly in terms of preparing the university to compete successfully, thirteen years later, for the ORNL management contract.

GROWTH OF MANY JOINT ACTIVITIES

In the 1980s, many new forms of the UT-Oak Ridge partnership started, centered on research and education in science and engineering. While the Science Alliance was the largest of the Centers of Excellence grants awarded by the state of Tennessee, two others formed at UT: the Center for Materials

Processing and the Waste Management Research and Education Institute. ORNL scientists were involved in both of these activities, for research and teaching.[36] The Waste Management Institute was formed in 1985 and located in the Energy, Environment, and Resources Center, which ORNL physicist John Gibbons had started in 1973.[37] This institute addressed a wide range of research relating to the plethora of waste challenges in the United States. For example, UT-ORNL distinguished scientist David White and UT microbiology professor Gary Sayler developed sensors to detect light emitted by genetically engineered bacteria that glow (bioluminesce) as they metabolize toxic substances in soil. These sensors were demonstrated to President George H. W. Bush during his visit to UT on February 2, 1990.

Two centers based in the UT College of Engineering, in conjunction with ORNL researchers, developed robust participation programs with industry. The Measurement and Control Engineering Center attracted eighteen industrial members, each of which contributed to its financial support to enable research projects in process control, sensor development, and pattern recognition. Dan McDonald, director of the ORNL Instruments and Controls Division, was the co-technical director, along with UT chemical engineering professor Charlie Moore. In 1986, this center won support from the National Science Foundation as a University/Industry Cooperative Research Center.

The UT Center for Materials Processing likewise involved ORNL researchers from three divisions and participants from industry. The center sponsored research on the mechanics of composite structures, including ceramics, metals, and polymers.

The Tennessee Center for Research and Development (TCRD) was located along the Pellissippi Parkway connecting Knoxville and Oak Ridge.[38] This center was funded by TVA, Martin Marietta Energy Systems (the DOE managing contractor in Oak Ridge), UT, and the state of Tennessee. TCRD lead many technology projects and was the home of two major centers, the Power Equipment Application Center and the Laser Technology Center. The Electric Power Research Institute, which many of the country's electrical-power companies sponsor, established and funded the former.

Also located adjacent to TCRD at UT's Pellissippi Laboratories were centers led by David White (Institute for Applied Microbiology) and Gary Sayler (Center for Environmental Biotechnology), both focused on research on microorganisms to destroy pollutants in the soil. These research activities helped initiate a UT graduate degree in environmental biotechnology. The shortage of research space on campus led to the opening of this laboratory complex roughly halfway between UT and Oak Ridge.

This wide variety of joint centers and institutes developed in the 1980s demonstrated that the UT-ORNL partnership was rapidly growing in many technical areas. This was part of the natural evolution of the relationship, where UT and ORNL learned how to collaborate in many areas of mutual benefit. The breadth of this relationship contributed to the success in forming the UT-Battelle partnership that assumed management of ORNL in 2000.

COMPETITION FOR THE SUPERCONDUCTING SUPER COLLIDER

The highest energy accelerator in the world for some years was the Tevatron at Fermilab near Chicago, sited there in the 1960s as the result of a national competition. It was a landmark particle accelerator with a ring structure of 3.9-mile circumference (more than a mile and a quarter in diameter), able to collide two beams of protons each at an energy of 500 GeV, producing proton-proton collisions with energies of up to 1.6 TeV (trillions of electron volts). One of its most important achievements was the 1995 discovery of the top quark, one of a family of six quarks that are thought to be the most fundamental building blocks of nature.

Twenty years after the construction of Fermilab, physicists lobbied for construction of an even larger proton collider to give more energy to probe in more detail the quark structure of matter, including whether there is an even more fundamental particle that lives inside the quark. The Reagan administration supported this initiative, and a national competition was opened for the siting of the Superconducting Super Collider (SSC).[39] This would be a proton accelerator at least ten times larger than the Tevatron, with superconducting magnets used for the first time to give higher magnetic fields that would bend the beam in this giant ring structure. Energy Secretary John Herrington announced that the first funds would be requested in the 1988 budget and that the large federal outlay would come from new funds for DOE, so that other science programs would not suffer.[40]

On August 21, 1987, Tennessee Governor Ned McWherter submitted an SSC site proposal to DOE. The proposed Tennessee site was located in the middle part of the state, south of Nashville, lying in four counties. A team of thirty professionals from across the state worked to characterize and choose this site. The accelerator would be built in limestone up to 350 feet below the surface. New faculty positions in high-energy physics were pledged by UT and other universities in the state. ORNL was an important part of the proposal, although it did not have a significant history of research in this field.

Late in 1988, the state of Texas won the nod for the SSC site, to be located at Waxahachie, south of Dallas. The Tennessee proposal was viewed as finishing second in this fierce competition.[41] However, much good came from this siting battle for the proposed $4.4 billion laboratory. The SSC would be housed in a fifty-three-mile oval tunnel in Texas, and the next competition concerned which teams from around the world would be chosen to build two huge detectors to observe and record the thousands of particles emitted when two protons collide with 20 TeV of energy for each. Construction of each huge particle detector would cost at least $600 million.

One of the proposed SSC detectors was called L-STAR, which MIT's Sam Ting, who won a Nobel Prize in physics in 1976, led. To design and hopefully build L-STAR, Ting assembled a large team of twenty universities, three national laboratories, and participants from more than fifteen countries. UT and ORNL were at the center of this large team. Led by Department of Physics head William Bugg, UT had been increasingly prominent in high-energy physics in recent decades, while ORNL was new to this field. Alvin Trivelpiece was director of the DOE Office of Energy Research and was instrumental in convincing the Reagan administration to fund the SSC. He became ORNL director in 1989 and offered laboratory support for L-STAR, including creating the Oak Ridge Detector Center, led by Tony Gabriel, to help design and build this giant device. Ting felt it was important to have ORNL and UT deeply involved in L-STAR.

ORNL staff were reassigned to the Detector Center, and new UT faculty were hired into joint positions with ORNL to develop greater expertise in this field. An April 1990 workshop at ORNL brought together L-STAR partners from around the world.[42] This 50,000-ton detector monolith would include several ten- to twenty-ton magnets constructed in Russia and five million silicon diode detectors as central components of a 3,000-ton calorimeter[43] for measuring energies of hundreds of protons and neutrons emitted in the high-energy collision. The Detector Center at ORNL and UT received a grant from the SSC project to develop these detector ideas in advance of an early 1991 decision on which detectors would be built.[44] After the SSC had made its selection, the chosen teams would have a year to write a detailed DOE proposal for funding the construction of its detector concept.

L-STAR was indeed one of two giant detectors chosen to be built at the SSC. In 1991, funds started to flow to ORNL for the Detector Center and to UT to support scientists and engineers working on the project at the university and various other institutions, including a rather large group in Russia. Money also came to UT from the Texas National Research Laboratory Commission (funded by the state of Texas as a matching component to the federal funding for the SSC) to build a silicon calorimeter as part of

L-STAR. Those were good years for the high-energy physics group at UT, as its annual spending exceeded $1 million for this work.

Problems arose for this huge project as SSC construction began in Texas. Primary among them were the project's continuing cost overruns, its lack of significant foreign contributions, and the end of the Cold War.[45] The projected SSC cost rose as the project proceeded and escalated from $4.4 billion to $12 billion when Congress canceled it in October of 1993, after $2 billion had already been spent. Work on the SSC and on L-STAR stopped amid much disappointment. The cancellation led to the decline of U.S. leadership in this field and opened the door for Europe to take the lead with the construction of the Large Hadron Collider (LHC) at the European Council for Nuclear Research (CERN) laboratory in Geneva. It was at the LHC where a large team of physicists discovered the long-sought Higgs boson, one of fundamental building blocks of matter.[46] The Nobel Prize in Physics 2013 was awarded to François Englert and Peter Higgs for their theoretical prediction of this entity that is apparently the origin of mass of subatomic particles.

In spite of this setback for high-energy physics, long-term benefit did occur for Tennessee. Twenty-five years later, UT physics faculty hired for the SSC initiative took the lead on innovative neutrino scattering experiments at the Spallation Neutron Source (SNS) at ORNL. The Fundamental Neutron Physics Beamline instrument on beam line thirteen at the SNS is operated by the ORNL Physics Division and has a close relationship with the university community in studying cosmology, nuclear and particle physics, and astrophysics. The SNS also emits neutrinos, which are fundamental particles that have no electric charge, almost no mass, and are therefore extremely difficult to detect and study.

Joint UT-ORNL faculty member Yuri Efremenko took the lead on building a special detector that enabled him and his colleagues to measure the rare coherent scattering of a neutrino from an argon nucleus.[47] This development has important implications for fundamental physics and serves at least as a small reward for the large effort mounted by UT, ORNL, and the state of Tennessee relating to the SSC. Science often works in this manner—delayed payoffs that arrive in unexpected ways based on earlier efforts.

A SECOND U.S. PRESIDENT ARRIVES ON CAMPUS

In his 1990 State of the Union speech, delivered on January 31, President George H. W. Bush articulated his commitment to improving science research and education in the United States and announced that he would double the budget for NSF. Two days later, he visited North Carolina State

University before traveling on to the UT campus. As one might expect, a visit by a sitting U.S. president is a big deal, and, reflective of Bush's commitment to bolstering research, he seemed particularly interested in learning about UT and ORNL's programs in science. It is likely that the friendship between Bush and then UT President Lamar Alexander was a key factor in Bush's decision to visit Knoxville.[48]

On Tuesday morning, three days before Bush's planned visit, *The Knoxville Journal* ran an editorial with the headline "Bush can learn from the Science Alliance." Cause and effect were not clear, but, on the morning the editorial ran, UT Vice President Homer Fisher called Lee Riedinger to say that the Science Alliance would be one of the highlights for the Bush visit, in addition to the president's public address. Initially, the segment on the Science Alliance was to be brief, but, by Wednesday, the session had been expanded to twenty

SCIENCE ALLIANCE BRIEFING: (FROM LEFT) DAVID JOY (DISTINGUISHED SCIENTIST), JOHN QUINN (UT KNOXVILLE CHANCELLOR), PRESIDENT BUSH, LEE RIEDINGER (SCIENCE ALLIANCE DIRECTOR)—FEBRUARY 2, 1990. COURTESY OF BETSEY B. CREEKMORE SPECIAL COLLECTIONS AND UNIVERSITY ARCHIVES, UNIVERSITY OF TENNESSEE, KNOXVILLE.

minutes, and the pace of work in the Science Alliance office suddenly picked up. On Thursday, the eve of Bush's visit, at 1:30 p.m., Riedinger, the Science Alliance director, was called to attend a briefing with the White House Advance Team concerning logistics for the session with the president. Later in the afternoon, reporters started calling, since word had gotten out that the Science Alliance would be featured prominently in the president's visit.[49]

At the Thursday briefing, the orientation of the room was discussed, including the three seats at the head table. President Bush would be seated at the center of the head table, UT President Lamar Alexander seated to his left, and Science Alliance Director Riedinger to his right. When he learned of the exclusive seating arrangement, Riedinger fully grasped the unique opportunity of spending twenty minutes sitting by the president's side to discuss UT-ORNL science programs. That afternoon, Riedinger and his team created two standup poster boards featuring relevant Science Alliance photos as a backdrop for the pending discussion.

On Friday, the day of the visit, the Science Alliance group—Riedinger plus eight UT-ORNL Distinguished Scientists—gathered in a room at the University Center. They heard that the president was running behind schedule and feared that their presentation might be cut from the program, but word soon arrived that the president had landed at McGhee-Tyson Airport. Riedinger and his colleagues quickly made their way to room 221 at the University Center, where the meeting would take place. The room featured tables positioned in a horse-shoe configuration, with seats at the side tables reserved for arriving state and federal dignitaries. The posters were set up behind the table, in front of a blue curtain. Shortly before the president's arrival, the door opened and twenty still and video photographers entered the room, and increasingly more chairs were added to accommodate the expanding audience.

Riedinger, who had managed to keep his emotions in check, experienced a surge of adrenalin and a slight case of the jitters when he heard the president in the hallway speaking to UT Chancellor John Quinn. It was show time, and all appeared to be going precisely as planned. U.S. Secretaries of Energy, James Watkins, and Education, Lauro Cavazos, were seated at the end of the head table, and chairs were brought in for U.S. Representatives from Tennessee: Jimmy Quillen, Marilyn Lloyd, Jimmy Duncan, and Don Sundquist. Knoxville Mayor Victor Ashe also settled into a seat.

Tension built as the Science Alliance team members stood behind their chairs at the side tables and the photographers' glaring lights illuminated the room. As the president entered the room and greeted the attendees,

Riedinger, who was positioned at the head table, next to where the president would be seated, realized that there might be a problem, and one that involved protocol, a key consideration whenever a U.S. president pays a visit.

Specifically, Riedinger realized that the head table was short one seat. The one vacant seat to the president's left had been reserved for UT President Alexander, but, then Tennessee Governor Ned McWherter, who was not scheduled to attend the session, showed up. Someone—McWherter, Alexander, or Riedinger—was about to lose his place at the head table. Riedinger, clinging tightly to the back of his chair, made it clear to McWherter and Alexander that he had no intention of yielding his place in the limelight, and

SCIENCE ALLIANCE BRIEFING: (FROM LEFT) LEE RIEDINGER (SCIENCE ALLIANCE DIRECTOR), PRESIDENT BUSH, GOVERNOR MCWHERTER, LAMAR ALEXANDER (UT PRESIDENT), AND THREE DISTINGUISHED SCIENTISTS: JACK WEITSMAN, GERALD MAHAN, BERNHARD WUNDERLICH—FEBRUARY 2, 1990. COURTESY OF BETSEY B. CREEKMORE SPECIAL COLLECTIONS AND UNIVERSITY ARCHIVES, UNIVERSITY OF TENNESSEE, KNOXVILLE.

he avoided Alexander's questioning expression. The standoff resolved quickly when Alexander noted Riedinger's resolve and ceded his own seat to the governor. As Alexander moved aside, for a second time, Riedinger realized that his tenure might be at risk, but, to Alexander's enormous credit, despite the unceremonious unseating, he allowed Riedinger, one of his employees, to retain his academic credentials.

With the momentary dilemma resolved, Riedinger delivered his presentation. He covered the key points and, on the fly, altered his rehearsed remarks to accommodate the unexpected presence of Governor McWherter. Some of the praise he had planned to bestow on Alexander, who had preceded McWherter as the state's chief executive, he redirected at the governor. In particular, Riedinger noted McWherter's generous ($800,000) support for ORNL's purchase of a Gammasphere detector array, a sophisticated tool critical for conducting physics experiments. Right on cue, McWherter leaned toward the president and, in a loud whisper, said, "Yes, I did that, Mr. President, but please don't ask me anything about what it is!" McWherter's audible aside brought a chuckle from the audience and lightened the mood considerably.

Riedinger finished after about ten minutes, then the president asked a few questions about UT's recruitment of students majoring in science. Alexander took the lead in responding, providing the relevant statistics. Chancellor Quinn noted that Science Alliance's Distinguished Scientists had been assigned to teach undergraduate-level science courses for the university. Gerald Mahan, the first Distinguished Scientist, hired in 1984, spoke for a few minutes about how much he enjoyed teaching freshman physics. The session ended, the president shook hands all around, and the entourage left the room for the next stop. On the way out, Bob Uhrig, also a Distinguished Scientist, had a friendly conversation with Education Secretary Cavazos. The two had roomed together thirty-five years earlier as graduate students at Iowa State University.

LASTING IMPRESSIONS FROM ONCE-IN-A-LIFETIME SESSION WITH A U.S. PRESIDENT

Riedinger recalls his impressions of the president: he looked rather young, was in good physical shape, was tall (six-foot-two), and appeared to listen well in the briefing. He also seemed genuinely interested in science education, as evidenced by his questions and remarks.

President Bush was taken to Alumni Gym for his address. Before the president came on, Lamar Alexander set the stage by announcing the introduction of Summer School of the South—a new summer program for math and science teachers.[50] Through the program, scientists from the university and ORNL would instruct two hundred teachers each summer. The State of Tennessee, Martin Marietta Corporation (managing contractor of ORNL and other DOE Oak Ridge facilities), and DOE supported the program through $1 million grants. In his remarks, President Bush discussed the need for excellence in science and math education, applauded the new summer program for science and math teachers, and marveled at the fact that the program's funding had been cobbled together in little more than one week. (Clearly, pending presidential visits can spur action and get results.) Bush also called for quick congressional approval of his plans to double the NSF budget and to increase the Eisenhower Math and Science Education Program by 70 percent.[51]

SUMMARY AND LOOK FORWARD

The decade of the 1980s ended with some critical gains and new critical connections in the partnership between UT and ORNL. In particular, the partnership evolved from one driven by individual UT faculty working with Oak Ridge researchers in specific fields to a higher level as leaders of the two institutions teamed their efforts and pooled their resources for a common good. A program to broaden the partnership, the Science Alliance, was born with state of Tennessee funds early in the decade, resulting in the initial hiring of ten Distinguished Scientists by the decade's end. The JIHIR was constructed with funds from the state and operated with funds from UT, ORNL, and Vanderbilt University. Though UT ultimately decided not to bid for the contract to manage four DOE facilities including ORNL, the sweeping and consequential events of the 1980s had prepared the university for a successful bid for the management contract for ORNL more than a decade later.

7

THE DECADE OF THE 1990S

After the Cold War Thaws, UT and ORNL Launch Ambitious Joint Projects

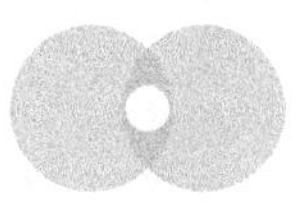

THE BERLIN WALL came down in late 1989 as the Soviet Union was in the process of unraveling. By the summer of 1991, various republics were ready to declare their independence from Moscow. On December 25, Soviet President Mikhail Gorbachev resigned and handed over the powers of government to President Boris Yeltsin of Russia, and the Russian flag was raised over the Kremlin. This marked the end of the USSR and, with it, the end of the Cold War. This brought a big change for the United States and its national laboratories, as the Cold War, a decades-long driver of R&D programs, was now gone or at least greatly diminished.

New opportunities came and evolved in this decade. In accordance with a DOE plan established in the 1980s,[1] ORNL would be in line for a new research reactor (the Advanced Neutron Source) to replace the High Flux Isotope Reactor that had been built in the mid-1960s. However, this proposal ran into trouble with the Clinton Administration in 1996, and some quick decision-making by ORNL Director Alvin Trivelpiece led to the replacement of this reactor proposal by a successful plan to build an accelerator-based source of neutrons. Ground was broken for the SNS as the decade of the 1990s was closing.

For advancement of the UT-ORNL partnership, the big opportunity came late in the decade, when UT teamed with Battelle Memorial Institute to win the contract to manage ORNL, a concept unthinkable even a decade earlier. But, numerous precursor developments over the years and especially during the 1990s facilitated the later award of the coveted ORNL management contract. One was the formation of the joint National Transportation Research Center on a plot of farmland halfway between Oak Ridge and Knoxville. Another was expansion of the Science Alliance program for jointly hiring top scientists and engineers.

As often the case, the significant developments of the 1990s resulted from the leadership and innovation of forward-thinking people. Five of the more important of them were UT's Joseph Johnson and Homer Fisher, ORNL's Bob Honea and Al Trivelpiece, and Battelle's Bill Madia.

On the world front, the decade of the 1990s is often remembered as a time of relative peace and prosperity, especially in the wake of the collapse of the Soviet Union. The rise of the Internet began a radical new era in communication, business, and entertainment. Peace came as the Cold War ended, but a localized Gulf War would prove pivotal in shaping the future of the Middle East. The U.S. and Allied forces launched aerial attacks on Baghdad, Iraq on January 18, 1991. These world events were reflected in various ways in developments in the UT-Oak Ridge partnership.

Tiger Teams Visit Oak Ridge

President George H. W. Bush appointed Admiral James Watkins as the Secretary of Energy on March 1, 1989. Three months later, on June 6, a convoy of about thirty vehicles carrying more than seventy armed agents of the FBI and the Environmental Protection Agency raided DOE's plutonium-processing plant at Rocky Flats, Colorado, for suspected environmental crimes.[2] This plant manufactured fissionable plutonium cores for U.S. nuclear bombs. Watkins was upset not only with the health and safety issues at this facility but also over the fact that his managers in the local DOE office had not alerted him to these dangers. In 1989, Watkins decided to take the initiative and order inspections of all DOE facilities by a special force known as Tiger Teams.[3] These teams, as large as sixty persons, visited each DOE site, including ORNL and Y-12, for about a month and identified violations of safety and environmental protocols.

These visits followed self-assessments and inspections by managing contractors and preliminary inspections by local DOE offices.

This elaborate inspection process resulted in a 1,400-page report on ORNL operations, and the laboratory responded with a 1,000-page document. The DOE inspectors identified 413 deficiencies at the laboratory whose remediation would cost nearly $1 billion. Subsequent analysis revealed that DOE had requested long before the Tiger Team visited all but $60 million of those funds, but ORNL had received none of those funds, presumably due to budgetary constraints at DOE.[4]

Some improvements were certainly called for, but the notion that DOE facilities would ignore major risks to workers or public health and safety bordered on the insulting. ORNL dutifully implemented many of the suggested improvements, but the predominant and quietly held view at the national laboratory at that time was that these investigative intrusions were a huge waste of time and money. Nevertheless, this exercise, although painful at the time, led to an increased focus on compliance and safety that improved the operational environment at DOE facilities in Oak Ridge and elsewhere.

JOINT HIRING OF TOP RESEARCHERS

The state of Tennessee's funding of the Science Alliance in 1984 led to the initial joint hiring of distinguished scientists and engineers in fields of research mutually important to UT and ORNL. Tennessee Senator Howard Baker kicked off that program with a ceremony in the Mansfield Room of the U.S. Capitol after then Tennessee Governor Lamar Alexander had orchestrated funding for the project. Ultimately, the Science Alliance helped both institutions greatly by attracting the top-tier talent that neither could have recruited on its own.[5]

Initially, these distinguished scientists were officially based at the university, but ORNL covered half their salaries, benefits, and research-support packages. The rationale for this arrangement arose from the award of academic tenure, which UT was in a position to offer while ORNL could not. Leaders of the two institutions believed that they could not attract high-level talent without the promise of university tenure.

The success of this program of jointly hiring research stars led to the idea of making joint hires at all academic levels, in part to attract rising stars in research areas of joint interest to the partnering organizations. Members of this new cadre were called collaborating scientists (the name changed to joint faculty a decade or so later). The first hires were made around 1990, with the participating scientists based at UT.[6]

The idea later evolved to "share the risk" by having some collaborating scientists based at UT and some at ORNL. In this way, each institution would assume the long-term risk and commitment for supporting the researchers, should the other institution decide to opt out of the cost-sharing arrangement. This equal sharing of risk was and is a fine concept, but it proved difficult to implement.

This difficulty started to become apparent in 1992 when the UT Department of Physics agreed to cost share and appoint as collaborating scientists two researchers hired earlier into ORNL's Physics Division. The agreement was made, and the two individuals started to spend half of their time teaching courses in the UT physics department. However, the problem resulted from difficulties in transferring funds from UT to ORNL to cover half of the costs for these two participating physicists. In particular, the university and the national laboratory had vastly different structures for assessing indirect costs—essentially overhead. This vexing problem persisted for three years, while Jim Ball, the director of ORNL's Physics Division, patiently waited for the checks from UT to arrive.[7]

Overhead costs at all DOE laboratories, including ORNL, are at least twice as high as those at a university. DOE does not provide central operating funds to its national laboratories, so each facility has to impose hefty overhead costs on every research program that operates under its roof. By comparison, a state university receives an annual operating budget, and it charges its enrolled students for tuition. As a result, these state and tuition revenues partly cover university infrastructure costs (e.g., costs for facilities operation, support staff, maintenance, accountants, lawyers, etc.). A research grant to the university is charged for overhead, but these costs are lower than those at a national laboratory.[8]

The UT Department of Physics was perfectly willing to send half of salary and benefits to ORNL for these two collaborating scientists but could not afford the *de facto* doubling of the costs of these packages due to the laboratory's high overhead rate. In 1991, Lee Riedinger shifted from his position as director of the Science Alliance to serve as UT's interim vice chancellor for research when the incumbent in this position suddenly took another job. Normally, an interim appointee serves for only a short time, but, in this instance, Riedinger served in this capacity for four years. In this central role, he took the lead in trying to draft an agreement that would allow UT partially to escape the financial burden of ORNL's steep overhead costs as the university moved forward to reimburse the laboratory for half of the salaries and benefits of the two ORNL-based collaborating scientists.

Numerous meetings with lawyers and accountants at both institutions ensued, and this process dragged on for two years. During that time, Riedinger and William Bugg, head of the Department of Physics, tried to avoid talking with ORNL's Jim Ball because they knew that Ball's first question would be, "When will UT remit to ORNL two years of promised support for these two shared physicists?" The problem seemed intractable until a possible solution emerged in spring of 1994.[9]

Alvin Trivelpiece

Alvin Trivelpiece received his doctorate from Cal Tech in 1959 and began a career in plasma/fusion physics. He served on the faculties at the University of California, Berkeley, and then the University of Maryland for a total of eighteen years before entering the private sector. He was corporate vice-president of Science Applications, Inc. in La Jolla, CA, for three years until 1981. Then, he spent a crucial six years as director of the DOE Office of Energy Research, where he led the planning for a suite of new major scientific research facilities at national laboratories.[10]

AL TRIVELPIECE. COURTESY OF OAK RIDGE NATIONAL LABORATORY.

The Reagan administration was friendly to the idea of a major investment in new facilities emphasizing fundamental research, while reducing investments in energy strategies that the private sector should pursue. A major planning process by an esteemed panel resulted in the 1984 Seitz-Eastman report[11] to the National Research Council articulating a prioritized list of large experimental facilities for the next generation of research in materials science. Trivelpiece skillfully used this report (and others) to make the case and build a consensus for new facilities at five national laboratories: Argonne, Berkeley, Brookhaven, Jefferson, and Oak Ridge, with construction budgets that would be supported over the next twenty years. This ambitious plan did eventually succeed.

After DOE, Trivelpiece spent two years as the executive officer of the American Association for the Advancement of Science and, in January 1989, became ORNL director, a position he held for eleven years until UT-Battelle LLC assumed management of ORNL in April 2000. Trivelpiece served with distinction as the ORNL leader, laying the groundwork for a large supercomputing program and steering the lab through the difficult waters leading to the Spallation Neutron Source. And, he did much to promote the UT-ORNL partnership.[12]

President Bill Clinton appointed Martha Krebs to Trivelpiece's former position as director of the Office of Energy Research in late summer of 1993. Krebs, who held a doctorate in physics from Catholic University of America in Washington, D.C., left her position as associate director at Lawrence Berkeley National Laboratory to assume the DOE post. Trivelpiece welcomed Krebs, an old friend, to ORNL for her first on-site review of the laboratory's programs in early 1994. Conducted annually at all Office of Energy Research national laboratories, such reviews were attended by DOE sponsors of the laboratories' programs, featured presentations by project leaders, and provided a forum for guiding and directing the laboratories' future research activities.

Riedinger was asked to make a presentation on the strong partnership between the laboratory and the university in general and on the joint hires facilitated by the Science Alliance program in particular. After the talk, Krebs asked politely if there were any problems in executing this innovative partnership. Riedinger gulped and explained that it was impossible to remit payment to ORNL for the university's share of the compensation packages provided to laboratory-based collaborating scientists because of the big difference in overhead rates and the inability to negotiate an agreement that would provide a waiver of some of the laboratory's steep overhead costs.

Laboratory officials in the room visibly cringed when Riedinger raised this nagging issue, because, in so doing, Riedinger was airing a bit of "dirty laundry" that was perhaps best not revealed to the new DOE official responsible for funding the laboratory. Al Ekkebus, one of this book's authors, worked on Trivelpiece's staff at that time. He was in the room and felt the tension when this issue was raised. Krebs, though initially taken aback by the frank assessment of the issue, quickly pledged that she would resolve it. And, she made good on her commitment.

Upon Krebs' return to Washington, she ordered her staff to solve this problem and to solve it quickly. She shook the administrative tree and meetings were hastily called up and down the command chain. Ultimately, a number of DOE officials in Washington, D.C. and Oak Ridge collaborated in working toward a solution. The result, a University Research Agreement, was negotiated and signed in 1995, allowing the university to pay only part of the laboratory's high overhead costs, and, soon enough, university funds were on the way to ORNL to support half of two collaborating scientists' compensation packages.[13]

For a time, a few ORNL administrators, who would have preferred to avoid the direct engagement of the director of the Office of Energy Research in resolving the issue, took Riedinger's name in vain. Nevertheless, the problem was solved, removing any barriers to future hires of collaborating scientists based at ORNL.

OAK RIDGE'S NEXT NEUTRON SOURCE

From the beginning, Oak Ridge was regarded as a "reactor town." The Graphite Reactor went critical in November of 1943. The 1950s were the decade of reactor development at ORNL, and this tradition resulted in the construction of HFIR, which first achieved criticality in 1965. HFIR still runs today and remains one of the best and highest flux research reactors in the world.

HFIR has been the country's best producer of actinide elements for a host of R&D purposes. Actinide elements, which account for consecutive atomic numbers 89 to 103, are all radioactive, and many (those beyond uranium, element 92) were synthesized in the laboratory and do not exist in nature. In addition, HFIR was essential for a recent discovery of a new element beyond the actinides, element 117 (tennessine, Ts), which was produced in a year-long experiment at a synchrotron in Dubna, Russia using a radioactive target of the actinide element berkelium (Bk, atomic number 97) produced and prepared at HFIR.[14] Although HFIR's primary mission was isotope production, it was also equipped with four beam lines so that some of the generated neutrons could be used to learn about the properties of materials, building on neutron-scattering work done by Ernest Wollan and Cliff Shull at the Graphite Reactor in the late 1940s and early 1950s.

ORNL maintained a strong program in neutron scattering centered on HFIR starting in 1966. By the late 1980s, a successor to HFIR was being sought. The Advanced Neutron Source (ANS) would produce the world's highest flux for neutron scattering, provide isotope production facilities at least as good as those of HFIR, and provide tools and techniques for irradiating various materials as good as or better than those of HFIR.[15] ANS would continue the long tradition of neutron-scattering experiments for a host of research projects in engineering, biology, physics, chemistry, and material science.[16]

When Ronald Reagan became president in early 1981, a new source of neutrons for research was not constructed in the United States since the 1960s. Meanwhile, European countries continued to open new neutron-based research centers and assume world leadership in this field. When Reagan was elected president, he had pledged to increase funding for fundamental research in the United States. This led to the well-regarded Seitz-Eastman panel, co-chaired by Dean Eastman of IBM and Frederick Seitz, retired head of the National Academy of Sciences and director of the Clinton Training School in the late 1940s. This high-powered panel recommended the construction of a set of billion-dollar research facilities at national laboratories to ensure the future health and competitiveness of the U.S. scientific community.

At the time, Alvin Trivelpiece was the director of DOE's Office of Energy Research and took the lead in deciding how to distribute these large new science facilities equitably among the national laboratories so that each would boast a new research facility poised at the forefront of science and technology. The new jewel for ORNL would be ANS, a 300-MW research reactor with three times the power of the existing HFIR. Trivelpiece's later appointment as director of ORNL in 1989 greatly increased the laboratory's emphasis on this big new research reactor project.[17]

Unfortunately, issues with the ANS project began to surface just a few years later, long before the facility could be built. One issue was the reasonable decision to design and build this research reactor to use highly enriched uranium (with a concentration of at least 20 percent fissionable ^{235}U), which had always been the isotope of choice for research reactors capable of delivering the high flux of neutrons needed for sensitive experiments. By comparison, nuclear reactors used in industry for generating electricity use only 3- to

5-percent enrichment but are built on much larger scales (e.g., 1,000 MW, compared with 100 MW for HFIR and 300 MW for ANS) and have robust sets of safety and security systems. The early decision was to build ANS with this same extensive set of safety measures, which increased the cost estimate well beyond the early $1 billion projection.

This was also a time of heightened concern among politicians over construction of any civilian device that relied on highly enriched uranium. The United States was urging other countries to refrain from using this fuel source for their reactors, due to the fear that the highly enriched uranium could be diverted from its use in a research reactor to being a key part of nuclear weapons. Still, DOE was moving ahead with plans to build ANS with highly enriched uranium fuel rods. Experts discussed building ANS with low-enrichment uranium, but doing so would make it difficult to attain sufficient neutron flux in the reactor to carry out the desired research programs. There were also perceived management problems associated with the ANS project, and these problems strengthened the hands of those opposed to building a reactor using highly enriched uranium. However, the final flaw came with the 1994 estimate that the cost of ANS construction would be nearly $3 billion and that proved to be a bridge too far. The Clinton administration subsequently removed funding for ANS from the president's fiscal year 1996 request to Congress. Thus, ANS died before ground could be broken.[18]

Out of the ashes of ANS came the drive to fund a promising accelerator-based source of neutrons for research—and one that would produce a pulsed source of neutrons as opposed to the steady flow of neutrons issuing from a typical reactor. The Seitz-Eastman panel had recommended that both types of neutron sources—pulsed and steady—be built for a host of research purposes. A competition was held for the siting of a pulsed neutron source, and four laboratories submitted excellent proposals: Argonne, Brookhaven, Los Alamos, and Oak Ridge.[19] ORNL won this competition in part because it was Oak Ridge's turn for a big new facility, particularly in the wake of the canceled ANS. Accelerator-based sources of synchrotron radiation used for materials research were funded at Berkeley and Argonne; a large accelerator-based source of neutrons were being expanded at Los Alamos; and Brookhaven was building the Relativistic Heavy Ion Collider for nuclear physics research.

This led to funding for the Spallation Neutron Source at ORNL. Conceptual design of SNS was funded in fiscal years 1997 and 1998, and $130 million came in 1999 for a more-detailed design. Vice President Al Gore visited ORNL on January 21, 1998, to announce the Clinton Administration's request for funding to initiate SNS construction.[20] SNS, with a projected price tag of $1.4 billion, would be composed of a high-energy and high-intensity accelerator for making a proton beam that would strike a massive target and produce neutrons by the spallation process.[21] Through such a process, fragments of the target material—so called spall—are ejected from a body due to impact or stress.[22] In the case of SNS, high-energy protons hit the target's nuclei and spall off various nuclear particles, including neutrons.

SNS began with high expectations and much excitement, but, by early 1999, problems started to arise. A late January DOE review of the project resulted in criticism in part due to the outside feeling that ORNL did not have a team of people experienced in design and construction of large accelerators.[23] And, some members of the U.S. Congress expressed concerns about the project. The House of Representatives was unhappy that the state of Tennessee would charge sales tax on items procured for SNS construction at a rate higher than that in other states that could host this facility. There was even a threat of canceling the SNS project before ground was scheduled to be broken in late 1999.

In response to these issues (real or perceived), ORNL director Trivelpiece removed the current head of the SNS project, Bill Appleton, a highly respected materials scientist and laboratory leader, and replaced him with David Moncton from Argonne National Laboratory.[24 25] To his great credit, Moncton was an accomplished accelerator builder and previously had served as leader of several large projects.[26] At Argonne, he took the lead in the successful construction of the laboratory's new jewel, the Advanced Photon Source. He had the background and credentials that ORNL and DOE leadership were looking for and so Moncton came to ORNL in March 1999 as an Argonne employee assigned to the SNS project.

Moncton had a stellar background. He received DOE's prestigious E. O. Lawrence Award for his research with synchrotron x-rays. He attended graduate school at MIT, starting in 1970. Cliff Shull, who served as one of his research mentors for his doctoral research conducted at the reactors at MIT and at Brookhaven National Laboratory, recruited him. Shull and Ernest Wollan were the originators of using the scattering of neutrons to study material properties in their pioneering work at the ORNL Graphite Reactor in the late 1940s, until Shull's departure for MIT in 1955.

The leadership of Al Trivelpiece at ORNL through the decade of the 1990s was crucial to capturing the next generation neutron facility. He wrote the DOE plan for large new research facilities (all at the billion-dollar level) at five national laboratories, and, then at ORNL, he guided the rapid transition from pursuing a reactor (ANS) to capturing a pulsed neutron source (SNS). Ground was broken for the Spallation Neutron Source on December 15, 1999.

BEGINNING OF MAJOR SUBCONTRACTING TO THE UNIVERSITY

Another important outcome of the UT-Oak Ridge partnership has been increased subcontracting awards from major laboratory programs to the university. Through these subcontracts, ORNL programs can, as needed, tap the specialized UT expertise necessary for the laboratory's programs to meet their deliverables. Small ongoing subcontracts were written for twenty years between ORNL's research divisions and their disciplinary counterparts at UT, but, early in this decade, the award of such subcontracts surged.

Subcontracting is the practice of assigning, or outsourcing, part of the obligations and tasks under a contract to another party (the subcontractor).[27] Subcontracting is especially prevalent in areas where complex projects are the norm, such as construction and information technology, or in cases where specialized talent needs to be added quickly. Defense agencies utilize this process commonly, to allow their prime contractors to issue subcontracts to more efficiently achieve the goals and timetables of the contracted work. National laboratories use the subcontracting mechanism to get help on projects brought to the organization. The proximity and strong partnerships in numerous areas make UT an attractive subcontractor for many ORNL programs.

After Ronald Reagan became the U.S. president in January of 1981, funding for research programs in ORNL's Energy Division started to decline significantly due to a change in priorities for federal funding of R&D. ORNL Associate Director Bill Fulkerson was deeply concerned about the loss of research funding for this part of the laboratory and asked Bob Honea, a group leader in the Energy Division, to find new programs and new funding for this division. Honea succeeded, and one of the new connections he developed was with the Military Traffic Management Command in the DOD. This command contracted with Honea's group to develop a new system that came to be known as the Mobility Analysis and Planning System (MAPS). This started a new avenue of funding for ORNL programs that would develop greatly in future decades.[28]

ORNL had to overcome two problems, speed and adequate staffing, in order to deliver promised outcomes for this and other defense-related projects. It was (and is) a challenge for a national laboratory rapidly to ramp up expertise and meet deadlines in the environment created by a constrained budget. The high overhead costs at any national laboratory (at least twice that of most universities) heighten these challenges, as discussed earlier in this chapter. At ORNL, a solution came by partnering with UT and engaging in subcontracts to help meet deliverables. DOD is able to send money for a project rather quickly and simply to a DOE laboratory through a Military Interdepartmental Purchase Request (MIPR).[29] Then, ORNL could rapidly subcontract some of those funds to UT to engage additional personnel.

To enable the increased level of needed subcontracting, the Pellissippi Research Institute was formed at UT around 1990 and led by Don Alvic, who received a PhD degree in geography from UT a decade earlier while working to develop geographic information systems (GIS) in Bob Honea's group at ORNL. Ample space was available for construction of Alvic's institute at UT's Dutchtown Road facility, completed in 1988 roughly halfway between Oak Ridge and Knoxville. At the time, UT Executive Vice Chancellor Homer Fisher took the lead in contracting with a private developer to build and lease to the university much-needed laboratory space for newly hired distinguished scientist David White. First one and then a second building provided space (20,000-square-feet each) for other joint UT-ORNL programs, including Alvic's Pellissippi Research Institute.

The Gulf War began in August 1990, in response to Iraq's invasion and attempted annexation of Kuwait. This sudden war presented many problems for the United States, including a dearth of the logistics necessary to deploy troops, weapons, supplies, and housing quickly to a remote and distant place. In response, DOD funded ORNL's Joint Flow and Analysis System for Transportation (JFAST).[30] The Air Mobility Command Deployment Analysis System (ADANS) was another ORNL program conceived at this time. Through these programs, ORNL and UT researchers collaborated to develop software that significantly advanced the state of the art in airlift scheduling and provided the Military Airlift Command with new tools for schedule analysis. ADANS was used to analyze how best to organize the logistics for the massive Operation Desert Storm airlift.

Staffing and expertise were needed for JFAST and ADANS, and Alvic had the background and contacts at the university to help with recruiting personnel, drawing on graduate students, faculty, and newly hired researchers. As Alvic bolstered staffing, the number of subcontracts increased to provide the rapid implementation and improved performance of R&D essential for each funded project. However, at the time, the concept of quick approval and rapid turnaround was largely foreign to the often-plodding UT Office of Research, which historically was underfunded and not fully acquainted with the concept of "customer service."[31]

As a result, Alvic had to walk each subcontract proposal through the slow-moving process and endeavor to persuade the Office of Research staff to put his proposal on top of the large stack of paperwork to be reviewed before being sent out for approval. Being persistent and creative did help expedite this process, as Alvic and his ORNL contact convinced the laboratory's head of purchasing to set up a simpler task-order mechanism[32] in hopes of breaking up the UT log jam. Utimately, Alvic and UT's Vice Chancellor for Research Tom Collins succeeded in convincing the university's comptroller to accept this new way of doing business.

This situation became Riedinger's in the summer of 1991, when he was appointed as interim vice chancellor for research after Collins left UT to assume a post at Oklahoma State University. The Pellissippi Research Institute was awarded a $25-million, five-year contract from ORNL, executed rather hastily by way of task orders. UT's rapid approval of the contract reflected a decided shift from the old (rather laggard) way of doing business to a new operating procedure that allowed the university to capitalize quickly on emerging research opportunities.

THE CONSOLIDATION OF TRANSPORTATION RESEARCH

Research relating to transportation grew and thrived through the 1980s for both the university and the national laboratory. At UT, the research was funded primarily by Tennessee's Department of Transportation; DOE's Office of Energy Efficiency and Renewable Energy was the primary funding source for ORNL. In 1993, ORNL Associate Director Bill Fulkerson once again summoned Energy Division group leader Bob Honea and asked him to conduct an inventory of all transportation research underway at the three DOE installations (ORNL, Y-12, K-25) on the Oak Ridge Reservation.

To everyone's surprise, Honea discovered a combined $60 million of ongoing transportation research programs, with no one in charge of overseeing and directing the then piecemeal research effort. In response, Fulkerson appointed Honea to coordinate the overall transportation research program, which Honea subsequently named the Oak Ridge Transportation Technology Program (ORTRAN). This became the umbrella program that included transportation-related research at the three Oak Ridge facilities managed by Martin Marietta and funded by different federal agencies. Presumably, the coordination and consolidation of these separate research programs would lead to new synergies and opportunities.

Bob Honea

Bob Honea grew up in Athens, Georgia and attended the Georgia Military College on a football scholarship for a year before transferring to the University of Georgia for bachelor's and master's degrees in geography. In 1968, he relocated to Tennessee to teach at East Tennessee State University (ETSU).[33] He divided his time between the university and the Marshall Space Flight Center in Huntsville, Alabama and played a major role in the start-up of NASA's Landsat program, an ambitious project that utilized satellites to study and map the Earth's surface.

In 1971, Honea left ETSU to earn a PhD in geography at the University of Florida and then joined the research staff at ORNL in 1975, where he later led programs in GIS transportation modeling, many of them in support of the U.S. military's Air Mobility Command Deployment Analysis System. In 2006, after twenty-seven years at the laboratory, Honea transferred to the University of Kansas to direct the Transportation Research Institute. In 2011, Honea was awarded the Anderson Medal, the highest honor given by the Applied Geography Specialty Group of the Association of American Geographers.[34]

BOB HONEA AND AL TRIVELPIECE. COURTESY OF OAK RIDGE NATIONAL LABORATORY.

Perhaps as important as Honea's research accomplishments were his creativity and entrepreneurship, both of which proved crucial when he was put in charge of the Oak Ridge transportation program. At the time, organizational structure and space were two problems significantly hindering future growth. Honea responded by convening a meeting at the Tellico Yacht Club of transportation research leaders from the three sites in Oak Ridge and from UT. The meeting focused on what to do about space constraints for the various transportation research projects. To enhance collaboration and grow the transportation program, meeting participants realized that it was essential to secure a building or facility where they could co-locate the various components of the then scattered transportation research effort. They had not been able to find an existing facility on the Oak Ridge Reservation, so they decided to explore alternative sites located close by.

Six years earlier, in 1988, Honea convinced UT Executive Vice Chancellor Homer Fisher to have a developer build the second 20,000-square-foot research center on Dutchtown Road to house both UT and ORNL staff related to his research programs at the laboratory. Honea explored the possibility of a similar arrangement whereby UT would take the lead on building a transportation research center, but, unfortunately, the new vice chancellor for research, Mike Devine, who succeeded Riedinger (after Riedinger returned to the UT Department of Physics) in 1995, was not willing to take a chance and pursue this promising possibility. Honea followed other possible leads, none successful.

A new strategy surfaced when UT executive assistant Billy Stair got involved and called a crucial meeting. Stair grew up in Kingsport, Tennessee, attended college at Penn State University, and returned to Tennessee for a career in politics and public policy. He worked in Tennessee state government for eighteen years, serving in both the legislative and executive branches. From 1987 to 1994, he served as Governor Ned McWherter's senior policy advisor. In 1994, Stair came to UT to work as executive assistant to President Joe Johnson, focusing on government relations, communications, athletics, and various budgetary initiatives. In 2000, he would move to ORNL as director of communications, when UT-Battelle assumed management of the laboratory.

Stair knew of the challenges in building the greatly needed new space for the two institutions' transportation research. In 1995, he called a meeting of Bob Honea, Pat Wood (a Knoxville developer), and Melissa Ziegler, director of the Knox County Development Corporation (KCDC) and an Oak Ridge native. They learned that KCDC was willing to provide land for this new facility, but could not build, own, or lease any future structure to ORNL and UT.

Honea went to see Katie Kates, a real-estate contract specialist in DOE's Oak Ridge Operations office. Together, Honea and Kates creatively set up a mechanism by which a private developer, Pellissippi Investors LLC, could construct this transportation facility and maintain ownership for a fixed period of time (a similar arrangement to a twenty-five-year home mortgage). Honea's funded transportation programs would pay lease fees to the developer over this period, and ORNL would lower the overhead charges on this off-site space by requiring only a 10-percent "space charge" rather than the usual 43 percent for normal government-owned buildings on the national laboratory site. This innovative deal was approved (rather surprising at the time), and the developer committed to build a roughly $25-million facility, which probably would have cost $40 million as a DOE construction project. This facility would occupy KCDC-donated farmland adjacent to Pellissippi Parkway, roughly halfway between Knoxville and Oak Ridge.[35]

Ground was broken for the new center in 1999 in a ceremony including Representative Jimmy Duncan and U.S. Transportation Secretary Rodney Slater.[36] In 1998, DOE Secretary Federico Peña announced a grant award to support purchase of required research equipment and christened the facility the National Transportation Research Center (NTRC). NTRC opened in October 2000, after six persistent years of effort. This 85,000-square-foot facility features twelve research laboratories, 175 offices, a conference center, and reception area. Nearly two hundred research staff from ORNL, UT, and private industry soon occupied NTRC, which had an annual funding base in excess of $90 million. Unfortunately, faculty from the UT College of Engineering (where transportation research was conducted) never fully engaged in NTRC projects. The prevailing justification was that NTRC was too far from campus—all of thirteen miles.

The construction and opening of NTRC was important not only from the standpoint of uniting disparate Oak Ridge transportation research efforts under one roof, which prompted synergy and growth, but also for spurring additional facility developments. UT-Battelle would use a similar partnership model, approved by DOE's Oak Ridge Operations office, to construct critically needed research space on the laboratory's campus. Two new buildings would later be constructed adjacent to NTRC to house DOE's first Manufacturing Demonstration Facility, devoted to R&D in additive manufacturing. This time, UT faculty became deeply involved in this joint research facility, and some of them would go on to assume leadership roles there.

UT, ORNL BEGIN TO BUILD COMPUTING PROWESS

A substantial increase in computing capabilities occurred at ORNL and at UT in the decade beginning in 2000, but groundwork for that growth was laid in the 1990s. Indeed, during this decade, across the country, there was an increased focus on distributed and parallel computing and on the then nascent Internet. Twenty years later, ORNL and UT would each boast one of the largest and most powerful supercomputers in the world. As in most success stories, capable individuals led these advances through pursuing bold visions.

Alvin Trivelpiece, though an experimental fusion physicist by training, understood the need for strong computing and computational science as an underpinning for most every modern scientific discipline, and many credit Trivelpiece with placing ORNL on the road to becoming a major supercomputing power. Around 1992, he recruited Ed Oliver, who created ORNL's new Office of Laboratory Computing. One of Oliver's more consequential

JACK DONGARRA, RIGHT, IS CONGRATULATED BY ORNL DIRECTOR THOMAS ZACHARIA FOR WINNING THE A. M. TURING AWARD. THE SUMMIT SUPERCOMPUTER AT ORNL IS IN THE BACKDROP. COURTESY OF OAK RIDGE NATIONAL LABORATORY.

decisions was to appoint computational materials scientist Thomas Zacharia to his first major administrative job, director of the laboratory's Computer Science and Mathematics Division.

Zacharia's 1998 appointment began the rapid rise in computing capability at ORNL and marked the start of Zacharia's impressive move up the administrative ladder. In 2002, he would become ORNL's first associate laboratory director in the new Computing and Computational Sciences Directorate, and, in 2017, he would be appointed laboratory director. UT-Battelle, by then the new ORNL managing contractor, would likewise play a major role in the laboratory's evolution into a globally respected computational powerhouse.[37]

Jack Dongarra was hired from Argonne National Laboratory in 1989 to become one of the dozen UT/ORNL distinguished scientists,[38] with Dongarra specializing in numerical algorithms in linear algebra, parallel computing, and the development, testing, and documentation of high-quality mathematical software. In 1993, Dongarra led the creation of the list of the world's TOP500 Supercomputer Sites, based on his LINPACK benchmark, a mathematical tool that makes it possible to calculate precisely a supercomputer's power and speed. Semi-annually for the past twenty-five years, Dongarra's benchmark has been used to identify the globe's five hundred top-performing computers.

In the years since, Dongarra has received numerous professional awards for his leadership in designing and promoting tools for evaluating the mathematical software used to solve numerical problems common to high-performance computing. At UT, Dongarra built a strong team of researchers in his Innovative Computing Laboratory, and, in the latter half of the 1990s, he won almost $4 million in research awards from the NSF, a significant portion of the cumulative $42 million in NSF funds awarded to UT during that time. In early 2022, the Association of Computing Machinery announced that Dongarra had won the 2021 A. M. Turing Award for his pioneering contributions to the evolution of high-performance computing over the past forty years.[39] The award is respected as the Nobel Prize of computing and is named for Alan Turing, the British mathematician and cryptanalyst, who many consider to be the father of theoretical computer science and artificial intelligence. Along with the recognition, Dongarra received a $1 million prize, with financial support provided by Google Inc.

Zacharia, Dongarra, and others played essential roles in building the reputations of the two institutions in parallel computing. Additionally, major advances after 2000 would propel both institutions toward the top of the list of leading U.S. universities and laboratories in the computing field.

Calculating Supernovae

Supercomputers at ORNL and at UT have been used for many calculations in a host of fields of science and engineering. One of the most novel uses has been to understand the dynamics of a supernova.

When a large star (maybe ten times the size of our Sun) nears death, the nuclear fusion that has fueled the star for its lifetime peters out due to the consumption of most of the light elements that drive fusion. Gravity then takes over and the large star rapidly collapses over a few hours of time to such a high density that a massive explosion occurs—a supernova. This is the most violent event that regularly occurs in the universe and is important for several reasons. One is that the high temperature and density of nuclear particles during the explosion rapidly result in the production of many heavy elements that are ejected and spread throughout the galaxy in which the supernova occurs. Eons later these heavy elements comprise key parts of a new star that forms, along with planets later on. Indeed, most of the elements that formed the Earth and everything on it have come from one or two supernovae that exploded in our part of the Milky Way galaxy billions of years ago. The small remnant of a supernova is a neutron star or a black hole, some of the more mysterious objects in the universe.

Astrophysicists have been trying to understand the dynamics of a supernova, and supercomputers are necessary to perform large-scale calculations of the many physics processes taking place in a huge collapsing and then exploding star. ORNL/UT joint faculty Anthony Mezzacappa and his team have been leaders in this area using the ORNL supercomputer.[40] They achieved a breakthrough in making the simulated supernova explode by incorporating the strategic role of a tremendous flux of neutrinos emanating from the core of the collapsed star.[41]

Neutrinos are formed as a byproduct of the forced merger of electrons and protons in the collapsing star. Neutrinos have no electric charge, almost no mass, and interact rarely with material through which they pass. However, the large number of neutrinos formed in the explosion is apparently responsible for providing a huge boost to the expanding cloud of elements formed in the supernova. That expanding cloud seems to stall in the computer simulations, but then the large flux of neutrinos expanding outward provides heat and gives a needed boost to the stalled sphere of ejected elements. Then, the supernova explodes in full force. Understanding the physics of this process was a big step forward in the field of astrophysics.

The ORNL/UT team (including collaborators from other universities) learned another important fact about supernovae in their supercomputer simulations.[42] While the original expansion of the ejected material in the exploding star is spherical, an instability occurs in the shockwave fueled by neutrino heating. This instability results in a momentarily elongated, rather than spherical, ejected cloud, which then can begin rotating. The rotation is rapid, which explains why the remnant product of the explosion (a neutron star) has an extremely high rotational speed.[43] Neutron stars are small (around 10 km in radius), and, on average, they rotate once per second—amazingly fast. This important result from the team's simulations was possible only because the ORNL supercomputer was large enough to enable three-dimensional calculations of the physics involved. Smaller, less powerful computers could handle only two-dimensional calculations, in which case this instability would not be apparent.[44]

Mezzacappa was an ORNL Corporate Fellow and director of the UT/ORNL Joint Institute for Computational Sciences before moving to the UT Department of Physics and Astronomy as the Newton W. and Wilma C. Thomas Chair. In 2022, he received the prestigious Alexander Prize, named for former UT president and U.S. Senator Lamar Alexander and his wife, Honey. It recognizes superior teaching and distinguished scholarship.

An interesting addendum to this story involves one of the other team members, Bronson Messer. He was a UT student with undergraduate degree in physics in 1991 and a PhD in 2000 with a dissertation on core-collapse supernovae. In addition to his research using supercomputers, Messer found time to appear on the popular TV show *Jeopardy* and, in fact, was a three-time champion in 2003, winning around $30,000.[45] Now, he is the Director of Science at the Oak Ridge Leadership Computing Facility at ORNL.[46]

RESEARCH AT ORNL AND UT LEADS TO TWO BREAKTHROUGHS IN FORENSIC SCIENCE

The broad expertise of ORNL and UT in various areas of science and engineering led to forefront research programs in conventional fields but also impacted areas related to public affairs. One example is forensic science. Director Alvin Trivelpiece recognized an opportunity for ORNL to respond to a local community need grounded in that discipline. In July 1993, Knoxville Police detective Art Bohanan was perplexed because he could not find the fingerprints of a 3-year-old child who had last been seen entering a car and later found murdered. Bohanan found the adult suspect's prints in the car but none belonging to a child. In his search for answers, the detective queried the FBI, the National Institute of Justice, and Scotland Yard. He even asked a colleague in Russia what he knew of the characteristics of children's fingerprints. However, Bohanan received no useful information.

Later, during a tour of his forensic laboratory by a group of community leaders hoping to learn more about the Knoxville Police Department, Bohanan told them of his problem. One of the attendees happened to be Al Trivelpiece, ORNL director. Trivelpiece invited Bohanan to the national laboratory to discuss the problem in greater detail. In a meeting with ORNL scientists Trivelpiece had assembled to discuss Bohanan's dilemma, Bohanan met Michelle Buchanan, an analytical chemist specializing in organic mass spectroscopy.

Mass spectrometers have a long history in Oak Ridge, with some of the biggest units in the world housed at the Y-12 Plant during the Manhattan Project. These particular versions of the mass spectrometer were called calutrons and were used to separate uranium isotopes. More broadly, a mass spectrometer is an analytical tool to measure the mass-to-charge ratio of one or more molecules present in a gaseous sample that has been ionized (electrical charge removed from the molecule) and passed through a magnetic

MICHELLE BUCHANAN (RIGHT), ORNL ANALYTICAL CHEMIST AND FUTURE DEPUTY DIRECTOR FOR SCIENCE AND TECHNOLOGY, WORKING WITH KNOXVILLE POLICE FINGERPRINT SPECIALIST ART BOHANAN. COURTESY OF OAK RIDGE NATIONAL LABORATORY.

field. These measurements can be used to identify unknown compounds by determining their molecular weight.

Buchanan selected some volunteers—adults and children—to provide swabs of the chemicals present on their fingers. The results were unequivocal: the chemicals from the children's fingers contained rapidly evaporating volatile chemicals, such as free fatty acids, while adult prints display longer-lasting compounds.

To get a sense of the differing rates of evaporation for these two compounds, imagine following tracks made from walking in highly volatile gasoline compared to those made from walking in motor oil. After a short period of time, the tracks from the gasoline would disappear while those from the motor oil would linger. While unable to recover the missing child's fingerprints, at least Bohanan now understood what had happened to them.[47]

The search for this first identifying compound opened a door to a second. The researchers detected a wide variety of substances present in fingerprints, including cholesterol and nicotine. They then investigated the potential of identifying trace components in fingerprints that could distinguish among individuals. Such components in fingerprints could yield a profile of a suspect that provided key identifying characteristics like, "female, smoker, diabetic, cocaine user."[48]

Meanwhile, another new UT-ORNL line of forensic research would assist law-enforcement officials in locating the hidden graves of human murder victims. While an undergraduate student at Virginia Tech, Arpad Vass joined a diving team conducting research in Antarctica to study the physiological adaption of life present in lakes. Vass enjoyed the biological sciences and had begun to explore forensics after witnessing a knife fight between computer operators in the hospital where he worked. A friend from his Antarctica studies recommended that Vass conduct graduate work with David White, the UT/ORNL distinguished scientist and microbiologist. White, in turn, introduced Vass to William Bass, UT's noted forensic anthropologist.[49]

Vass earned his PhD in 1991 under Bass' guidance, and his dissertation research focused on time-since-death studies. A body can decay as quickly as two weeks or more slowly over two years, depending on four factors. Temperature is important, as every 18°F increase in the body's environment doubles the rate of chemical reactions and, thus, the rate of decomposition. Other factors include water or humidity and the pH (acid or base) of the environment. Extremes of acidity or alkalinity speed up the degradation. Meanwhile, anything that blocks oxygen from the body, such as burial in soil or water or high altitude, slows the degradation process. The interaction of all these factors determines the speed of the body's decay through chemical degradation that begins with enzymes, followed by bacteria and fungi, and then insects.[50]

Another path of Vass' research focused on the different chemical vapors formed and released from the body over time as it decays. Many of the experiments to investigate these vapors were conducted at UT's Anthropological Research Facility, better known as the Body Farm. Vass' work on the decomposition odors released by human corpses buried in clandestine graves provided a chemical fingerprint that can help law-enforcement officials find these burial sites and gather evidence that ultimately points to the victim's killer. Vass, then a forensics expert in ORNL's Biosciences Division, started by identifying 478 specific volatile compounds associated with burial decomposition and narrowed these down to the top thirty in order of perceived importance for locating buried bodies.[51]

Vass and colleagues took the research a step further and developed a dedicated tool to assist in searching for buried crime victims. Named Lightweight Analyzer for Buried Remains and Decomposition Odor Recognition (LABRADOR), the device detects volatile organic chemical compounds relevant to human decomposition. While not as sensitive as a canine's nose, LABRADOR has the advantageous ability to detect and alert the operator to the precise amount of odor present, which is a key factor in pinpointing the location of a grave or in looking for victims following natural disasters. Vass left ORNL and, in 2013, joined avaSensor LLC as chief science officer. More recently, he has worked as an instructor at UT's National Forensics Academy.

FROM COLLABORATOR TO MANAGER: FORMATION OF THE UT-BATTELLE PARTNERSHIP

As discussed in the previous chapter, for decades there were suggestions that UT should manage ORNL. Indeed, discussions of a management role for the university began in the mid to late 1940s, when there were real questions about the continuation of the Oak Ridge facilities following completion of the Manhattan Project and the end of World War II. However, despite the speculation, it would be many years before UT would be positioned to vie for such a management role.

Consider, for instance, that, following the war, UT's level of research expertise in most fields—particularly the sciences and engineering—was not close to being competitive with that of the nation's top-tier universities or with the talent and skill resident at ORNL. Consider, too, that Tennessee's political leaders generally were unwilling to assume the financial risk and legal liability associated with operating large and complex facilities for the federal government.

But, over the decades, as this book documents, UT made significant progress, evolving into a more research-focused university, and, meanwhile, the UT-ORNL partnership continued to grow through such joint initiatives as the Science Alliance, which brought high-level national talent to East Tennessee. Additionally, by the mid-1990s, when the management contract for the Oak Ridge facilities would once again be open to bids, UT was prepared to enter the process as a strong competitor.

Starting in 1984, Martin Marietta Energy Systems managed the three federal facilities in Oak Ridge—the Y-12 plant, ORNL, and the K-25 site. In early 1995, Martin Marietta managed these three sites as well as most of the other properties on the 34,700-acre Oak Ridge reservation and programs at both the facility in Paducah, Kentucky and the Portsmouth plant in Piketon, Ohio. In March of 1995, Lockheed and Martin Marietta merged to form the Lockheed Martin Corporation, and different business units assumed responsibilities for Oak Ridge contract management. On August 15, 1995, DOE announced that ORNL would no longer be operated as part of the contract with Y-12 and K-25 under Lockheed Martin. Instead, Lockheed Martin Energy Research Corporation (LMER) was formed with the sole mission of operating ORNL for DOE. Lockheed Martin Energy Systems maintained a separate management contract for Y-12 and K-25.[52]

These significant changes with Lockheed Martin led to speculation that DOE would soon announce a bidding process for the ORNL management contract. On June 13, 1995, UT President Joe Johnson was quoted in a *Knoxville News Sentinel* article, saying that the university would consider getting involved in a bid process if Energy Secretary Hazel O'Leary should decide to open the management contract to competition.[53] In recent years, these contracts were typically awarded for five years, and Martin Marietta received one renewal, marking ten years managing ORNL. Johnson also said in the article that Lee Riedinger, then UT interim vice chancellor for research, would lead the effort should the process start. On July 11, Riedinger and UT Senior Vice President Fisher sent to Johnson and UT, Knoxville Chancellor Bill Snyder a plan that the university should follow if DOE were to open bidding for the contract. However, the plan was shelved when O'Leary announced in August 1996 a two-year contract extension to Lockheed Martin for management of ORNL and Y-12, valid until the end of March 2000. In addition, she said DOE intended to work towards separate contracts for management of the three sites in Oak Ridge—ORNL, Y-12, and cleanup and revitalization of K-25[54]—which likely necessitated the two-year contract extension.

In May of 1998, ORNL nuclear physicist Alex Zucker (retired from leadership roles at ORNL) paid an unexpected visit to Fisher and Riedinger, now head of the UT Department of Physics, essentially confirming rumors that DOE would soon open the competition for the ORNL management contract. DOE had just granted a second two-year extension to Lockheed Martin, but DOE made it clear that this would be the last such extension and that a competition would occur during this next two-year period. This set in motion a chain of events that would alter the university forever and lead to substantial changes at ORNL.

Alex Zucker

Alex Zucker came with his family to the United States after Nazi Germany invaded Austria in the spring of 1938; his grandparents remained in Austria and later perished in the Holocaust. Zucker attended high school in New York City, joined the Army and stormed Omaha Beach on D-Day, and, through funding from the GI Bill, pursued his college education following the war.[55] After Zucker earned a PhD in nuclear physics from Yale, ORNL recruited him in 1950. His first assignment was to build a cyclotron using one of the enormous magnets from a Y-12 calutron that had been used to enrich uranium during the Manhattan Project.[56]

ALEX ZUCKER. COURTESY OF OAK RIDGE NATIONAL LABORATORY.

This cyclotron project was the first to measure the effects of nuclear reactions on the element nitrogen. At the time, President Truman authorized an ambitious project to build and test a hydrogen bomb in a rapidly escalating arms race with the Soviet Union. The extremely powerful hydrogen fusion bomb—whose detonation was triggered by a "little" uranium fission bomb akin to the ones dropped on Japan in 1945—would produce locally high temperatures upon explosion in the atmosphere. Because the Earth's atmosphere is 71 percent nitrogen, there was fear that the explosion's high temperature and neutron emissions would cause an unintended but powerful reaction among nitrogen nuclei, perhaps even causing the atmosphere to ignite and burn. Such an outcome would have potentially catastrophic consequences for life on the planet—and consequences that extended well beyond the bomb's direct impact.

This frightening possibility was discounted by theoretical nuclear physicists, but, while theorists project probable outcomes, in this situation—with no margin for error—someone needed to measure these actual reaction rates on nitrogen nuclei. And, thus, Zucker fielded his first assignment at ORNL and spent a year building the cyclotron and an associated ion source to allow him to accelerate electrically charged nitrogen atoms. The eventual experiment succeeded and proved that nitrogen nuclei in the atmosphere would not fuse and burn during an explosion of a hydrogen bomb. This was a sound and important scientific result and following it was a decade of testing in the atmosphere of increasingly powerful hydrogen bombs by the United States and the Soviet Union.[57]

Over Zucker's career at ORNL, he built a reputation as an innovative and well-respected nuclear physicist and was asked to take on increasingly important leadership roles at the laboratory, eventually becoming interim laboratory director in 1988. Upon his retirement in 1993, Zucker joined the UT physics faculty. Throughout his long career at ORNL, Zucker, though officially an employee of the laboratory, had always been a "friend of the university," and he demonstrated that friendship when he alerted UT Senior Vice President Fisher and Riedinger that DOE would soon open the bidding on the Oak Ridge management contract and, further, when he strongly suggested that UT enter the process.

Two meetings of note occurred during May, 1998. During the first, on May 3, Zucker discussed with Riedinger the pros and cons of UT competing for the ORNL management contract on its own, with a partner, or as a favored subcontractor for whoever the next managing entity might be. On

May 26, these two met with Fisher to discuss possible courses of action for the university.

By then, many at the university and in the local community felt that the time had come for UT to compete for a lead management role at ORNL, as Zucker had suggested. However, others felt strongly that UT was not ready for this big step and did not possess the expertise or the resources to make a competitive bid. In fact, some at DOE's Oak Ridge Operations office arrived at that very conclusion. Among them, Joe Lenhard, a senior office official, was tapped to convey that message to UT leadership. Pete Craven, an Oak Ridge business leader and UT supporter, arranged for Lenhard to meet with Fisher. At the meeting, Lenhard argued that UT was not strong enough to mount its own bid to manage ORNL but should instead skip the competition and play a subordinate role in supporting the organization DOE would ultimately select to receive the management contract.

Fisher replied forcibly that UT was indeed ready to choose a partner—an equal partner—and compete and noted that no other organization knew the laboratory as well as UT or could better help build its programs. Fisher impressed both Craven and Lenhard by citing the broad range of joint programs that had been established and the number of faculty and graduate students engaged in research at the laboratory. Prior to the meeting, neither Craven nor Lenhard was fully aware of the extent of UT's existing engagement with ORNL. Fisher stated that a secondary role for UT could risk the loss of the existing partnership programs—including the joint appointments facilitated by the Science Alliance—depending on which contractor ended up chosen. Lenhard apparently went back to the local DOE office and reported that UT was serious about competing this time.[58]

Soon thereafter, in early June, Fisher met privately with Jim Hall, former Oak Ridge Operations manager on leave for another DOE assignment, to discuss (off the record) the potential for UT to compete for the contract if rebid. In a small conference room at the Oak Ridge DoubleTree Hotel, they discussed the potential merits and liabilities of UT's involvement. It was a positive meeting, and Fisher asked Hall to meet with him again, but this time with UT President Johnson present. They held that meeting the following week in the same location. The important takeaways from those two sessions were: 1) Battelle Memorial Institute was, at the time, DOE's most effective laboratory contract manager; but 2) UT would be a formidable competitor if it teamed with the right partner; and 3) other universities should be engaged in a bid in some capacity. Following that meeting, Johnson instructed Fisher to assemble a group to begin preliminary discussions about how UT would respond to the opportunity, were it to arrive.[59]

MOMENTUM BUILDS FOR A UT MANAGEMENT ROLE

Strong encouragement for UT to compete came from Richard Genung in a private meeting with Homer Fisher. Genung, director of the ORNL Chemical Technology Division from 1988 to 1994, was named ORNL deputy director in 1997. In the early summer of 1998, Genung asked Barry Goss, president of Oak Ridge-based Pro2Serve (a technical and engineering company) and UT consultant, to arrange a meeting with Fisher. Genung, who earned his master's and doctorate degrees at UT, urged Fisher to push the UT administration to move quickly to announce its intention to vie for the contract, insisting that an early announcement from the university would likely cause other potential bidders to opt out of the competition, if and when the contract was rebid. Genung noted the great need for upgrading facilities at ORNL and

CARTOON BY CHARLIE DANIEL, *KNOXVILLE NEWS SENTINEL*. COURTESY OF BETSEY B. CREEKMORE SPECIAL COLLECTIONS AND UNIVERSITY ARCHIVES, UNIVERSITY OF TENNESSEE, KNOXVILLE.

strengthening the computational sciences, biological sciences, and energy and environmental sciences. This was strong and important encouragement from a well-respected ORNL leader, who died far too young in October 2002 at age 55.

By this time, DOE decided to award separate contracts for its Oak Ridge operations.[60] This provided UT with the opportunity to pursue only the contract for ORNL, without taking on either Y-12 with its nuclear weapons mission, or the massive environmental cleanup effort being executed across DOE's Oak Ridge Reservation and at the gaseous diffusion plants in Kentucky and Ohio. Local community leaders started to hear about this and take positions. Frank Cagle, *Knoxville News Sentinel* editor, wrote that UT should indeed compete but questioned whether university leadership was up to the challenge.[61] Political satirist Charlie Daniel included a cartoon about what the ORNL sign would become if the Tennessee Vols were to manage the lab.

In June 1998, a small university team began work on various aspects of the effort to compete for the ORNL management contract: Fisher, Riedinger, Zucker (retired from ORNL, now an adjunct faculty in the UT physics department), and research vice chancellor Mike Devine. On August 17, President Johnson sent a letter to Martha Krebs, head of the DOE Office of Energy Research, saying that UT supported Lockheed Martin's continuing role in managing ORNL, but, if a contract competition occurred, the university would be involved in some manner.

Others read the political tea leaves, believing that UT would have a favored status should a competition occur. On August 27, the small UT team hosted a visit by Dennis Barnes, then president of the Southeastern Universities Research Association (SURA). SURA was the managing contractor of the Thomas Jefferson National Accelerator Laboratory in Newport News, Virginia since its opening in 1984. SURA was a well-respected university consortium, and Barnes and his leadership team were interested in pursuing an ORNL management contract, perhaps as a partner of UT. Everyone assumed that this time the successful bidder would have to be a partnership that involved a university or universities, as well as a company experienced in overseeing the complex operational issues of a large national laboratory. At the time, DynCorp expressed interest in such a partnership, so Barnes and colleagues visited with DynCorp's Oak Ridge officials the next day. Another possible suitor—IIT Research Institute in Chicago—talked with Fisher in July. The maneuvering had begun.

Maneuvering was also underway in Oak Ridge, centered on stated concern by ORNL director Trivelpiece about getting the Spallation Neutron Source funded and any possible disruption that might result from a change in the managing contractor.[62] The laboratory nervously awaited the first allocation of construction funds for the SNS, and Trivelpiece publicly worried that DOE bringing in a new ORNL management team at this critical juncture would harm the process. Clearly, he was lobbying for a continuation of the Lockheed Martin contract. In an interview in October,[63] he further made the point that DOE could cancel the contract competition in order to insure "smooth sailing" for the SNS in its early construction stages.

In early September, Fisher and Riedinger talked with UT President Johnson and laid out a detailed plan of committees, people, and needed resources to mount a serious bid. On September 28, Fisher and Devine visited two important officials—Oak Ridge Operations office manager Jim Hall (back from a temporary assignment) and Pete Dayton, later named to lead the office's contractor selection process—to express the university's interest in mounting a bid and to probe them on their view of this possibility. In the meeting, Fisher and Devine learned that the draft request for proposals would be published in November, the final request (after a comment period) in March, and a contract decision announced in December 1999.[64] This suddenly became a serious—not to mention time-sensitive—business. The two Oak Ridge Operations officials fully supported a UT role in the next contractor partnership but would not elaborate on what exactly UT's involvement might or should look like. Did Hall and Dayton envision a minor role for the university or a major one? That remained to be determined.

A key step in UT's preparation to enter the bidding process, should it unfold as Hall and Dayton indicated it would, was for the university to get early approval from the UT Board of Trustees and Tennessee Governor Don Sundquist to engage in this process. Joe Johnson quickly took care of these two important briefings and soon secured the crucial authorization, though more formal approval of a plan and a partnership would be needed later.

Joe Johnson

Joseph Johnson served as president of the UT statewide system from 1990 to 1999 and again as interim president from 2003 to 2004. Prior to joining UT, Johnson worked in Tennessee state government from 1960 to 1963 as chief of the Budget Division, executive assistant to the governor, and deputy commissioner of finance and administration. At UT, he served as the executive assistant to the university president from 1963 to 1969, vice president for development from 1969 to 1973, chancellor of the Center for the Health Sciences from 1970

JOE JOHNSON. COURTESY OF BETSEY B. CREEKMORE SPECIAL COLLECTIONS AND UNIVERSITY ARCHIVES, UNIVERSITY OF TENNESSEE, KNOXVILLE.

to 1973, and vice president for development and executive vice president from 1973 to 1990.[65]

Johnson had a long and rich career at UT and continued to serve the university in retirement. He was a UT and Knoxville legend for his grace and finesse in working with people.[66] No one was better at remembering names, knowing details of the lives of friends and colleagues, and sending hand-written thank-you notes. Johnson, not particularly partial to computers but a skilled multitasker, was well known for drafting notes on yellow legal pads on one topic while fully engaged with visitors on another. His support was unwavering along every step in the development of UT-Battelle management team and the partnership between UT and the federal facilities in Oak Ridge. Johnson died in 2023.

PROPOSAL PREPARATION BEGINS IN EARNEST

On October 14, 1998, President Johnson and Chancellor Snyder announced a set of teams to investigate different aspects of a potential bid for the ORNL contract. Fisher and Devine would coordinate the process, while four committees would explore various aspects of a proposal (science and technology; finance and legal; technology transfer; and safety, environment, and security). A steering committee would take input from these four tightly focused committees. Riedinger was named the UT team leader. The working group drafted a budget and hired a handful of advising consultants to contribute background, flesh out new information, and identify emerging needs. Team members scheduled visits to other national laboratories, including Brookhaven (on Long Island) and Berkeley. It became clear early on that just the process of competing for such a large and strategic award would be expensive, with a price tag likely as high as several million dollars.

Rumors spread rapidly about the process. In one newspaper account, ORNL director Alvin Trivelpiece denied that he had encouraged the University of Texas to consider a management bid.[67] The same article noted that UT's Fisher impressed many with his and the university's level of preparation for the unfolding process. Indeed, through October and November, numerous potential partners for the management contract visited UT to make their pitch. Among them were IIT Research Institute, DynCorp, Bechtel, Midwest Research Institute, Parsons, CH2M Hill, Morrison Knudsen, Teledyne, Duke Engineering, Babcock and Wilcox, Thermo Electron, and Science Applications International Corporation. Different proposed roles for UT included a general operating partner; an infrastructure operating partner; or a subcontractor focused on nuclear operations, technology commercialization, and/or information technology. Three university consortia visited to propose their inclusion in the arrangement as partners: SURA, ORAU, and the University Research Association (URA). Part of the intrigue involved whether some of these entities would mount their own bids and, thus, directly compete with UT.

Encouragement for UT to compete for the ORNL contract came from various sources, including the managing editor of the *Knoxville News Sentinel*.[68] He argued that it is important for UT to compete, as a win would lead to increased prestige and possibilities for expansion of its programs.

Homer Fisher

Homer Fisher earned a bachelor's degree in economics and a master's of business administration at Auburn University before serving thirty-eight years in higher education administration—sixteen years at Florida State University and twenty-two at UT. Fisher worked as the vice chancellor for business and finance at UT, Knoxville before then President Lamar Alexander appointed him senior vice president of the state-wide UT system. In all of his roles, he was extremely supportive of faculty research initiatives, often finding creative solutions to vexing problems.

HOMER FISHER. COURTESY OF BETSEY B. CREEKMORE SPECIAL COLLECTIONS AND UNIVERSITY ARCHIVES, UNIVERSITY OF TENNESSEE, KNOXVILLE.

Before most people in UT's central administration recognized the importance of the Oak Ridge federal facilities to the university's faculty and students, Fisher vocally advocated, leading or supporting nearly every joint initiative between UT and ORNL. Additionally, he played a critical role in guiding the university in its competition for the ORNL management contract. After retiring from the senior vice president position in 1999, Fisher managed UT's Oak Ridge office, where he served as a liaison between the state university and the Oak Ridge community.[69]

On October 26, Fisher wrote a letter of thanks to Justin Wilson, deputy to Governor Don Sundquist for policy. Fisher and Johnson met the day before with Wilson to lay out the UT strategy and to explain the university's goals in managing ORNL. This rather long list of goals included "increasing the involvement of UT and other research universities in determining and carrying out the research agenda of the laboratory" and "teaming with others who can contribute to the strengthening of the research mission of ORNL."[70] The letter recounted a recent meeting of UT and Battelle Memorial Institute[71] leadership for early discussions of how Battelle might join an academic partnership of UT and SURA.

Through this fall flurry of activity, the UT committee leaders waited for an official and more serious visit from leaders of Battelle, based in Columbus, Ohio. Battelle is a long-standing nonprofit technology development company and has been heavily invested in managing federal laboratories. Battelle has managed Pacific Northwest National Laboratory (PNNL) since 1965 and won the contract to manage Brookhaven National Laboratory in 1996 in partnership with Stony Brook University. There was much speculation at the time that Battelle was, itself, considering entering the bidding process for the ORNL contract and further speculation that the Ohio-based company was intently watching how UT was navigating the process to determine how, ultimately, Battelle might contribute. An important person in this strategic dance was Pro2Serve's Barry Goss.[72]

Goss worked for years in circles relating to DOE and Battelle and understood the landscape well. Fisher and Goss talked in early fall about a potential UT partnership with Battelle but agreed that these discussions should, for the moment, be deferred. Then, one day in mid-November, Fisher and Riedinger met with Goss and stressed that the university was near the end of its process, would soon have to pick a bidding partner, and needed to enter serious discussions with Battelle posthaste. "OK, I understand," Goss responded. "I will get Madia here next week." Bill Madia, a nuclear chemist, was PNNL director and a longtime member of Battelle's leadership team. Fisher and Riedinger questioned whether Goss could deliver on this pledge to summon Madia, but, in fact, he did.

Bill Madia came to UT on November 30 (the Monday after Thanksgiving) and was ushered into President Johnson's office along with some of the UT team leaders, including Fisher and Riedinger. It was clear to all after thirty minutes of discussion (guided by Johnson) that Battelle was the university's needed partner. UT and Battelle got engaged, so to speak, in that meeting and agreed to mount a bid together. After having met with so many other potential partners, it was clear to the UT people present that Battelle possessed the right set of experiences and expertise, and Bill Madia had the right mix of knowledge, charisma, and drive to make this partnership succeed, not only in winning the bid but in the future management of ORNL. Later that day, Madia spent two hours laying out his strategy, plans, and concerns to Fisher, Devine, Goss, and Riedinger. The die had been cast.

The next day, Fisher, Goss, and Riedinger—joined by UT team members Billy Stair and Frank Harris (director of the biology program at UT)—met with Ron Townsend, president of ORAU, about which universities to bring into the UT-Battelle partnership. ORAU would be part of this team and would represent its affiliation with a hundred or more degree-granting institutions. However, the UT leaders felt it was important to have a small number of strong southeastern universities join the effort as well, all to be awarded seats on the management partnership's eventual board of directors. The problem was that SURA had decided to compete on its own for the ORNL contract and was trying to recruit some of the same universities to be part of its team. Dennis Barnes, SURA's then president, initially warmed to the idea of joining with UT, but he was ousted at the end of October, and the association's new leadership team decided that SURA would mount its own bid.

This started the competition to sign up key universities to be part of the UT-Battelle team or the SURA team, and Townsend took the lead on these negotiations. Virginia Tech joined the UT-Battelle team quickly, and Townsend felt that Duke and North Carolina State would soon follow. Georgia Tech and Florida State were on the fence. Virginia was an unknown. Vanderbilt was another obvious partner but proved unwilling to sign up with either team. The issue was that Joe Wyatt, Vanderbilt's chancellor, was then the chair of the board of presidents of URA, another prominent university consortium, which was contemplating its own bid to manage ORNL.

December was in part consumed by jockeying to see which universities would join which bidding team. By the end of December, Townsend worked his magic, and the set of committed UT-Battelle core universities included Duke, Florida State, Georgia Tech, North Carolina State, Virginia, and Virginia Tech. Their presidents would not send official letters of commitment until the spring of 1999, but these universities were definitely part of the UT-Battelle team. On January 20, UT publicly announced its intent to mount a bid for the ORNL contract, teamed with Battelle, ORAU, and a set of core universities.[73]

William Madia

Bill Madia received a PhD in nuclear chemistry from Virginia Tech and, in 1975, began a long career with Battelle. He started a nuclear fuel-cycle analysis group at the company, developed nonproliferation procedures, and managed the plutonium fuel-fabrication laboratory.[74] In 1985, he was promoted to director of the Columbus (Ohio) Laboratories and later became president of Battelle Technology International, managing laboratories in Frankfurt, Germany, and Geneva, Switzerland. He was appointed director of Pacific Northwest National Laboratory in 1994, where he served until 2000. In April 2000, Madia was appointed director of ORNL and CEO of UT-Battelle LLC, where he served until 2003.

BILL MADIA. COURTESY OF OAK RIDGE NATIONAL LABORATORY.

Following his tenure with ORNL, Madia returned to Battelle in Columbus as executive vice president, retiring in 2007. Failing at retirement for the first time, he joined Stanford University in 2008, where he was appointed vice president for the Stanford Linear Accelerator Center (now SLAC National Accelerator Laboratory). He retired (again) in 2019, but those who know Madia expect that he will continue to pursue professional activities. Madia is an expert in science and technology policy and management and knows DOE well. He was a key leader in Battelle's expansion of its role in managing national laboratories from one (Pacific Northwest National Laboratory) to now eight for DOE plus one for the Department of Homeland Security.

THE UT-BATTELLE TEAM ASSIGNS ROLES

Fisher and Madia signed a memorandum of understanding between UT and Battelle on December 4, establishing the agreement principles for UT-Battelle LLC, a fifty-fifty partnership. UT would take the lead on providing scientific direction, metrics, strategies, and key personnel. Battelle would oversee ORNL's management and chart the laboratory's technology direction. The partnership was now official.

The kickoff meeting of the UT-Battelle team occurred January 11–15, 1999, at the ORAU facilities in Oak Ridge. There were numerous issues to discuss, many eventual decisions to make, and a proposal to write. Bill Madia directed PNNL for a decade and led Battelle's successful bid with Stony Brook University to manage Brookhaven National Laboratory. He clearly had the most experience and was easily in charge. Homer Fisher, Madia's UT counterpart, was assigned to lead the university in its efforts to win this contract.

Various committees were established to begin drafting various parts of the UT-Battelle proposal to submit to DOE. Jeff Smith of Battelle was tapped to chair the laboratory operations committee, and UT's Riedinger was named as the co-chair of the science and technology committee, along with Ray Stults, who served as an associate director for PNNL. Other committees would focus on technology transfer, human resources, fees and costs, legal issues, and the transition plan to succeed Lockheed Martin and assume the reins of laboratory management. No one really knew when the proposals would be due since DOE had not yet issued the final request.

The science and technology committee was the biggest, with seventeen members, because the largest part of the proposal would address a range of issues necessitated by ORNL's breadth in its science and technology enterprise. Most of the members were faculty from the partnering universities, and all were accustomed to writing proposals. However, this proposal differed, not the usual scores of pages of text proposing the scientific research to be undertaken, but rather a set of viewgraphs—transparent images to be projected on a screen, in the days before PowerPoint software—to convey succinctly the partnering organizations' core strengths and capabilities. And, of course, the viewgraphs would not directly address the research to be undertaken but rather propose how to manage and foster the science behind it. Academics are not really trained in these management issues, so the science and technology committee had much to learn, and, as it turned out, the members faced quite a steep learning curve. The idea was to read and understand the request for proposals and to address and answer each point faithfully. It sounds easy, but it was not.

Another big task facing the team was to decide on whom to propose as the UT-Battelle leadership team to manage ORNL. First was the director. It was not clear if Lockheed Martin would submit a bid to continue its management tenure, so UT-Battelle decided not to name Trivelpiece, the incumbent laboratory director, as its candidate. After some discussion, Bill Madia agreed to be the laboratory director designate. Next, Madia and Fisher decided that a Battelle person should be named as the deputy director for operations and a UT person as the deputy for science and technology, reflecting the primary expertise and interests of the two partnering institutions. By February, the UT-Battelle team decided that Jeff Smith would be proposed for the former role and Lee Riedinger for the latter. On February 19, Riedinger first saw his name displayed on the proposed organizational chart (for the team to consider) and characterized his pending appointment as "a heady and scary experience, with all the unknown joys and trials to come."

In February, an important meeting occurred in Nashville. UT President Johnson, Vice President Fisher, and Johnson's executive assistant Billy Stair met with Governor Don Sundquist and Deputy Governor Justin Wilson about a substantial commitment of state funds to back the UT-Battelle proposal. And, there was clear precedent for such state-backed financial support. In 1982, then Governor Lamar Alexander committed state funds ($350,000) for the construction at ORNL of the first joint institute, JIHIR. The facility, owned by UT but operated on the ORNL campus, focused on joint programs in nuclear physics. Now, UT-Battelle proposed to construct three additional facilities to house UT-ORNL joint institutes, in computational, biological, and neutron sciences. Each was intended to be larger than the original 6,000-square-foot heavy ion research facility, so a substantial outlay of state funds would be needed to finance construction. A pledge of financial support from the Tennessee governor would make the UT-Battelle proposal all that more attractive to DOE reviewers. Governor Sundquist concurred and pledged $12 million in support of the proposed projects.[75]

Political intrigue continued through the spring. On March 16, SURA dropped its plans to submit a proposal to bid for the ORNL contract. By April 1999, only UT-Battelle announced its intent to compete for the ORNL management contract, and DOE was worried.[76] Having an important competition with only one entrant was not a good sign that this was a desirable contract or a well-run process for procurement. To allow more time to find a second competitor, DOE extended the deadline to submit proposals by forty-five days to July 12.[77] Two weeks later, the University Research Association's president, Fred Bernthal, expressed possible interest in bidding.[78] This set up an interesting dynamic, as Bernthal and Riedinger were long-time nuclear physics colleagues and the former brought the latter into Senator Baker's office in 1983, and now they might compete for the ORNL contract.

On April 6, Madia picked up intelligence that URA would mount a bid, with Lockheed Martin as its partner. This was announced a month later on May 4.[79] DOE was likely happy to have a second bidder, but this new bidder now posed a direct threat to the UT-Battelle plan. And, this meant that URA member Vanderbilt University would be on the opposing team.

Vanderbilt's and UT's decades-long rivalry in the sporting arena now

STATE OF TENNESSEE

DON SUNDQUIST
GOVERNOR

February 24, 1999

Dr. Joe Johnson, President
The University of Tennessee
800 Andy Holt Tower
Knoxville, Tennessee 37996-0180

Dear Dr. Johnson:

As you know, we all strongly support the University of Tennessee's efforts at obtaining the Oak Ridge National Laboratory M&O contract. Being awarded this contract is not only important to the University of Tennessee, it is extremely important to the State of Tennessee as a whole.

As an indication of our support, this letter is to advise you that at the appropriate time we will support the necessary appropriations to construct a facility for housing the Joint Institute for Biological Sciences ("JIBS") and to construct a graduate program facility in Oak Ridge ("GPF). It is estimated that the JIBS facility will cost Eight Million Dollars ($8,000,000) and the GPF facility is estimated to cost Four Million Dollars ($4,000,000). Although we cannot guarantee the General Assembly will appropriate these funds, absent unforeseen circumstances, we are confident we will be successful.

Please feel free to include this letter in your proposal to the Department of Energy. If we can be of any further assistance, do not hesitate to call upon us.

Best regards,

Don Sundquist
Governor

John S. Wilder
Lt. Governor and
Speaker of the Senate

Jimmy Naifeh
Speaker of the House

State Capitol, Nashville, Tennessee 37243-0001
Telephone No. (615) 741-2001

LETTER OF SUPPORT FROM TENNESSEE GOVERNOR, SPEAKER OF THE SENATE, AND SPEAKER OF THE HOUSE. IN POSSESSION OF THE LEAD AUTHOR.

moved into a different realm. To many at UT, Vanderbilt's entry into the bidding contest raised the stakes considerably. Vanderbilt University was not a member of UT-Battelle's core university group for the ORNL management proposal, and Vanderbilt's decision not to join the UT-Battelle team and its role as competitor for the ORNL management contract heightened tension between the two Tennessee universities—tension that previously largely had been confined to athletic contests.

Shortly after, a rumor surfaced that Martha Krebs, director of the DOE Office of Energy Research, was being considered for a vice provost position at Vanderbilt. This would be viewed as a conflict of interest since she was defined as the final decision maker on the contract award, and, of course, Vanderbilt was part of the competing Lockheed Martin-URA team. In July, former ORNL Director Herman Postma discussed this rumor in a public talk in Oak Ridge.[80] Homer Fisher and Barry Goss also heard this rumor and appealed to the officials at the DOE Oak Ridge Operations office to have Krebs removed from the decision-making process. She was and, as it turned out, she did not accept the position at Vanderbilt.

Pressures of Preparation

Creation of the viewgraphs continued through the spring, with the science and technology committee composing the storyboards that would explain, in detail, how UT-Battelle would manage science at ORNL. June 11 was a day of reckoning. On that Friday, Riedinger presented his committee's draft sales pitch, which comprised fifty viewgraphs, to the partnership's "pink team"—internal reviewers who would evaluate how thoroughly and precisely the viewgraphs addressed the expectations and demands expressed in the request for proposals, which a one-inch thick book of double-sided paper contained.

Riedinger recalls that, when he concluded his presentation, the reaction was far from what he had anticipated. "Every professor understands that it's important to be gentle and supportive when a young graduate student proudly presents the initial draft of his or her first paper," says Riedinger. "The worst thing a professor can do is to wad it up, throw it in the waste-paper basket, and tell the kid to go and start over. But that is more or less what happened on June 11." However, in this instance, Riedinger served as the dressed-down and demoralized student.

Indeed, the pink team, far from complimentary of Riedinger's storyboards and presentation, essentially told him to go and start over. As a widely published professor who had given many scientific talks, Riedinger recalls feeling stunned by the response. Battelle's Jeff Smith noted Riedinger's long face on that late afternoon and wondered if his UT counterpart would even bother to return on Monday to continue work on the proposal. But, as Riedinger recalls, after consuming a few stiff beverages on Friday evening, he returned to his office on Saturday to mightily revise (i.e., start over on) most of the storyboards to address the pink team's chief criticism: the density of words and concepts presented on Riedinger's viewgraphs made them difficult to follow, even for a group of individuals well acquainted with the UT-Battelle team's overarching strategy and various management capabilities. To a DOE reviewer, they would be utterly indecipherable.

On July 7 and 8, Riedinger presented his significantly revised viewgraphs to the partnership's "red team"—a second group of internal reviewers—and received a more favorable response. The red team determined that the team's presentations were overall in good shape for full proposal submission on August 2.

After a long seven months of preparation in 1999, the proposal was done. This was a 1.5-inch-thick loose-leaf binder filled with documents, including 182 viewgraphs of what UT-Battelle proposed in operations, science, management, and new facilities. Of these, forty-four pertained to future directions in science and technology. One explained how UT-Battelle would expand ORNL expertise in four areas by adding new initiatives (including the proposed Joint Institutes for Neutron Sciences and Biological Sciences), leveraging the National Transportation Research Center, and building new strength in such important areas as neutrino research.

In addition to the viewgraphs, the final documents submitted to DOE included the credentials of each partnering institution, profiles of leadership candidates, financial records, past performance forms, and plans for a transition to a UT-Battelle management structure at ORNL, should the partners clinch the contract. These seven large boxes of proposal material were submitted to DOE on August 2.[81] The package included a new July 22 letter from Tennessee Governor Sundquist and the speakers of the Tennessee House (Jimmy Naifeh) and Senate (John Wilder) with expanded financial support (now $18 million, rather than the earlier commitment of $12 million) for the construction of joint institutes at ORNL should UT-Battelle win the contract.[82] UT-Battelle's competitors—Lockheed Martin and URA—asked for the same commitment from the state but were turned down.

A few days later the Lockheed Martin-URA proposal was submitted and the name of their proposed laboratory director surfaced: Mike Knotek, formally of Pacific Northwest National Laboratory, as was Madia.[83]

The presenting team traveled to Battelle's headquarters in Columbus, Ohio for rehearsals during August 3–6. Each presentation was filmed and critiqued, revised accordingly, and presented again—not a familiar process for a university professor, but an effective one. The team was ready to make the presentation to the DOE Source Evaluation Board in Oak Ridge. The stakes were extremely high but losing was not in Bill Madia's lexicon.

A new UT president, Wade Gilley, assumed the office on August 1. Joe Johnson, who had supported this endeavor from the beginning and understood well the importance of ORNL to UT, retired. Gilley therefore opened the UT-Battelle presentation to the Source Evaluation Board on August 16, followed by a two-hour presentation of storyboards by various members of the designated leadership team. The rehearsals at Battelle headquarters paid off, and the presentation came off without a hitch.

Bill Sansom, a local business leader and vice chairman of the UT Board of Trustees, made concluding remarks and noted the strong support of UT and Governor Sundquist (who served as chairman of the UT Board). Gilley, though newly installed as UT president, capably delivered his introductory remarks, and everyone was optimistic that he would be an engaged and supportive president in the future, as Joe Johnson had been in the past. Ultimately, this would turn out *not* to be true, but it would take almost two years before the fireworks started to detonate and scandal beset the Gilley administration.

Following the August presentation, DOE's selection process went dark. Rumors circulated about which team would win the contract, but, until the official announcement came, they were only rumors. The UT-Battelle team members did their best to return to their normal duties, but it was a time of heightened anxiety. On September 13, the nine members of the Tennessee delegation to the U.S. House of Representatives sent a letter of support for UT-Battelle's management proposal to DOE Secretary Bill Richardson. However, still no definitive word arrived from the energy agency. Finally, on a day in October, Madia and Fisher received a telephone call from DOE telling them to be ready at 10 a.m. the following morning to receive some important news.

At the appointed time on October 20, they settled into their offices at UT-Battelle headquarters on New York Avenue in Oak Ridge, and, on schedule, the phone rang. The DOE official on the other end of the line announced that UT-Battelle had been awarded the ORNL management contract. UT President Gilley and Tennessee Governor Sundquist had been informally advised of DOE's decision the night before but were asked not to reveal the good news. Later, DOE made the official announcement that UT-Battelle LLC was chosen as the next managing contractor of ORNL; the partners would assume responsibility for management of the laboratory on April 1, 2000.[84] The success of this competition would officially open a broad new critical connection between UT, Battelle, and ORNL.

The award of the ORNL management contract to UT-Battelle LLC brought much acclaim within Tennessee, both locally and throughout the state. DOE's Milt Johnson, the official selection official, said that one reason for the win by UT-Battelle was that it had accepted all terms and conditions set down in the request for proposals, whereas Lockheed Martin-URA did not.[85] In other words, the winning contractor agreed to play by all the DOE rules.

Government Owned, Contractor Operated: The GOCO Model at Oak Ridge

To execute the Manhattan Project, the U.S. Army Corps of Engineers quickly mobilized both universities and private industry to secure the scientific and engineering talent needed to achieve the ambitious goal of building the world's first atomic weapons. At Oak Ridge, Kellex (a subsidiary of the M. W. Kellogg Company) designed, J. A. Jones Construction built, and Union Carbide Corporation operated the K-25 plant. Stone and Webster, with scientists and engineers from the University of California at Berkeley playing a supporting role, designed and constructed the Y-12 plant; Tennessee Eastman Corporation operated the plant. DuPont designed and constructed, while the University of Chicago operated, the pilot-scale plant for production and separation of plutonium, known as both X-10 and Clinton Laboratories.

This arrangement of government-funded facilities built and operated by private contractors with the necessary expertise continued after the Manhattan Project. The government-owned, contractor-operated (GOCO) model was formally adopted in the Atomic Energy Act of 1946 (Public Law 79-585), under which Congress authorized "contracts for the operation of Government-owned plants so as to gain the full advantage of the skill and experience of American industry."

At X-10, soon to become ORNL, the Monsanto Chemical Company took over from the University of Chicago in July 1945 but decided in May 1947 not to renew its contract. After a period of extreme uncertainty for ORNL, AEC selected Union Carbide as the laboratory's operating contractor, beginning in March 1948. Union Carbide had already replaced Tennessee Eastman as the operating contractor for Y-12 in May 1947, and it would continue to operate all three of these facilities for AEC and its successors, the Energy Research and Development Administration and the U.S. DOE, until 1984. The Paducah Gaseous Diffusion Plant in Kentucky was added to Union Carbide's contract with AEC in 1952.

In 1982, Union Carbide notified DOE of its intent to withdraw from the contract. Union Carbide's decision came during a turbulent period for DOE.

In his 1980 election campaign, Ronald Reagan called for abolishing the department, and several bills proposing the realignment or closure of some DOE facilities were introduced in Congress. In addition, contractor compensation was relatively low; Union Carbide received an annual fixed fee of about $8 million for its management and operation of the facilities at Oak Ridge and Paducah.

A competition for the contract was held, and, in December 1983, DOE announced that Martin Marietta Corporation had been selected to manage and operate ORNL, Y-12, K-25, and the Paducah Gaseous Diffusion Plant. Martin Marietta Energy Systems, Inc. assumed full responsibility for these facilities on April 1, 1984. (The Portsmouth Gaseous Diffusion Plant in Piketon, Ohio, would be added to the contract in November 1986.)

Although DOE had elected to continue its practice of "bundling" the management and operation of these sites, this new contract included a performance-based fee. This change from the old cost-plus-fixed-fee model reflected concerns raised in a number of high-level inquiries and reports on the status of DOE facilities and contracting practices.

The fall of the Berlin Wall in 1989, presaging the end of the Cold War, amplified these concerns. Attention was drawn to the fact that many DOE management and operation (M&O) contracts had been extended multiple times with no competition. In 1990, the U.S. General Accounting Office (now the U.S. Government Accountability Office) designated DOE contracting "a high-risk area vulnerable to waste, fraud, abuse, and mismanagement."[86] In response, DOE implemented a number of changes to the M&O contracting model.

These changes would soon impact Oak Ridge, where missions were evolving to match national needs and priorities. Uranium enrichment operations at K-25 ceased in 1985, and the site was permanently shut down in 1987 and slated for environmental remediation. The entire Oak Ridge Reservation was placed on the Superfund program's National Priorities List in 1989, the same year that DOE created an Office of Environmental Management to oversee the cleanup of its nuclear legacy across the nation.

With the merger of Martin Marietta and Lockheed Corporation in March 1995, Martin Marietta Energy Systems became Lockheed Martin Energy Systems, Inc. (LMES). A two-year extension of the LMES contract, signed in August 1995, converted the terms for management of ORNL from an award-fee basis to a fixed-fee arrangement, similar to that in place at DOE's other multiprogram laboratories. The extension also provided for the possibility of an entirely separate contract for ORNL, held by a contractor other than LMES.

Later that year, DOE elected to request the establishment of a separate contractual relationship for ORNL. On January 1, 1996, a new Lockheed Martin operating unit, Lockheed Martin Energy Research Corporation (LMER), became the M&O contractor for ORNL.

In August 1996, DOE announced a set of contract changes for its Oak Ridge facilities. The M&O contract with LMER for ORNL would be extended for two years, through March 2000, and opened to competition, with the new contract to begin in April 2000. The M&O contract with LMES for Y-12 would also be extended for two years and competed. DOE also implemented a new approach to environmental cleanup, awarding a management and integration contract to Bechtel Jacobs Company LLC in December 1997. This contract covered environmental remediation not only at K-25, but also at Y-12, ORNL, and the gaseous diffusion plants in Kentucky and Ohio.

The M&O contracts for ORNL and Y-12 would soon be awarded to new teams: UT-Battelle LLC, a partnership of the University of Tennessee and Battelle, began managing and operating ORNL on April 1, 2000, on an award-fee contract. BWXT Y-12, a joint venture of BWX Technologies and Bechtel National Incorporated, began managing and operating Y-12 on November 1, 2000.

Thus, at the beginning of the 20th century, three different contractors once again managed DOE's three major Oak Ridge facilities after decades of a management by a single contractor.

SUMMARY AND LOOK FORWARD

The award of the ORNL management contract to UT-Battelle significantly impacted the laboratory (including construction of a new campus), Battelle (this was the largest and most diverse management contract yet won by the organization), and UT (a big advance in national prestige and a parallel growth in cooperative programs). At ORNL, UT-Battelle brought a new way of thinking and operating, which led to construction of a large number of new facilities designed to enable 21st century scientific research, all on budget and on time—outcomes that were, of course, important to the sponsor (DOE). ORNL's strengths in neutron science and supercomputing gained international prominence.

At UT, the partnership with ORNL expanded more than anticipated. Indeed, in future decades the number of joint faculty would grow from a handful to several hundred. New areas of research, including additive manufacturing, would flourish. Then, there were the financial rewards. Using an elaborate annual grading system, DOE awards a management fee to UT-Battelle based on performance, and, as a result of the partnership's consistently high ratings, significant new financial resources infused university coffers. After covering corporate expenses, UT-Battelle gives half of the remainder to UT and half to Battelle.

This financial boost has amounted to a few million dollars to each partner annually.[87] At the university, this financial reward has been reinvested in collaborative programs with ORNL, including high-level joint hires and operation of joint institutes. And, the management contract brought UT into a small group of other universities that participate in the management of national laboratories, including the University of California, Berkeley; the University of Chicago; and the State University of New York, Stony Brook (also called Stony Brook University).

The decade of the 1990s ended with a momentous ceremony on December 15, 1999, to break ground[88] for the Spallation Neutron Source, a scientific facility that fires high-speed protons at a target, making neutrons that scientists use to scatter off a material and study its structure and properties. In the 1940s and 1950s, Oak Ridge ranked as the world's premier site for neutron-scattering research, based on the experiments conducted first at the Graphite Reactor and later at the Oak Ridge Research Reactor. HFIR became the centerpiece of ORNL's neutron scattering work following its completion in 1965. Once completed in 2007, the $1.4 billion SNS would provide capabilities complementary to those of HFIR and re-establish Oak Ridge as the world's premier site for neutron-scattering research. In addition, the SNS would expand a critical connection to allow tremendous new research opportunities for UT faculty and students.

The 1990s began with UT generally being regarded as a respectable university with increasing depth in the sciences and engineering but not yet equipped to manage a large federal research facility. Indeed, in the early years of the decade, UT's research strengths paled in comparison to those of ORNL and many of the nation's premier research universities. However, during ten years of extraordinary growth and increased collaborative engagement—particularly through joint UT-ORNL research endeavors—UT, with its partner Battelle, silenced the doubters and won the management contract. The award bestowed on the university considerable prestige and welcomed revenues for reinvestment in joint programs, but, more than that, it set the stage for significant further advancements that the UT-Oak Ridge partnership would achieve over coming decades.

8

THE DECADE OF THE 2000S

UT Enters the Big Leagues after Sixty Years of Steady Progress

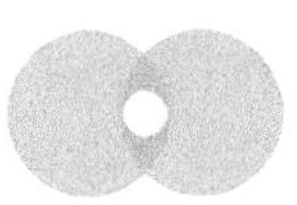

THE DECADE OF the 2000s began soon after the October 20, 1999, announcement that UT-Battelle LLC had won the competition to manage ORNL. The resulting changes at ORNL and UT were profound—perhaps greater than those anticipated by even the most optimistic appraisers of the management partnership. In the 1980s and 1990s, concerns over full funding of the DOE national laboratory system led to fear that a laboratory might have to be closed, causing questions about ORNL's future. However, by the end of the decade of the 2000s, ORNL advanced substantially on the unofficial list of the best national laboratories and was positioned to assume the top spot (in the eyes of many) in the following decade. This rapid rise in accomplishment and reputation at ORNL resulted from the leadership of UT-Battelle and the innovation and hard work of several key individuals.

UT also ascended in rankings and successful programs during the 2000s. As this book's opening chapter notes, in the early 1940s, UT had no doctoral programs, but, by 2000, not only had UT substantially bolstered its number of terminal degree programs, it also had joined the short list of universities involved in managing and operating national laboratories.

To a large extent, the institutional advances in this decade resulted from the effective leadership of a number of high-level individuals. Bill Madia (Battelle) and Homer Fisher (UT) took the lead on the capture of the contract to manage ORNL. Madia then served as laboratory director for three years at the beginning of this decade. The contributions of these two stellar individuals to the UT-ORNL partnership were greatly supplemented in the decade of the 2000s by many other key individuals, including Thom Mason (ORNL director from 2007 until 2017), ORNL's Thomas Zacharia (the chief architect of supercomputing in East Tennessee), Loren Crabtree (UT, Knoxville chancellor and key supporter of UT-Battelle), and Phil Bredesen (governor of Tennessee from 2003 to 2011).

Administrative turnover through the decade affected both the university and the laboratory. Over the course of ten years, three different directors served at the helm of ORNL, but these transitions in leadership were smooth and harmonious, with no loss of institutional momentum. By contrast, the changes in leadership at UT were chaotic and disruptive, with three presidents leaving early or forced to resign by the Board of Trustees.

MANAGEMENT RESPONSIBILITIES AT THE NATIONAL LABORATORY CHANGE HANDS

UT-Battelle LLC was given ninety days to be on site at ORNL and be prepared to assume the management of the laboratory on April 1, 2000. The trains had to keep running on time, so to speak, on the first day of the new administration. The transition from the Lockheed Martin Energy Research leadership team to the new UT-Battelle team had to be effected gracefully, and these ninety days starting on January 3, 2000, were crucial.

The winning management proposal by UT-Battelle was built around the concept of simultaneous excellence in science, in operations, and in community outreach. The proposal's 182 viewgraphs laid out the expectations for ORNL under the new management structure, but now was the time to put the plan into action. One of the challenges was that ORNL's physical campus was old, rundown, and not befitting a modern research institution. Nearly two-thirds of the laboratory's buildings were at least forty years old.[1] Operating costs were high, at least in part due to old and expensive-to-maintain facilities and equipment. The safety record of the laboratory fell short of DOE's rigorous expectations.

Meanwhile, employee morale was low after some years of rumors that

ORNL was not highly ranked in the constellation of federal research laboratories and might in fact close if DOE had to jettison a national lab for budgetary reasons. This reduced ranking was not a result of ORNL declining in research expertise and productivity, but more a function of other national laboratories jumping ahead. Large new scientific user facilities were opened at other laboratories: among them, the National Synchrotron Light Source at Brookhaven National Laboratory in 1982 and the Advanced Photon Source at Argonne National Laboratory in 1995. ORNL lacked a major new experimental facility of such scope over those two decades, which made the success of building the SNS so important. However, at some level, this new project struggled. Even as ground was being broken for the new facility in December 1999, there was risk that the SNS project might be canceled.[2] Clearly, there was much work to be done as UT-Battelle prepared to assume management duties.

National Laboratory Difficulties

As UT-Battelle assumed operation of ORNL, many challenges beset the national laboratories. These administrative difficulties were documented by advisory and oversight groups to DOE and other government agencies. Almost thirty reports from 1978 to 1998, as identified by the U.S. General Accounting Office, focused on management weaknesses at DOE and its laboratories.[3] These reports highlighted concerns about the laboratories' missions being unfocused, DOE micromanaging the facilities, and the laboratories not operating as an integrated system. A scholarly article (one author was former Ford executive John McTague) discussed these difficulties and made proposals on how to improve the national laboratory system.[4] Discussions were so intense that Congressional hearings were held to discuss the future of the DOE facilities and potentially the establishment of the laboratory-closing commission along the lines of the Base Realignment and Closing Commissions (BRACs).[5]

In the 1980s and 90s, research opportunities at various national laboratories leaped into the forefront as synchrotron-based light sources[6] were being planned, constructed, and commissioned (at Argonne, Brookhaven, and Berkeley). By comparison, ORNL had no comparable construction of a large experimental facility. ORNL suffered due to the 1993 cancellation of the SSC in Texas, as it lost large associated research programs. ORNL's proposed ANS, a new reactor, ran into political difficulties in 1996 and was canceled;[7] an accelerator-based source of neutrons replaced ANS. Adding to ORNL concerns were safety issues at the HFIR, which resulted in fines and a shutdown until they were resolved.[8] Against this backdrop, UT-Battelle assumed management of ORNL, feeling an urgent need to improve the performance and reputation of ORNL.

DOE required the ninety-day transition period, and the current contractor (Lockheed Martin) was obligated to provide temporary space for the incoming contractor's transition team. The UT-Battelle team was given space in old Quonset huts at ORNL, and the temporary inhabitants dubbed these structures the "Winter Palace," with appropriate sarcasm.[9] (The eighteenth-century Winter Palace in St. Petersburg, Russia, was constructed as the ruling family's winter residence, decorated in fine Baroque style.) UT-Battelle's temporary quarters were part of a complex that had been constructed in the late 1940s as a metallurgical laboratory and research shop.

The structures were never intended for long-term use, but they survived for more than fifty years. The final residents were the UT-Battelle transition team, whose members suffered through peeling paint, a leaking roof, and unreliable heat. To some on the team, the term sour grapes, on the part of the departing management organization, came to mind. However, the poor conditions of the Winter Palace actually increased the team's *esprit d'corps* and led to a highly successful transition. And, of course, being housed in this old and rundown

BUILDING 2000 IN 1949. COURTESY OF OAK RIDGE NATIONAL LABORATORY.

BUILDING 2000 IN 1999. COURTESY OF OAK RIDGE NATIONAL LABORATORY.

building reinforced UT-Battelle's commitment to renovate the aging campus. Advantages of working in the Winter Palace were that: the transition team could be located together, no ORNL staff had to be relocated, and it was away from ORNL administrative areas where incumbent staff or local DOE officials might want to pry into what was happening with the new team.

UT-Battelle: A Partnership Focused on Excellence

The U.S. DOE relies primarily on contractors to carry out its missions at its national laboratories and other facilities. This practice, which originated during the Manhattan Project, has made it possible for the agency to draw on the expertise of the private commercial sector and the academic community while providing general oversight, review, and programmatic direction to the contractor at each facility.

Since April 1, 2000, UT-Battelle has held the management and operation (M&O) contract for ORNL. Formed as a fifty-fifty limited liability partnership between UT and Battelle Memorial Institute, UT-Battelle LLC is the legal entity responsible for accomplishing the mission of the laboratory. (For the story of how UT and Battelle decided to team up and compete for the ORNL M&O contract, see chapter 7 on the 1990s.)

The two partners brought complementary strengths to the new company.

The university had strong connections to DOE's Oak Ridge facilities dating back to the Manhattan Project. Battelle, a global research and development organization headquartered in Columbus, Ohio, had extensive experience in managing national laboratories and other major research facilities, including DOE's Pacific Northwest National Laboratory, National Renewable Energy Laboratory, and Brookhaven National Laboratory.

The UT-Battelle partnership also includes ORAU, a nonprofit consortium of one hundred colleges and universities, and among them are seven leading southeastern universities: Duke, Florida State, Georgia Tech, North Carolina State, Vanderbilt, the University of Virginia, and Virginia Tech. The UT-Battelle Board of Governors, which oversees UT-Battelle's activities and assures stakeholder satisfaction, now represents each of these institutions.[10] The positions of board chair and vice chair are exchanged between the president of the UT system and the CEO of Battelle every two years. The corporate officers of UT-Battelle form the core of the ORNL Leadership Team, with the company's president and CEO serving as director of ORNL.

UT-Battelle's proposal for the M&O contract for ORNL established ambitious goals, guided by a philosophy of simultaneous excellence in three areas: science and technology, laboratory operations, and community engagement. UT-Battelle developed detailed plans for each area.[11] In science and technology, emphasis was placed on expanding ORNL's research capabilities, building stronger university partnerships, and ensuring that ORNL was equipped to support DOE's missions. In laboratory operations, UT-Battelle committed to improve safety and environmental protection, operate ORNL more efficiently, and revitalize the laboratory's aging facilities. In community engagement, UT-Battelle focused on supporting science education, investing in the communities in which its employees live, and encouraging staff at all levels to volunteer their time and talent to build thriving communities.

A contract of six hundred pages outlines UT-Battelle's responsibilities in its management and operation of ORNL and sets the amount of the available annual performance fee that the company can earn. DOE assesses UT-Battelle's performance each year, using a Performance Evaluation and Management Plan that sets out goals and objectives for seven areas. The grades in each area are compiled in an annual "report card" and determine the amount of performance fee paid to UT-Battelle. Once corporate expenses are paid, UT and Battelle divide the remaining fee.

The success of the UT-Battelle partnership is evident in DOE's decision to grant four five-year extensions of the original contract, signaling the department's satisfaction with the management and operation of ORNL by UT-Battelle. As of this writing, the M&O contract between DOE and UT-Battelle is due to expire on March 31, 2025.[12] The decision on whether to extend the contract yet again or put it up for competition rests with the Secretary of Energy. All five prime contracts between DOE and UT-Battelle are publicly available.[13]

SPALLATION NEUTRON SOURCE

The SNS project was designed in a complicated way to be a close collaboration among five national laboratories: Oak Ridge, Argonne, Berkeley, Brookhaven, and Los Alamos.[14] Part of the purpose of the collaborative structure was to bring the unique expertise of these five laboratories to bear on what would be the largest science construction project that DOE had yet undertaken. Another part, of course, was to spread the resources to other laboratories and thereby earn their buy-in and support—not to mention the support of eight additional members of the U.S. Senate.

Everyone realized that this would be a difficult partnership to maintain, with an emphasis on delivering parts of the project on budget and on time. One of the early issues involved the design and delivery of the linear proton accelerator that would initiate the spallation process to produce the neutrons. Los Alamos possessed recognized expertise in linear accelerator (linac) technology, but there were early concerns about the specifications and delivery date of its design. Previous ORNL Director Alvin Trivelpiece brought David Moncton from Argonne National Laboratory to ORNL to lead the $1.4 billion SNS construction project. Under Moncton's leadership, the project was substantially modified, including the choice of a superconducting linear accelerator provided by a new partner, the Thomas Jefferson National Accelerator Facility (otherwise known as Jefferson Laboratory) in Virginia.[15]

SNS Production of Neutrons

Neutrons produced in a nuclear reactor come from the fission of uranium (^{235}U) and are steady state in character, i.e., produced continuously. Another way to produce neutrons for various experiments uses an accelerator-based approach, with pulses of high-energy protons striking a target and producing neutrons by the spallation process.

The linear accelerator (linac) produces a beam of negatively charged hydrogen atoms (hydrogen with one extra atomic electron added, H^-) at high energy (1 GeV or 1,000 MeV). The linac consists of normal conducting components and also superconducting accelerator cavities that give the beam its high

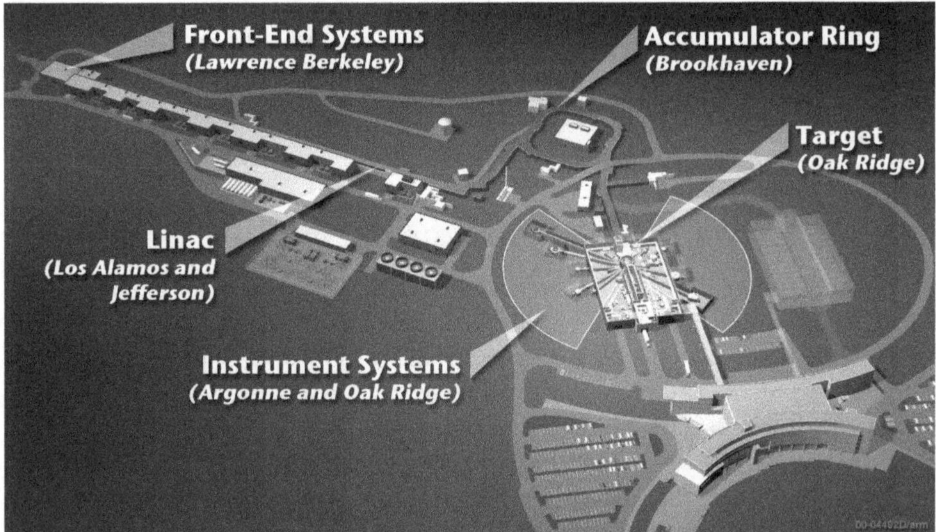

LAYOUT OF ORNL SPALLATION NEUTRON SOURCE COMPLEX. COURTESY OF OAK RIDGE NATIONAL LABORATORY.

energy and pulsed time structure (at radio frequencies).[16] Next the H^- beam is sent into an accumulator ring, which starts with a diamond foil that removes the two electrons from each H^- atom to produce a beam of protons (with a positive electrical charge). The proton beam goes around the accumulator ring approximately 1,200 times to produce an extremely short, sharp bunch of protons that is directed to the target.

This bunched proton beam emerges from the accumulator ring in short pulses (1-millionth of a second in duration) sixty times per second and is directed to a large vat of recirculating liquid mercury (the target) to produce a host of nuclear particles, including neutrons. The produced neutrons are emitted from the mercury target in short pulses sixty times a second and sent to one or more instruments for performing pulsed neutron-scattering measurements on various materials under study. It is advantageous for some physics measurements to combine the neutron-scattering results from a reactor (steady-state beams of neutrons) with those from the SNS (pulsed beams of neutrons). ORNL uniquely offers both types of facilities for neutron-scattering experiments.

By the time of the management contract transition, the SNS project faced several challenges. One troublesome issue was the sales tax to be charged in Tennessee for items purchased in the state by contractors for the project or even for items purchased elsewhere and brought to Tennessee. The sales tax in Tennessee has always been high, in part because there is no state income tax. DOE estimated that these Tennessee taxes would add about $30 million to the SNS facility's total cost. In March 1999, House Science Committee Chairman Jim Sensenbrenner, a Republican from Wisconsin, termed this tax situation as unacceptable—the sales tax bill would be higher in Tennessee than in other states that could provide alternate sites for the facility.[17] His committee's authorization of funds for the SNS, included in H.R. 1655, had seven conditions to continue funding for SNS beyond February 1, 2000:

that senior management positions be filled by qualified individuals;
that an external review validate the project's cost baseline and project milestones;
that the duties and obligations of each participating laboratory be defined in legally binding terms;
that the project director have direct supervisory responsibility over the SNS staff based at the collaborating laboratories;
that the Secretary of Energy delegate primary authority for the project to the project director;
that the Tennessee sales tax for the construction of the SNS not exceed taxes in states where SNS could have been constructed (i.e., California, Illinois, New Mexico and New York); and
that the DOE secretary report on the project's progress annually.[18]

The sales tax issue was a difficult problem to solve because it would require the action of the Tennessee legislature. On November 1, 1999, Tennessee Governor Don Sundquist proposed a tax plan that included an exemption from the state's sales tax for the construction of SNS until 2009. However, seventeen days later, on November 18, the Tennessee legislature voted to adjourn the legislative session, without addressing the SNS sales-tax issue.[19] This left SNS and its support on Capitol Hill in limbo.

In January 2000, the UT-Battelle team came to the rescue on the difficult sales tax conundrum.[20] UT's leadership garnered the support of the governor and the two speakers for rapid passage of legislation that would provide a sales-tax exemption for SNS procurement items, and, miraculously, this legislation passed on January 28, at a time when the legislature was supposed to be adjourned.[21] This was a huge boost for the beleaguered SNS project, and it soon gained solid footing with the U.S. Congress and DOE. In fact, Congressman Sensenbrenner stated that all of his concerns were addressed and his House Science Committee now supported full funding of the SNS project.[22]

ORNL hired Thom Mason in 1998 to direct the SNS Experimental Facilities Division and take the lead on construction of the complicated and expensive detectors that would sit on various beam lines and allow a host of experiments using the pulsed neutrons generated by SNS. Mason was a young Canadian science standout, only 34 years old when he came to Oak Ridge from the physics faculty at the University of Toronto. Mason's management expertise had come to the attention of ORNL scientists when he chaired the 1997 International Conference on Neutron Sciences in Toronto. Unfortunately for Mason and for the other conference attendees, the conference banquet was held on a boat on Lake Ontario. A rain and wind storm arose and, according to many attendees, the dinnerware slid off the tables and a few bruises and even broken bones resulted for the participants. In spite of that bad luck, Mason was hired at ORNL in 1998 with a strong background in neutron scattering and condensed-matter physics. Once in place in Oak Ridge, Mason achieved many successes and, later in this decade, became ORNL director.

The SNS construction project settled onto an even keel under the leadership of David Moncton. The other national laboratory partners were generally delivering key pieces of the project on time, and DOE was happy. However, as time progressed, project managers encountered challenges in maintain-

THOM MASON. COURTESY OF OAK RIDGE NATIONAL LABORATORY.

ing the momentum of this large multi-laboratory project and keeping it on budget and on time (of supreme importance to DOE), in part due to the size and complexity of the project. In addition, Moncton found it difficult to maintain some of his responsibilities at the Argonne Advanced Photon Source while leading the SNS project.[23] The directors of ORNL and ANL asked Moncton to choose one or the other to focus upon fulltime, and he chose to return to Argonne.[24] Suddenly, in February 2001, the SNS project needed a new leader. And it needed one quickly.

ORNL implemented a rapid selection process to hire an accomplished and respected accelerator builder to take the helm at the SNS. Such a qualified person was judged essential, since a critical part of the project was to design and construct the linear proton accelerator and the accumulator ring. After weeks of talking to various accelerator gurus, no one candidate emerged who could move to Oak Ridge quickly to head up the SNS project.[25] Finally, the ORNL leadership team decided to hand the keys to Thom Mason and elevate him to SNS director.[26]

Mason certainly possessed the credentials to design and build instruments for experiments at the end stations of the pulsed neutron facility, but he had no background or experience in building accelerators. Putting Mason in charge of the entire project was in some ways a big gamble, but nearly everyone who encountered him acknowledged that he was probably the smartest person they had ever met and had a smooth and even approach to leading people. By the time the UT-Battelle Board of Governors met on May 23, 2001, Mason was tapped to head the SNS project. The board reviewed details of the project, thanked Moncton for his service, and certified that the project was in good shape.

Thom Mason led the successful completion of the project, and SNS delivered its first beam on target in 2006 and opened as a user facility in 2007, after having been built on scope, on schedule, and on budget. Although not a trained accelerator expert, Mason is one of those rare humans who can absorb, understand, and remember the details of any complex system, and he applied those skills to learning the nuts and bolts of accelerators in short order. His success in leading the SNS project was among the deciding factors in his appointment as ORNL director in 2007. Following his departure from ORNL in 2017, Mason became a Battelle vice president and, in 2018, was named director of Los Alamos National Laboratory.

The Joint Institute for Neutron Sciences (JINS) was dedicated on December 3, 2010, and Governor Phil Bredesen was present for the ceremony.[27] The $10.5 million of state funds toward construct JINS was the last piece of the $32.1

million provided by Tennessee to construct three joint institutes at ORNL, as promised in the UT-Battelle proposal.[28] JINS has since been renamed the Shull Wollan Center, in honor of the early 1950s inventors of neutron scattering at ORNL.

UT Faculty Support the Spallation Neutron Source

The proximity of the SNS to the UT campus profoundly impacted the university's faculty, students, and programs. In fact, several UT faculty received NSF grants that facilitated the development of SNS instruments. Linda (Lee) Magid, a professor in UT's Department of Chemistry, coordinated the preliminary workshop for the U.S. neutron scattering community in 1998,[29] which set priorities for the initial suite of instruments and formally established the group of users of ORNL's neutron facilities, the SNS and HFIR. Magid also obtained NSF funding for two workshops related to developing the SNS's large complicated detectors, which now number twenty.[30] [31]

NSF also supported the conceptual design and engineering study of the VISION spectrometer through a grant to John Larese of the chemistry department. The SNS VISION instrument characterizes molecular vibrations in materials and is useful in studying hydrogen storage, hydrogen photovoltaic materials, drugs and pharmaceuticals, polymers, proteins, catalysis, and batteries.[32]

Expanding the community of researchers interested in neutron scattering has been an important objective for the university. Peter Liaw, a professor in UT's Department of Materials Science and Engineering, received an NSF Integrative Graduate Education and Research Training Award covering nine years and involving nine UT departments, ORNL, and other institutions.[33] The goal was to enable PhD students to study the complex properties of existing structural materials and to design new ones. Through UT's NSF-funded International Materials Institute, a network was developed to advance the science, education, and training for neutron scattering in materials research. Liaw also received an NSF grant for development of loading and furnace facilities for use at the SNS VULCAN instrument, allowing for neutron-scattering analysis of real-size samples to measure stress and strain on large pieces of industrial equipment.[34]

EAST TENNESSEE'S RAPID RISE IN SUPERCOMPUTING

For generations, scientists and engineers maintained that research was built on the twin pillars of experiment and theory. By the end of the twentieth century, computing emerged as a third pillar of science. For example, new airplane designs are now modeled and tested using elaborate computer codes, with a minimum amount of far more expensive wind-tunnel testing. Likewise, computer-aided drug design reduces the cost and time associated with bringing new pharmaceuticals to market.

However, this stout and sturdy third pillar of science, at least initially, slowly developed, and early computing machines were often sluggish and limited in scope and capability. Built in 1954, ORACLE was ORNL's first computer, and it had a remote-controlled magnetic-tape data-retention system, which bestowed on it the largest computer memory to date. ORACLE was used, among other applications, to solve design issues associated with the development of nuclear aircraft, as part of an ambitious program of the 1950s that, notably, never resulted in creation of a functioning nuclear airplane.[35]

The speed with which digital computing has advanced is breath-taking—within an average human lifetime, computers have become ubiquitous and extraordinarily powerful. Once ORACLE became obsolete in 1962, ORNL

FROM LEFT: JESSE POORE (HEAD OF UT COMPUTER SCIENCE), JACK DONGARRA (UT/ORNL), LEE RIEDINGER (DIRECTOR OF SCIENCE ALLIANCE) AT ORNL IN 1990. COURTESY OF BETSEY B. CREEKMORE SPECIAL COLLECTIONS AND UNIVERSITY ARCHIVES, UNIVERSITY OF TENNESSEE, KNOXVILLE.

purchased computers from commercial vendors or used systems extant at other Oak Ridge facilities. In the early 1980s, Alvin Trivelpiece, then director of DOE's Office of Energy Research, set in motion the idea of having high-performance computing resources available for use by DOE's energy research programs.

Parallel computing, in which many calculations are carried out simultaneously, came into its own in the 1980s. In 1985, ORNL installed its first distributed memory parallel computer, an Intel device with sixty-four processors capable of conducting 2 million calculations, or floating-point operations, per second (also expressed as 2 megaflops).[36] By 1990, this system was upgraded to a 128-processor device with 7.7 gigaflops (billion flops) of computing speed. In 1994, the Intel Paragon XP/S-150 was installed, performing at 150 gigaflops.[37] Another big event of 1994 was the introduction of ORNL to the internet and World Wide Web.[38] A key advancement in parallel computing was the development of a software package called Parallel Virtual Machine (PVM), led by UT/ORNL distinguished scientist Jack Dongarra and his research team.[39]

The upsurge in computing power continued as the number of parallel processors increased along with the speed of each processor.[40] In 2001, UT was operating a 20-gigaflop IBM computer that had been the forefront device at ORNL before the laboratory replaced it with a 1-teraflop (trillion flops) IBM parallel computer. At about that time, Thomas Zacharia, ORNL associate laboratory director for computing and computational sciences, came up with a novel plan that would economize computing conducted by the two institutions and, in the process, contribute to the ever-maturing UT-ORNL partnership.

UT was then spending $800,000 per year to operate its IBM. Zacharia proposed that UT send to his organization that amount plus an additional $500,000 per year to acquire 20 percent of ORNL's 1-teraflop computer, (i.e., 200 gigaflops), a substantial increase in computing bang for the university's buck. And, he pledged to invest most of that annual sum in hiring graduate students and postdoctoral scientists to conduct research in the UT/ORNL Joint Institute for Computational Sciences. UT could not possibly refuse this deal.

Thomas Zacharia

Thomas Zacharia grew up in Kerala, India and came to the United States to earn a master's degree in materials science from the University of Mississippi in 1984 and, in 1987, a doctoral degree in engineering science from Clarkson University. Zacharia joined ORNL as a postdoctoral researcher in 1987 and, two years later, became a research staff member of the Metals and Ceramics

THOMAS ZACHARIA. COURTESY OF OAK RIDGE NATIONAL LABORATORY.

Division.[41] He started and led the Materials Modeling and Simulation Group until appointed director of the Computer Science and Mathematics Division in 1998. Zacharia was named ORNL's associate laboratory director for the newly formed Computing and Computational Sciences Directorate in 2001 and later the deputy director. In 2017, he was named director of ORNL.

Zacharia is an idea man, known to move boldly and swiftly in pursuit of goals that others might believe are unattainable. He visited new laboratory director Bill Madia in 2000 to pitch his idea that ORNL could be the best in the world in computing. He then articulated his vision of what could become the Oak Ridge Leadership Computing facility. At that time, DOE was investing a mere $7 million annually into ORNL computing, but Zacharia believed that this investment could eventually grow to $40 million.

The ORNL leadership team was somewhat skeptical whether Zacharia's bold vision was achievable, but Zacharia endeavored to turn the dream into reality and ultimately made ORNL one of the world's top supercomputing sites. Henceforth, Zacharia's associates learned never to underestimate him or his drive.

The creation of new dedicated space for a large computing facility as part of UT-Battelle's ambitious plan to modernize the aging ORNL physical plant, much of which had been constructed in the 1940s and 1950s, strengthened

Zacharia's plan to turn ORNL into a supercomputing powerhouse. Within the first three years of UT-Battelle's management of the laboratory, eleven out-of-date buildings were torn down, with more slated to be razed over the next decade, led by Bechtel Jacobs Co., at that time the DOE-supported environmental restoration company in Oak Ridge.[42] With many of the old structures out of the way, UT-Battelle would create a new campus complex situated in the ample space that had served as ORNL's parking lot for fifty years.

THE MODERNIZATION OF ORNL'S PHYSICAL PLANT

In many ways, life in America changed with the terrorist attacks on September 11, 2001. The United States government created the Department of Homeland Security, Directorate of National Intelligence, and National Counterterrorism Center. Support for intelligence and law enforcement agencies also increased.[43] Airplane travel especially became more complicated and time consuming due to the rigorous pre-flight checking of passengers.[44] Information gathering on perceived terrorists greatly increased. These new and expensive security measures had a direct impact, no repeat of a foreign-bred terrorist attack on the United States since 9/11 has occurred.

The attacks of 9/11 immediately and permanently affected security at DOE laboratories. Before that time, the ORNL parking lot was situated outside the security fence, while the laboratory itself was contained mostly inside of it. At the time, Bethel Valley Road, which runs east-west right past ORNL, was unrestricted and open to public traffic. After the terrorist attacks, the security risks became abundantly obvious. Indeed, something clearly needed to be done to prevent the possibility of a large truck filled with explosives driving on the public road to within one hundred feet of the northeast corner of building 4500N, the site of the laboratory director's office.

Fortunately, Bethel Valley Road was a DOE-owned byway that the public had been permitted to use after the fence around the city came down in 1949. With considerable help from Tennessee political leaders, UT-Battelle was able to convince DOE to close Bethel Valley Road to the public—which, of course, prompted a public outcry—and move the guarded entrance portal five miles to the east and a mile to the west, creating a large, secure buffer zone around the laboratory and parking lot.[45 46] This then allowed UT-Battelle to begin fulfilling a promise made in its proposal to DOE to transform ORNL from a down-in-the-heels remnant of the Manhattan Project to a more accessible and purposefully designed facility. The renovated campus' layout would be open and easy to navigate with ease of movement for personnel to and from their cars and between buildings, once visitors passed through the guard gates some miles to the east and to the west. Automatic badge readers at entrances would provide a convenient means of entry while ensuring that only authorized personnel are able to enter each building.

Three sources of funds were tapped to allow UT-Battelle to actualize its ambitious plan to construct the new campus on the capacious old parking lot. State of Tennessee funds were used to construct the Joint Institute for Computational Sciences (JICS), as promised in the UT-Battelle management proposal and backed by Governor Sundquist. This 25,000-square-foot structure faces the new quadrangle from the west side, looking east. DOE funds were used to construct the visitor center, which faces JICS and houses a modern cafeteria (replacing one constructed in 1948), an office for issuing guest badges, and a second-floor suite of meeting rooms. On the south side of the quad sits a privately funded 300,000-square-foot building containing offices and laboratories for a number of research divisions, as well as computing facilities for large supercomputers.

Jon Coddington: Architect Extraordinaire

While a UT professor of architecture and head of the college's graduate program, Jon Coddington led the design activity of the UT-Battelle master plan for the ORNL main campus. This role led to his special recognition in the gala 2000 Awards Night ceremonies by UT-Battelle.[47] [48] Coddington involved UT faculty and graduate students while working with ORNL staff to develop a plan to transform how staff and visitors view the national laboratory. This plan revolved around a lawn surrounded by buildings similar to a college campus. The campus' buildings reflected rigorous environmental and energy-efficiency standards in the selection of construction materials.

Coddington's other significant design schemes include Knoxville's James Agee Park in the Fort Sanders neighborhood and Nashville's Bicentennial Mall. Coddington has been honored with the Tennessee Society of Architects Presidential Award for Distinguished Service and numerous awards from the American Society of Landscape Architects and the Tennessee Society of Architects.

Coddington has also held positions at Ball State University and the River City Company in Chattanooga and has recently retired from Drexel University in Philadelphia.

UT-Battelle devised an innovative scheme to have a private developer secure funding ($72 million) to construct this large office, lab, and supercomputing facility and then lease it to ORNL for thirty years.[49] Political

ORNL PLANT IN 2002. COURTESY OF OAK RIDGE NATIONAL LABORATORY.

persuasion had to be employed in convincing a conservative DOE to allow this type of private-sector construction and ownership on a national laboratory site. Among other benefits of such a funding arrangement, the facility was constructed in half the time and at half the cost than it might otherwise have been, had DOE directly funded the construction project.[50] A South Carolina company was chosen for this construction project, the first such privately funded venture at ORNL.[51]

A 40,000-square-foot computing facility was built in a large research structure on the new campus at a time when there was no supercomputer to house. This represented Zacharia's challenge and reflected his confidence in the ultimate outcome—confidence effectively predicated on the belief that "if you build it, they will come." He took the lead in developing a program that would, soon enough, fill these large, raised-floor, well-cooled rooms with supercomputers. These computer rooms were initially constructed with 12 MW of power and 3,600 tons of cooling capacity. But, as supercomputers grew in size and capability, they also required greater amounts of electrical power and cooling. In 2007, the power was increased to 15 MW and the

ORNL PLANT IN 2007. COURTESY OF OAK RIDGE NATIONAL LABORATORY.

cooling to 6,500 tons to support the Cray XT-4 Jaguar supercomputer rated at 119 teraflops.

In 2007, Zacharia also took the lead on securing a $64.4-million award to UT-ORNL's JICS, winning a national competition for a supercomputer funded by the NSF.[52] NSF typically awards grants to colleges and universities, not national laboratories, but, at the time, UT did not have a big enough name in large-scale computing to head a credible NSF proposal. To enhance the university's clout, Zacharia was made a UT-ORNL joint faculty member (and later a vice president for science and technology[53]) so that he could serve as lead investigator on this proposal.

This approach worked, and, in August 2007, NSF announced that UT had won what was the largest federal grant in its history ($65 million).[54] A key selling point to the NSF was that the 1-petaflop (1,000 teraflop) computer—called Kraken, after the mythical sea creature that gobbled up large ships—would sit in the big computer room at ORNL next to the DOE-funded 1-petaflop computer, since this building had enough space, power, and cooling capacity for several supercomputers. In fact, a third petaflop supercomputer was also added later, funded by the National Oceanic and Atmospheric Administration. Zacharia was right: he had the capacious facility built, and the supercomputers did, indeed, come. Plus, this critical connection between UT and ORNL enabled UT to acquire a supercomputer.[55] By early 2009, Kraken was ranked as the second-largest supercomputer in the world for open scientific computing.[56]

There was an ancillary advantage to appointing Thomas Zacharia to a joint faculty position so that he could lead development of the NSF proposal. To make this proposal as credible as possible, then UT President John Petersen took the lead in getting a sacrosanct and long-standing rule changed—something that few UT people thought possible. The rule: the prohibition on awarding university tenure to a part-time faculty member. Zacharia was based at ORNL, with UT sending the laboratory an agreed-to portion of his salary and benefits. So, Zacharia received no direct UT paycheck, and, thus, he could not be considered for tenure in terms of the existing university rules.

To his great credit, Petersen understood the importance of awarding tenure to Zacharia, and so he took the lead in getting the UT Board of Trustees to change the rule to make it possible for a part-time employee to be eligible for the award. An academic department faculty still had to vote in favor of tenure for Zacharia, based on his credentials. Although the ensuing debate in the Department of Electrical Engineering and Computer Science was spirited, the vote was favorable, and Zacharia was awarded tenure. This had later benefit to the UT-ORNL partnership, as it would be possible to grant tenure via the soon-formed Governor's Chair program, regardless of whether a fifty-fifty shared faculty member was based at UT or at ORNL.

In this decade, ORNL achieved the status of a top supercomputer laboratory, and, in 2009, the ORNL's DOE-sponsored supercomputer Jaguar was ranked number one in the world in the TOP500 competition, after it had been upgraded to the Cray XT5 architecture with peak performance of 1.4 petaflops.[57] The UT Kraken (NSF) supercomputer was ranked third at 1.0 petaflops.[58] This is a high-stakes competition, but ORNL has remained at or near the top of the list in the years since. At the writing of this book, an exaflop (a million teraflops or a thousand petaflops) has been achieved at ORNL with an operation of 1.8 exaflops on the Summit supercomputer, which won the prestigious ACM (Association for Computing Machinery) Gordon Bell Award at Supercomputing in 2018.[59] The calculation performed to win this award was led by UT-ORNL computational biologist Dan Jacobson and involved comparison of millions of genomes in short periods of time to help understand the genomic basis of the opioid epidemic. The Summit supercomputer has brought many successes to ORNL, and the prestige and capability of the university have advanced greatly through its NSF award of Kraken in 2007.

A REPUTATION-BOOSTING STRING OF SUCCESSES

Success often breeds more success, and that was certainly the case at ORNL in the decade of the 2000s. ORNL's national reputation had not been particularly impressive in the 1980s and '90s, and that tepid standing had presented a major challenge for UT-Battelle as it began its tenure as ORNL's manager in April of 2000. In the twenty years before UT-Battelle, ORNL lost some

ORNL COMPLEX ON CHESTNUT RIDGE: SNS AND CENTER FOR NANOPHASE MATERIALS SCIENCES ON THE RIGHT. COURTESY OF OAK RIDGE NATIONAL LABORATORY.

competitive ground to other U.S. national labs in terms of not winning big awards for new scientific user facilities, like the synchrotron-based light sources at Argonne National Laboratory (Advanced Photon Source) and Brookhaven National Laboratory (National Synchrotron Light Source). Soon after UT-Battelle took the management reins, a series of ambitious projects paved the way for ORNL's success and, along with it, a boost in the laboratory's national standing. Building SNS on budget and on time by 2006 would be essential. Winning DOE and then NSF supercomputer awards happened as a result of strong leadership and new facilities suddenly available to house and power the sophisticated computational devices.

Also in this decade, DOE awards led to creation of the Center for Nanophase Materials Sciences, which sits next to SNS, and is important for the materials science research of many UT faculty and students, in addition to ORNL researchers and those from many other institutions. The Bioenergy Science Center (BESC) is another success, and this award was predicated partially on the availability of a new building constructed by the state of Tennessee to house the Joint Institute for Biological Sciences (JIBS), as promised by Governor Sundquist in the UT-Battelle proposal.

The anchor facility for BESC, a multi-institutional partnership, was the JIBS building. The UT-based Tennessee Biofuels Initiative, funded in part by the state, bolstered the proposal to create BESC. Both of these activities are focused on finding alternatives to petroleum-based fuels, specifically to make biofuels not from corn, which has a marginal energy-gain ratio, but from cellulosic plant material, e.g., switchgrass. Another win came to ORNL in 2010 with the selection of the Consortium for the Advanced Simulation of Light Water Reactors (CASL), a multi-institutional team led by ORNL, to establish DOE's first Energy Innovation Hub. CASL sought to use high-performance computers to calculate from first principles all elements and all modes of operation of a nuclear fission reactor to aide in the design of new reactors or to improve the performance of existing nuclear power plants.

This was a successful decade for UT as well. Clearly, the NSF supercomputer award was a game changer, as was the Tennessee Biofuels Initiative. And, growth continued in nuclear engineering (and computational science) as a result of CASL. Meanwhile, the nanoscience center at ORNL provided excellent new experimental capability to UT faculty and graduate students and strengthened the university's already well-established and -respected program in materials research. The number of joint faculty grew greatly in the 2000s and numbered well over one hundred by the end of the decade. Additionally, Phil Bredesen, Tennessee governor from 2003 to 2011 (who had an abiding respect for science, having earned a degree in physics from

Harvard University), instituted and funded a new program of high-level joint faculty.

Polymers and Nanoscience

Polymers are large molecules composed of many long chains of smaller molecules or chemical units repeated throughout the chain. Polymers are a common and necessary part of our everyday lives. Naturally occurring polymers include cotton, starch, and rubber, while there are thousands of synthetic polymers in common use in various broad areas, e.g., sportswear and sports items (skis, snowboards, rackets, tents), electronics (solar cells, television components), packaging (films, bottles, food containers), and construction (garden furniture, PVC windows, pipes, paints). Polymers are widely used due to their unique properties: low density, low cost, good thermal/electrical insulation, high resistance to corrosion, and rather easy processing into final products. In addition, combination with other materials to form composites can enhance the properties of a polymer.

Nanocomposites are formed by mixing nanoparticles with long-chained polymers, which can enhance the useful properties of a polymer in products found in automobiles, fire retardants, and drug-delivery systems. A scientific question has been what size nanoparticle is most effective in these composites to produce the most enhancement of polymer properties.

This question is one of many that has been effectively addressed in the ORNL Center for Nanophase Materials Sciences (CNMS). The Department of Energy has built a set of nano centers at its various national laboratories, typically located near major experimental facilities. The first one was the CNMS sited next to SNS to combine nanofabrication at the CNMS with SNS-based neutron scattering to understand the nano structure and properties of materials. The team of UT-ORNL Governor's Chair Alexei Sokolov and his colleagues at CNMS addressed this question of how small the nanoparticles in a composite should be.

The assumption had been that 10 to 50 nanometers[60] was the optimum size of such nanoparticles. However, their experiments showed that particles only 1.8 nanometers wide were far more effective in increasing the temperature above which the polymeric material starts to flow, which translates to maintaining normal viscosity at higher temperature and retaining good processability, resolving a major challenge in polymer nano composite applications.[61] The smaller nanoparticles move around the polymer chains faster than the larger ones and also have larger effective surface area (more smaller particles have more surface area than fewer large particles), which improves the chemical interaction with the chains. These and other CNMS experiments will increase the use of polymer nano composites in many applications.

UT-BATTELLE EXPANDS ITS TEAM OF CORE UNIVERSITIES

An important part of the UT-Battelle proposal to DOE for managing ORNL was the slate of core universities that would be part of the board of governors and would engage with ORNL in new ways. Under the leadership of Ron Townsend, president of Oak Ridge Associated Universities, six top universities in the Southeast signed on to the proposal and were part of the winning team: Duke, Florida State, Georgia Tech, North Carolina State, Virginia, and Virginia Tech. There were supposed to be seven, as the original intent was to include Vanderbilt, due to its proximity to ORNL and decades of partnership with the laboratory and with UT in various areas of science. However, that did not happen when the proposal was submitted in 1999, because Vanderbilt, part of the Universities Research Association, took the lead on developing a competing proposal submitted with the incumbent contractor, Lockheed Martin.

Late in 2000, ORNL director Bill Madia asked Lee Riedinger, ORNL deputy director of science and technology, to do the legwork necessary to bring Vanderbilt onto the UT-Battelle team as a core university. Riedinger talked to officials at the six core universities and UT about the desire to bring Vanderbilt into the fold, and all involved supported the effort. Everything seemed in order for a presentation to the UT-Battelle Board of Governors on May 23, 2001, seeking approval to add Vanderbilt as a core university, with a seat on the board. As the UT-Battelle manager of the university partnerships at ORNL, Riedinger made what he thought would be an easy pitch to the board to add Vanderbilt, but it did not turn out that way. The chair of the board at that time was Wade Gilley, the UT President since August 1, 1999.

After Riedinger's presentation to the board, Gilley called the proposal a terrible idea that he could never support. He alleged that Riedinger had lied in claiming unanimous support for Vanderbilt's inclusion and said that such a move would ignore the longstanding animosity between the two universities. Vanderbilt had backed the wrong team in the proposal stage, and, according to Gilley, there is no way that UT could forgive that grave mistake. This attack went on for what seemed to Riedinger like an eternity in a room full of distinguished board members and participating officials. Needless to say, the motion failed, and Riedinger was in a state of shock, never having been so chastised in public in his long career.

Two days later, on May 25, Riedinger was able to speculate on a reason for Gilley's ferocity. An article appeared in the Knoxville newspaper about a brewing scandal involving Gilley and a member of his staff, Pamela Reed.[62] The article discussed an allegedly embellished résumé that had enabled Reed to be hired in previous positions at other universities and then at UT. In March 2001, Reed was named executive director of a new UT Center for Law, Medicine, and Technology, although some had questioned her qualifications for holding such an influential position. The article precipitated a media storm about the conditions of Reed's hiring and even an alleged intimate relationship between Reed and Gilley. This brewing storm may have been on Gilley's mind when he lashed out so forcefully at the UT-Battelle board meeting only two days earlier.

The Pamela Reed story was a sordid one, and the details of this summer-long series of media revelations is well covered elsewhere.[63] As a result, on June 1, Wade Gilley resigned as UT President,[64] launching an unpleasant trend for the university. Including Gilley's disgraced resignation, UT would experience three consecutive failed or at least compromised presidents, all of whom had been hired from universities beyond UT.[65] On June 13, Reed resigned her UT position,[66] and newspaper articles through the summer provided more details of the alleged Gilley-Reed relationship.[67] [68] It was not a cheerful time at the university.

The issue of Vanderbilt's inclusion as a core university was finally resolved over the next two years. Joe Wyatt had retired as Vanderbilt's chancellor in 2000 after years at the helm and was succeeded by Gordon Gee, who served in that capacity until 2007. To try to repair the rift, Gee spent all day on January 9, 2002, in Knoxville talking to various officials about the partnership at a briefing session held at the National Transportation Research Center. Among those in attendance at the session were Madia and Riedinger from ORNL, UT President Eli Fly (a long-time UT leader appointed interim president following Gilley's departure), Knoxville Mayor Victor Ashe, and several Knox County officials. These interactions were productive[69] and ultimately led to Vanderbilt's inclusion as a UT-Battelle core university, effective on September 30, 2004.[70] What had looked like a quick and easy outcome ended up taking nearly three years to consummate.

Leadership Succession at ORNL (Seamless) and at UT (Fraught)

ORNL had three directors in the decade of the 2000s, and the torch passed seamlessly from one to the other: from Bill Madia to Jeff Wadsworth in 2003 and from Wadsworth to Thom Mason in 2007. This continuity of direction and principles of management set the standard for strong leadership at the laboratory at all levels. Likewise, UT's scientific enterprise thrived through this decade but almost in spite of rapid changes at the top: five presidents, including two interims and three disgraced executives, over the course of ten years.

Following Gilley's abrupt departure in June 2001, the UT Board of Trustees initiated a search for the new president and formed a search committee with ten of its members. Reporting to the committee was a state-wide advisory council with seventeen members, including ORNL's Lee Riedinger. The advisory council evaluated resumes of the one hundred or so applicants for the presidency, conducted interviews in Nashville on January 25, 2002, and chose one candidate to the final round of deliberations: Marlene Strathe, the vice president for academic affairs at the University of Northern Colorado.

Meanwhile, the firm Korn/Ferry International conducted a private search process, at the behest of the Board of Trustees, and settled on one candidate for final consideration: John Shumaker, president of the University of Louisville, who now became a public finalist for the job, along with Strathe. The search committee of the UT Board of Trustees then interviewed both candidates and settled on Shumaker for the post on February 27, 2002. He was offered the job, and, on March 1, Shumaker announced that he would accept the position as UT's twenty-first president.

The search committee and the advisory council were all impressed with John and Lucy Shumaker when they came for a visit in early March 2002. The spouse of the university president usually plays an important role as a host and a close partner of the president in fund raising, official receptions, and public events. It seemed to all who met the Shumakers that they would be an effective team in serving UT.

John Shumaker's first day on the job as UT president was June 5. He secured an attractive $735,000 financial package (salary, expense account, benefits)—a package that the *Chronicle of Higher Education* ranked as the second highest among U.S. public university presidents. However, Lucy Shumaker did not come to Knoxville with John as, in fact, he had filed for divorce in April.

Shumaker's first year on the job went well overall. Among his other innovations and administrative successes, Shumaker took the lead in forming the UT Research Foundation, which would aid in technology development and the patenting of the output of faculty research. This new organization succeeded the UT Research Corporation, which UT physicist Kenneth Hertel (and others), as noted in this book's chapter on the early 1940s, had started. However, on other fronts, trouble was brewing.

In the summer of 2003, allegations began to surface that questioned Shumaker's ability to keep university and personal finances separated. He had a university credit card, and he used it for all his needs, professional and private, and did not keep invoices that would allow accountants to easily decide how much he owed to the university for his personal purchases. In June, reports surfaced about Shumaker's frequent use of the UT airplane, some for official business and some for personal travel.

On July 17, 2003, the *Knoxville News Sentinel* reported that "University of Tennessee President John Shumaker said he would reimburse the university more than $24,600 for twenty-five flights involving personal business on the school's plane and cut up his university-issued credit card."[71] And, for the first time, the name of Carol Garrison, president of the University of Alabama, Birmingham, surfaced. She had been the provost at the University of Louisville and a good friend of Shumaker, and many of Shumaker's private flights on the UT airplane were to Birmingham. Subsequent newspaper reports alleged that the friendship had evolved to an intimate relationship and one that may have predated Shumaker's arrival at UT.

On July 27, court officials in Louisville released thirty-six hours of videotaped footage from the divorce trial of John and Lucy Shumaker. This led to an August 2 newspaper article[72] that fits in the category of the truly bizarre. During the difficult and protracted divorce proceedings, Lucy Shumaker seemingly fed damaging information about John to *Knoxville Journal* reporter Ashley Rhea. One piece of information was about the alleged extramarital affair John had had with former Louisville provost Carol Garrison (later admitted to by both parties). As reported in the *Knoxville News Sentinel*,[73] things got even more bizarre, when allegations surfaced that the reporter Ashley Rhea, whom Lucy Shumaker had talked to, was actually Pamela Reed, who had been forced to resign from UT the previous summer after Wade Gilley had resigned as president.

More state audits followed, and the Shumaker story on spending of university funds spiraled out of control.[74] The issue of Shumaker's lavish spending reached a climax at a meeting with Governor Phil Bredesen on August 7, 2003, at which time Shumaker agreed to resign, which he did the next day, with a severance package that was later voided.[75]

This debacle marked the second failed UT presidency in three years. Gilley's tenure had lasted twenty-two months, Shumaker's only fourteen. After Shumaker resigned, Joe Johnson was called back as the interim president (he had served an eight-year term as president prior to Gilley's arrival in August of 1999). Under Johnson's steady and proven leadership, the healing began. Interim President Johnson stopped another intended Shumaker renovation—the replacement of the orange carpet in the president's office with a $40,000 *red* carpet. This would have represented a sacrilege, since red is the school color for the long-standing UT football rival, the University of Alabama.[76]

JOHN PETERSEN ASCENDS TO THE UT PRESIDENCY

Subsequent to Shumaker's resignation, UT launched another national search, but this time it was conducted completely in the open, in reaction to the partially private search that had led to Schumaker's appointment. On April 5, 2004, the search advisory council (an eighteen-member body made up of near equal numbers of students, faculty, trustees, alumni, and UT staffers) whittled forty-seven candidates down to twelve finalists and later to six.[77] One candidate who was not advanced was ORNL Director Bill Madia.[78] Madia earned great respect, first as director of Pacific Northwest National Laboratory and then at ORNL; however, his career had been at Battelle and DOE facilities, and he had never worked at a university, which the search committee regarded as an impediment.

On April 21, the six finalists were interviewed in public on the UT campus, and the interviews resulted in three finalists who were voted on again several times until the selection of John Petersen. Petersen, a chemist by training, served on faculties at several universities before becoming dean of sciences at Wayne State University and then chancellor at the University of Connecticut. His base salary initially was $380,000 plus a $20,000 annual expense account that he could use any way he chose.

Petersen took office on July 1, 2004, and his administration was viewed as successful in its first two years, as evidenced by a positive Board of Trustees review in June 2006.[79] Rankings of UT's Knoxville campus rose in several categories among the nation's top public universities. Petersen served well as chair or vice chair of the UT-Battelle Board of Governors (these positions alternate between the CEOs of the two partnering institutions). The signs were generally positive, and the feeling was that stability at the top had been restored. However, a year later problems started to arise.

On September 8, 2007, Loren Crabtree, chancellor of the Knoxville campus, went public with his criticism of a new UT system mission statement issued by President Petersen on August 9.[80] This plan seemed to strip authority away from individual campuses in areas including human resources management, information technology, planning and construction management, and athletics. Crabtree contended that this arrangement would prevent UT, Knoxville from becoming a great public university and lead to "mediocrity."[81] Crabtree said that his criticisms reflected a near unanimous sentiment voiced among some 350 surveyed campus employees.

This was a period of intense tension between UT system and campus administrations. By this time, Riedinger had moved back to UT after six years at ORNL and was serving as the interim vice chancellor for research under Crabtree. Frequently, meetings of the chancellor's cabinet focused on the UT system's efforts to bulk up its personnel rolls and wrest responsibility for key functions away from the Knoxville campus. This administrative turf war influenced numerous campus functions and departments and impacted the UT-ORNL partnership.

At issue was who—UT, Knoxville, or the UT system—would assume administrative control of the UT-ORNL partnership and how UT-Battelle's management role ultimately would play out. In fact, by 2008, the UT system had grown to the point that Petersen had twenty-four administrators reporting directly to him, including fifteen vice presidents, three chancellors, and two athletic directors. All had high salaries, nineteen had university-provided cars to use, and most had expense allowances ranging from $2,000 to $25,000 per year.[82] This all seemed excessive to the Knoxville campus administration.

Loren Crabtree was well respected by the faculty on the Knoxville campus and was regarded as a man of high principle. He chose to disagree strongly with his boss, President Petersen, on each aspect of what Crabtree argued was overreach by the UT system administration. People on the chancellor's staff would advise him to choose his battles and not contest Petersen on his every administrative move. However, battles occurred constantly, as the system administration continued to expand. This tense standoff between president and chancellor continued through the fall of 2007 and ended abruptly on January 4. Petersen had called Crabtree to his office, and the result was that the chancellor resigned on the spot.

Loren Crabtree

Amid the turmoil at UT associated with five presidents (including two interims) in the decade of the 2000s, Loren Crabtree effectively led the Knoxville campus administration, kept the train on the tracks, and, in addition, effected significant advances at the university and in its partnership with ORNL. He was hired by the first of the compromised UT presidents (Gilley), promoted to chancellor by the second (Shumaker), and forced to resign by the third (Petersen).

Crabtree had been the provost at Colorado State University when he was hired into the same position at UT, Knoxville, effective July 1, 2001.[83] The prospect of leveraging UT's close connection to ORNL to increase its rankings attracted him. Crabtree, who would be elevated to the position of chancellor of UT, Knoxville in 2003, quickly understood the strong role of science and engineering on the Knoxville campus in the partnership with Oak Ridge. This was clear when Lee Riedinger, then ORNL deputy director for science and technology, hosted Crabtree on his first visit to the laboratory on August 22, 2001. This started a personal and professional relationship that later resulted in Riedinger ending his six years at ORNL and returning to UT to serve as Crabtree's interim vice chancellor for research for a year or so, starting in September 2006.

During his leadership tenure, Crabtree focused on the improvements necessary for UT, Knoxville to elevate its national ranking. One step forward was in top student recruitment, due to a November 2002 vote by the citizens of Tennessee to initiate a lottery, the proceeds from which have been primarily devoted to scholarships for Tennessee residents to attend universities in the state. These Hope Scholarships have significantly impacted UT, Knoxville,

FROM LEFT: LEE RIEDINGER (ORNL DEPUTY DIRECTOR), LOREN CRABTREE (UT CHANCELLOR), LEE MAGID (UT CHEMISTRY PROFESSOR), THOM MASON (SNS DIRECTOR). IN POSSESSION OF THE LEAD AUTHOR.

including a marked increase in the number of top academic performers from the state's high schools choosing to matriculate at the university, leading to a more accomplished student body and a better university overall.

Improving the quality of the faculty and their research awards was also a focal point for Crabtree. He supported the state-funded construction and operation of facilities to house the UT-ORNL joint institutes to expand faculty and student research opportunities. Three of these would be built at ORNL in this decade with state funds promised in the UT-Battelle bid to manage the national laboratory. The fourth would eventually be built at the UT Research Park at Cherokee Farm.

The number of joint faculty expanded greatly through the UT-ORNL partnership. This was important, since the university alone could not afford to hire a large cadre of new faculty to boost the number of PhDs produced each year. Efforts in this faculty arena were boosted greatly in 2005 when Governor Phil Bredesen funded a new Governor's Chair program to hire top researchers into joint UT-ORNL positions in research areas of common strength and interest.

After Crabtree's forced resignation on January 4, 2008, a well-attended going-away reception was held at the University Center on January 17.[84] Crabtree decided next to become academic vice president for the organization Semester at Sea, a study-abroad program housed on a ship. He had sailed on this ship several times as a Colorado State University faculty member, and now he decided to do one final semester-long voyage as the executive dean. Crabtree called Lee Riedinger to invite him and his wife, Tina, to be two of his thirty-five faculty on the spring 2010 voyage around the world, along with six hundred undergraduates. This remarkable voyage was enjoyed by all.

Crabtree's considerable popularity on UT's Knoxville campus set the stage for a contentious meeting, organized by the Faculty Senate, between President Petersen and faculty on January 22, 2008. Professors questioned Petersen about all the contentious issues at the heart of the Crabtree-Petersen battles. A survey showed almost three-fourths of the more than 1,100 faculty members who responded lacked confidence in Petersen's "ability to lead UT in the future."[85] The following week, the Faculty Senate formally issued a vote of no confidence against Petersen.[86]

Both positive and negative events and circumstances affected the faculty's view of the Petersen administration. A big positive was the $65-million NSF award for a one-petaflop supercomputer (Kraken). The win in the national competition brought more respect to the UT-ORNL partnership; the laboratory would house the computer. On April 3, 2008, Petersen, Governor Phil Bredesen, and many other dignitaries celebrated this sizable award at an event at the Joint Institute for Computational Sciences on ORNL's campus.[87]

The considerable negative at this time was the state's (and the country's) financial woes. On May 7, 2008, Governor Bredesen announced[88] 5-percent cuts in the state budget, the number of state employees, and the UT budget. These decreases in state allocations would escalate as the 2008 financial crisis deepened.

The pressure on John Petersen only intensified in the following months and was further increased by an unfortunate confrontation at the president's residence on October 17, 2008. The president's wife, Carol, hosted a reception that day honoring the accomplishments of the Alliance of Women Philanthropists, whose mission was to educate, empower, and inspire women to be philanthropic leaders at UT. The list of celebrants included women who had given or directed $25,000 or more to any UT program. As the reception was ending, somehow Carol Petersen publicly questioned the leadership of Laura Morris, who served as chair of the alliance, and a loud argument ensued. As a result of this incident, Morris resigned from her leadership role and complained to the UT Board of Trustees. Board Vice Chair Jim Murphy ruled that Carol Petersen was, henceforth, banned from all contact with UT donors and staff members.[89] Morris subsequently agreed to resume her role as chair of the alliance.[90]

In December 2008, Petersen announced that UT faced $100 million in state-funding cuts. In response, he and all members of his executive staff agreed to take voluntary 5-percent pay cuts and relinquish their UT-provided cars, saving the university about $400,000 a year. Soon thereafter, the Board of Trustees mandated a first-ever five-year review of the president as a prelude to the July 2009 five-year anniversary of his hiring. In late January 2009, Petersen submitted his own analysis of his tenure in the job, which he viewed as highly successful.[91]

In a surprising development, Petersen announced on February 18, 2009, his intent to resign as president, effective March 1,[92] a week before he was to present budget plans to the Board of Trustees and four months before the end of five years on the job. So ended another UT presidency, but this one did not crash and burn as the two previous ones, Gilley's and Shumaker's, had. Prior to Petersen's resignation, rumors swirled about whether he would survive his five-year evaluation and be renewed for another term. Nationwide, the average tenure for university presidents had decreased (down to six and a half years in 2017),[93] so perhaps Petersen's departure after less than five years was not too surprising. Still, he likely could have retained his position if not for some unfortunate aspects of his term, especially the increase in the size of the UT system administration and the system's efforts to wrest control of decisions and operations that had previously fallen under the purview of campus administrations.

Another interim president came to the rescue, Jan Simek, a long-time professor of anthropology at UT, Knoxville. This was Simek's second emergency appointment, as he had been called on to serve as interim chancellor after Loren Crabtree suddenly left early in 2008. Petersen had hired a new chancellor, Jimmy Cheek, who arrived in February 2009, allowing Simek to retreat to his department for at least a few months. Simek was and is an excellent professor and superb administrator devoted to the university and willing to assume these difficult jobs in a time of need.

Materials Sciences at UT

Materials science has long been a foundational strength of ORNL, and, in the past thirty years, UT has built considerable strength in materials science, in part by leveraging the UT-ORNL Distinguished Scientist program to make seminal hires. The most crucial of these was Ward Plummer, who moved from the University of Pennsylvania to assume a distinguished scientist faculty position in 1992. Plummer was an excellent experimental materials scientist, focusing on surface physics. He worked on the electronic and vibrational properties of crystalline surfaces and pioneered the use of surface science analytical tools (e.g., field emission spectroscopy and electron energy-loss spectroscopy) for his research. He was elected to the National Academy of Sciences in 2006, a profound honor for the best and brightest among the country's scientific community.

Probably Plummer's proudest legacy was the researchers he recruited to join him in UT faculty roles, ORNL staff positions, or both via joint faculty positions. He led the addition of six faculty to the Department of Physics and a like number in other university departments, usually in joint roles with ORNL.[94] He was perhaps the best model of what the two institutions wanted to achieve by hiring a distinguished scientist: Plummer served as a magnet for new programs and people at both institutions. In 2001, Plummer took the lead in starting UT's Tennessee Advanced Materials Laboratory, which arced across a number of departments and helped to attract funding for new research instruments, facilities, and people. This morphed into the Joint Institute for Advanced Materials (JIAM), which was initiated in 2005 (with Plummer a driving force and director) by $30 million in federal and state funding to construct a new building on UT's campus.

However, as sometimes happens, an intense "mover and shaker" can grow dissatisfied with the fact that the institutions are not moving ahead as quickly as he desires. This happened with Plummer, who left Tennessee in 2009 and moved to LSU to assume a chaired position in physics and serve as a special assistant to the vice chancellor for research. At LSU, Plummer was asked to repeat what he had accomplished at UT and ORNL and build materials science strengths in a variety of departments, which he did well until his sudden death in 2020.

Another distinguished UT-ORNL joint faculty, George Pharr, who had spent a decade in a dual research role between the two institutions, succeeded Plummer as JIAM director. In 2014, Pharr, who was elected to the National Academy of Engineering, is a man with many honors, but maybe the best known had nothing to do with materials science. In 2015, Pharr was featured in a thirty-second video clip, shown during televised UT football games, playing chess with Peyton Manning, then a quarterback for the Denver Broncos and a local icon after a record career at UT.[95] It was not clear who was the better chess player.

PHIL BREDESEN: A GOVERNOR WITH A KNACK FOR POLITICS AND A BRAIN—AND PASSION—FOR SCIENCE

Phil Bredesen was elected governor of Tennessee and served two terms from 2003 to 2011. He was an anomaly in this role in several ways. First, he was a twice-elected Democratic governor in a state dominated by Republican voters. Secondly, he was a physics major in college. Bredesen grew up in a lower middle-class family in the northeast, earned a scholarship to Harvard, and decided to major in physics after working for a summer as an undergraduate at Cornell University's High-Energy Synchrotron Source. After being installed as governor, Bredesen opened an accelerator-physics conference in Knoxville and spent most of his welcoming talk discussing the accelerator he had worked on at Cornell decades earlier. This greatly impressed the scientists in attendance.

After college, Bredesen dabbled in politics and worked on the campaign of Minnesota Senator Eugene McCarthy for the Democratic presidential nomination in 1968. He then started working in the healthcare field and moved to Nashville, Tennessee, a national hub of the healthcare industry. Bredesen founded the HealthAmerica Corporation, which he sold in 1986. He then served as mayor of Nashville from 1991 to 1999 and later as governor. Since 2011, he has chaired the Nashville-based Silicon Ranch Corporation, the U.S. solar platform for Royal Dutch Shell and one of the largest independent solar power producers in the United States. After an unsuccessful run for a U.S. Senate seat in 2018, Bredesen founded Clearloop, a renewable energy startup.

Bredesen's two terms as Tennessee's governor began in 2003 with a predicted state budget shortfall of $800 million due primarily to the overextension of the state's healthcare system (TennCare). His predecessor, Don Sundquist, had hoped to remedy the budget deficit by initiating a state income tax, but he narrowly lost that legislative vote. Thus, Bredesen had to start his tenure by cutting the TennCare rolls significantly. The Great Recession of 2008, which, again, necessitated drastic budget cuts, beset his second term.

In spite of difficult financial times, Bredesen was a friend to education. He especially promoted ways to enhance the partnership between UT and ORNL. In 2005, Bredesen pledged $10 million to initiate a program to hire top scientists into well-funded UT-ORNL joint positions—the Governor's Chair program.[96] This program became an important new critical connection that helped both institutions recruit top talent for joint research programs.

Governor Bredesen had other significant impacts on the UT-ORNL partnership. He provided seed funding for the UT-based Tennessee Biofuels Initiative,[97] which shared its scientific mission and direction with the DOE-funded Bioenergy Science Center at ORNL. In 2009, Bredesen took the lead on allocating federal American Recovery and Reinvestment Act funds to the state to build the 5-MW West Tennessee Solar Farm coupled to the Tennessee Solar Institute.[98] The solar farm and institute focused on job creation, education, renewable power production, and technology commercialization in the field of solar energy.

Perhaps Bredesen's last big idea as governor was to suggest that UT and ORNL cooperate to initiate a joint doctoral program in the interdisciplinary field of energy science and engineering. Lee Riedinger was tapped to initiate this new PhD program, and the UT-ORNL center that, in 2010, became the program's academic home and is called the Bredesen Center in tribute to the former governor. The story of these successes is told more fully in chapter 9.

A CADRE OF WORLD-CLASS RESEARCH SCHOLARS BOLSTERS THE UT-ORNL PARTNERSHIP

The Governor's Chair program was initiated in 2005 with the help of Governor Bredesen and provides funds to attract top research scholars into joint UT-ORNL positions in fields of common strength and interest. It succeeded the Distinguished Scientist program initiated twenty years earlier.

In 2006, Jeremy Smith, a British-born computational molecular biophysicist, was the first Governor's Chair hired.[99] He was attracted to this joint position largely because of the region's supercomputing assets, neutron scattering capabilities, and the increasing emphasis on molecular dynamics. Amid the COVID-19 pandemic that would dominate nearly every aspect of American life in 2020–21, Smith's computer simulations of biological macromolecules for early-stage drug discovery helped guide pharmaceutical companies on the dynamics of different protein targets important for the eventual treatment and prevention of this potentially deadly virus.[100] As of 2019, Smith has published well over four hundred peer-reviewed scientific articles.

By 2020, ORNL and UT, Knoxville appointed fifteen accomplished researchers to Governor's Chair positions, with two more appointed to the national laboratory and the UT Health Sciences Center in Memphis.[101] Half of these appointments fall in the broad science of "materials properties," which has been a research strength of ORNL and UT for decades. Each of the seventeen Governor's Chair appointees is an established leader in their

YILU LIU. COURTESY OF OAK RIDGE NATIONAL LABORATORY.

STEVE ZINKLE. COURTESY OF OAK RIDGE NATIONAL LABORATORY.

SURESH BABU. COURTESY OF OAK RIDGE NATIONAL LABORATORY.

own field and a stimulus for ongoing research at the university and the national laboratory. Among them is Yilu Liu, an electrical engineer specializing in smart-grid technologies in electrical power production and distribution. She focuses on developing new ways to monitor the flow of electrical energy through the nation's power grid and devising ways to make it more reliable.

Steve Zinkle, an expert on materials properties in high-radiation environments, is focused on developing high-performance radiation-resistant materials for advanced nuclear fission- and fusion-energy applications. Suresh Babu, a materials scientist, explores the design, production, and performance of materials used to create component parts for advanced manufacturing and additive manufacturing.

Avoiding Power Blackouts

The United States is a heavily electrified country and depends on a vast and a stable power distribution system. The electrical power grid is a sensitive system of millions of miles of power lines fed by thousands of power-generating stations and tapped by millions of customers using the electrical power. Instabilities occur regularly and sometimes cascade into major blackouts, like the one that started in northern Ohio on August 14, 2003.[102] This was the world's second most widespread blackout in history and affected an estimated ten million people in Ontario and forty-five million people in eight U.S. states. A generating plant in Eastlake, Ohio, went offline amid high electrical demand, which strained high-voltage power lines that later went out of service when they came in contact with overgrown trees. The cascading effect that resulted ultimately forced the shutdown of more than 250 power plants and a power blackout for twenty-four hours affecting twenty million people in the northeast.

A challenge is to quickly detect a problem on the power grid—the sudden loss of a generating power plant, for instance—and take rapid counter measures in minutes to keep the grid balanced and operating. In addressing that challenge, Yilu Liu and her colleagues at UT and ORNL developed FNET,[103] a frequency-monitoring network of small inexpensive sensors located in more than a hundred places in the eastern half of the United States. All sensors are GPS synchronized and connected to computers at the two institutions. These sensors send the current frequency and voltage of the local electrical power ten times each second to central computers at UT and ORNL. In this way, a sudden outage of a generating station in some remote location can be detected by the resulting sudden dip in the frequency of the alternating current, for example, from 60.0 to 59.8 cycles per second, enough to cause a major stability problem in succeeding minutes. FNET allows one to watch on computer screens a wave of grid fluctuations travel quickly through various U.S. states as the disturbance spreads rapidly. Such constant monitoring of the grid's stability allows utilities to see a disturbance initiated thousands of miles away and take action within minutes to ramp up local power generation and prevent that distant outage from escalating into a major blackout.

At UT, Liu established the Center for Ultra-Wide-Area Resilient Electric Energy (CURENT), funded by the National Science Foundation. At ORNL Liu leads the associated DOE effort—GridEye—of considerable interest to the North American Electric Reliability Corporation.[104]

David Millhorn

David Millhorn, a Chattanooga native, received a bachelor's degree from UT, Chattanooga and a doctoral degree from Ohio State University.[105] He was a physiologist who spent much of his career on the faculties of the University of North Carolina and the University of Cincinnati. In 2005, he joined the UT system administration as vice president for research and economic development, a role he served in until his death in 2017.

Millhorn had a profound effect on the university and the partnership with ORNL. He led implementation of the Governor's Chairs program with joint

UT-ORNL appointments and spurred the collaborative UT-ORNL team involved in winning a $65-million NSF grant awarded in 2007 for UT's Kraken supercomputer. He spearheaded planning, development, and construction of the UT Research Park at Cherokee Farm, intended as a mix of university and public-private partnership research and development ventures. The anchor tenant on Cherokee Farm is the Joint Institute for Advanced Materials,[106] the first such joint facility not built on the Oak Ridge reservation. That building opened in 2015, and a half dozen industrial research operations now occupy Cherokee Farm. The most recent (2020) was a Volkswagen of America innovation hub to be focused on development of electric vehicles.[107]

SUMMARY AND LOOK FORWARD

The decade of the 2000s began an entirely new era in the long relationship between the university and the national laboratory. In 2000, UT-Battelle began its tenure as manager of ORNL, an arrangement unimaginable twenty or more years earlier. ORNL scientists had first suggested the involvement of UT in managing the laboratory as early as 1946, but the arrangement would not be realized for more than fifty years. Step-by-step advances in the university's expertise and reputation occurred over this half century, spurred by dynamic leaders and by regular and strategic state-level investments.

Indeed, four consecutive Tennessee governors invested directly in the UT-Oak Ridge partnership over the thirty years leading up to 2010 (Alexander, McWherter, Sundquist, and Bredesen), and the financial support would continue with Governor Bill Haslam (2011–19) and with Governor Bill Lee (2020–present). Probably no other state can claim this substantial and consistent level of investment of state resources in collaborative efforts between a federally supported laboratory and the state's educational institutions.

In many ways, the decade of the 2000s marked a major turning point for ORNL and for UT. ORNL's reputation improved rapidly in this decade, in part due to UT-Battelle's leadership. Large projects were awarded to ORNL and completed on budget and on time. SNS was a $1.4 billion construction project that began operation in 2006. The laboratory also became a supercomputing powerhouse in this decade, in part because of UT-Battelle's extensive renovation of ORNL's physical plant, resulting in the removal of many old buildings and construction of new ones.

A large modern computer facility was built as part of this renovation, complete with abundant power and cooling capabilities. The availability of this facility played a significant role in ORNL's selection as DOE's first Leadership Computing Center. UT was able to compete successfully for an NSF-funded supercomputer housed at the ORNL facility. Supercomputing became a vital critical connection between UT and ORNL.

This decade also saw the addition of a hundred new joint faculty between the two institutions, building the research expertise at both locations. As promised in the UT-Battelle proposal, three new facilities to house joint institutes were constructed at ORNL with state funds, and each played a prominent role in winning new research grants (at both institutions) and attracting new faculty. In this decade, Governor Bredesen funded and intiated a new program of jointly hired scientific standouts (Governor's Chairs). And, at the end of the decade, UT introduced a new joint doctoral program related to energy, further strengthening the link between the university and the national laboratory.

This decade of progress succeeded in part due to continuity of leadership at ORNL but despite uneven leadership at the top of the UT system. Over the decade, three successive ORNL directors sustained a consistent and successful emphasis on excellence at the laboratory in science, operations, and community outreach. Meanwhile, five presidents, three of whom resigned under pressure, administered UT. Nevertheless, under the capable leadership of Chancellor Loren Crabtree—a committed supporter of the UT-ORNL partnership—UT's Knoxville campus continued to thrive through much of the 2000s, and the pace of success would only accelerate for the university and the laboratory in the decade to come.

9

THE 2010S

Joint Scientific Expertise and Critical Leadership Build New Degree Programs and Bolster a Shared Future

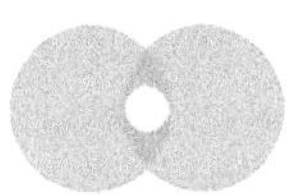

THE FORMATION OF UT-Battelle LLC and its success in winning the contract to manage ORNL led to many advances at both UT and ORNL through the decade of the 2000s and continuing into the 2010s. Strong, capable people in strategic leadership roles are usually the catalyst for success, and this certainly proved true for UT and ORNL through this decade. Joe DiPietro served as UT president from 2011 to 2018, bringing stability and spurring progress. UT, Knoxville Chancellor Jimmy Cheek had significant positive impact on the campus during his tenure from 2009 to 2017. He took the lead on a sorely needed campus building program and played a crucial role in the initiation of the UT-ORNL Bredesen Center and its two joint interdisciplinary PhD programs. He hired Taylor Eighmy as his vice chancellor for research in 2012, and Eighmy took the lead on achieving a national win in advanced manufacturing with creation of the Institute for Advanced Composites Manufacturing Innovation (IACMI)[1] in partnership with ORNL and with funding from the federal government and matching entities.

At ORNL, Thom Mason ended his decade as laboratory director in 2017; Thomas Zacharia succeeded him. Mason had led the unique SNS project, and Zacharia had put ORNL on the world map in supercomputing.

Transitions affect institutions in different ways. By the end of the 2010s, retirement came for many of the people who had spent years working to build the UT-ORNL partnership. The UT president (DiPietro) and chancellor (Cheek) retired and were replaced. Senator Howard Baker died in 2014, and Senator Lamar Alexander decided not to stand for reelection in 2020. These transitions seem to represent the loss of leadership for Tennessee at the federal level, which Baker fostered beginning decades earlier. A national descent into pitched political partisanship at the end of the decade certainly affected the country and the state, and its future impact on the university and the federal facilities in Oak Ridge is not yet clear. Transitions, even the more graceful ones, often bring uncertainty.

FORMATION OF THE CENTER FOR INTERDISCIPLINARY RESEARCH AND GRADUATE EDUCATION

Tennessee Governor Phil Bredesen worked in multiple ways to build UT's reputation and expertise and had considerable clout as the official chair of the university's Board of Trustees. As discussed in the previous chapter, Bredesen funded a strategic Governor's Chair program to bring leading researchers in science and engineering fields to high-level joint appointments between UT and ORNL, and the first was hired in 2006.[2] He also funded a biofuels R&D program[3] that linked well to the new DOE-supported BioEnergy Science center led by ORNL. He used funds from the American Recovery and Reinvestment Act of 2009 to initiate the Tennessee Solar Institute.[4] He understood that one way to raise the ranking of science and engineering departments at UT was to leverage even more strongly the people, programs, and facilities in Oak Ridge.

Bredesen's last big idea as governor was to enhance the partnership between UT and ORNL by establishing an academic connection between the two institutions that built on the research links that had steadily grown over the decades and the management link that had come via UT-Battelle's management and operation of the laboratory. In late 2009, Bredesen entertained various ideas on how to achieve this new academic connection, and he received different, sometimes conflicting, proposals to that end. UT System Vice President for Research David Millhorn envisioned the establishment of a new UT campus at ORNL operated by the UT system administration.

The new UT campus at the laboratory would be parallel to the existing UT campuses in Knoxville, Chattanooga, Martin, and Memphis.

Governor Bredesen was also hearing from UT, Knoxville, Chancellor Jimmy Cheek, who advanced his own vision for an educational presence in Oak Ridge: leverage the strong research overlap between his faculty and students and ORNL researchers and build new joint PhD programs relating to energy (one of the governor's important issues), as well as other scientific fields. This difference of opinion between UT, Knoxville and the UT system brought to mind Yogi Berra's fourth most quoted saying, "It's like *déjà vu* all over again"[5] Indeed, the previous chancellor, Loren Crabtree, had argued constantly with his boss, UT President John Petersen, over the issue of who controlled the UT-ORNL partnership. The frequent confrontations over this and other issues of control led to Petersen forcing Crabtree's departure in January 2008.

The governor's office was confused about the competing ideas coming from Knoxville on this topic. Finally, in November of 2009, someone from Bredesen's office called Margie Nichols, the UT, Knoxville vice chancellor for communications. She tracked down Jimmy Cheek at the University of Cambridge in England where he was attending "The Legacy of the Cold War," a conference jointly sponsored by the Baker Center at UT and the Churchill Archive Centre of Cambridge. Nichols relayed the confusion in the governor's office over which plan the governor should support and asked Cheek what Thom Mason, ORNL director, thought about these two conflicting plans. Fortunately, Mason was also in Cambridge to participate in this event, which allowed Cheek to quickly reaffirm that Mason supported the chancellor's ideas.

Over the next few days, Cheek and Mason worked with their staffs back in Tennessee to write a short joint plan for the new academic program they would work to build. The current UT president was Jan Simek, UT distinguished professor of anthropology, who had been pressed into service as interim chancellor after Crabtree's forced departure and then as interim president after Petersen left. Simek supported the Cheek-Mason plan, as did Jim Murphy, the vice chair of the UT Board of Trustees (and a friend of the governor). To quote another favorite Berra saying, "You've got to be very careful if you don't know where you are going, because you might not get there."[6] Now the two institutions and the governor's office had a plan.

Governor Bredesen decided to support the idea of UT, Knoxville and ORNL establishing a joint center and initiating some kind of interdisciplinary PhD program. He came to Knoxville in early December 2009 to discuss his intent to call a special session of the legislature in January to address key issues of higher education in Tennessee. He agreed with Cheek and Mason on a plan for a new interdisciplinary center linking UT, Knoxville and ORNL and offering a joint PhD program in energy science and engineering. As for the cost, Cheek and Mason argued for $6.2 million in one-time funding to kick-start this program and provide partial funding over five years, in addition to support that would come from UT's and ORNL's funded research programs. Bredesen agreed with this plan and moved forward with it.[7] [8] [9]

A special session of the Tennessee Legislature in January 2010 resulted in the passage of the Complete College Tennessee Act of 2010. The purpose of this act was not only to increase the number of college graduates in Tennessee but also to "elevate the status of the University of Tennessee, Knoxville, as a top-tier national research institution through expanded collaboration with the laboratory." Pertaining to the UT-ORNL partnership, two key directions are specified in the act:

> (a) The University of Tennessee is authorized to establish an academic unit of the University of Tennessee, Knoxville, for interdisciplinary research and education in collaboration with the Oak Ridge National Laboratory.
> (b) The chancellor of the University of Tennessee, Knoxville, and the director of the Oak Ridge National Laboratory are authorized to enter into an agreement concerning collaboration in interdisciplinary research and education designed to accomplish the purposes of this part. The provisions of this agreement shall address matters including, but not limited to, the appointment and oversight of graduate students, the appointment of ORNL staff as faculty, and the development of interdisciplinary curricula between the two institutions.[10]

To support this new center and doctoral program, Governor Bredesen inserted $6.2 million in his final administrative budget request, funding that the legislature approved and the governor signed;[11] the funding was made available in the UT budget of July 1, 2010.[12] The die was cast and the new program was on its way, but the details of the new center and doctorate program had yet to be discussed. Support for this new doctoral program was clear at ORNL, as Director Thom Mason spoke highly about how the greatly expanded number of graduate students working on research at the laboratory would benefit ORNL programs.[13]

In January 2010, Chancellor Cheek and ORNL Director Mason appointed a task force to design the new center and develop the new doctoral program. This thirteen-member group included representatives from four colleges at UT (agriculture, arts and sciences, business, and engineering); the dean of the graduate school; and four associate laboratory directors at ORNL. Wayne

Davis (dean of the UT College of Engineering) and Jim Roberto (ORNL associate laboratory director for partnerships) chaired the task force. Both were long-time members of their institutions, and both had worked for years in numerous ways to strengthen the partnership.

The job of the task force was to design the new PhD program relating to energy and the center in which it would be based and administered—and to accomplish these tasks quickly. The development of new doctoral programs usually takes years at a university bound by tradition. Indeed, creating a new interdisciplinary degree program in an overwhelmingly disciplinary institution like a university proves a real challenge, which is why such programs are rare. But, in this case, the governor wanted this done rapidly, while he was in his last year of office.

The task force worked hard through the spring of 2010, studying, for example, other universities that had created interdisciplinary graduate degrees. The task force talked to potential customers—among them, companies and research organizations that might have interest in hiring someone with an interdisciplinary doctorate in energy. The question of employability is legitimate, as universities are historically disciplinary, with colleges aligned along traditional departmental lines, usually unchanging, and guided by a strong tendency to hire faculty with degrees in those traditional areas.

By comparison, national laboratories are more often interdisciplinary, as the nature of the national problems they are charged to investigate usually cuts across traditional fields. Plus, national laboratories often redraw their divisional boundaries to attack the current national science and technology challenges and to respond rapidly to funding opportunities. Hiring people with interdisciplinary doctorates is generally easier in a national laboratory. The task force was encouraged by the private sector's expressed interest in hiring people with a breadth of doctoral education somewhat wider than normal. Still, the devil lurked in the details.

By late May 2010, the task force had a preliminary design for the Center for Interdisciplinary Research and Graduate Education (CIRE) and a doctoral program in energy science and engineering. Both marked new ground for UT. A doctoral program spanning four colleges was never contemplated before, and a center administrating a degree program was an entirely foreign concept (degrees are traditionally issued by academic departments and colleges). Davis and Roberto presented these plans to Cheek and Mason, who were pleased with the result of the task force's efforts.

The next big step was finding a director for this fledging center, someone who could develop these concepts into detailed plans and then lead them through the university's and state's long approval processes—and do all of this before Bredesen left office. An internal search (UT and ORNL) was opened for the CIRE director in early June. Soren Sorensen was a member of the task force and the head of the UT Department of Physics, a post he had held since Lee Riedinger transferred to ORNL as deputy director in 2000. On June 7, Sorensen called Riedinger and encouraged him to apply for the position.

Riedinger was initially wary of performing yet another administrative job, as he had enjoyed his return to full-time teaching and research in fall 2007 after holding a number of administrative posts over the previous thirty years. Maybe age sixty-five was the time to "coast to retirement." However, the opportunity and the sense of duty were difficult to ignore. To have a director who understood the people and the processes of both the university and the national laboratory was important, and Riedinger realized that he had the best such dual experiences and felt he really needed to pursue this new post. He applied, was interviewed on July 21, and, on August 12, Chancellor Cheek offered him the position.[14] Riedinger's new role as CIRE's first director officially began on September 1, but he had his first strategy session with Cheek on August 19.

A VITAL LESSON FROM SENATOR BAKER: ALWAYS "COUNT THE VOTES"

Governors expect rapid implementation of their policies and programs, and Phil Bredesen was no exception. It was his idea to initiate some kind of new doctoral program related to energy and involving UT and ORNL in some intimate way. He had provided start-up funds for this new program ($6.2 million), and he expected the center to be operational and fully staffed before he left office in January of 2011. This may sound straightforward, but the speed of new program implementation at universities is generally glacial, not gubernatorial (when a strong governor expects quick action).

There were many steps of approval along the way to implementing a new center and a new doctoral program, and the two in question were novel (and quite suspicious to some) and, therefore, more difficult to achieve with haste. The energy science and engineering PhD program had to be reviewed and approved at all levels so that a faculty could be assembled and the national recruiting of doctoral students could begin in January of 2011. Riedinger's first step was to take the task force's draft plans for the center and a doctoral program and work these into proper shape for going through the chain of approvals. The first such step was a meeting with the UT Graduate Council on September 23, 2010, and its members voted their approval of the new PhD program after a lengthy discussion of the details. The next step was sending the descriptive package to the UT Faculty Senate in preparation for the all-important meeting of that body on October 18.

Storm clouds appeared on the horizon as members of the Faculty Senate delved into the proposed interdisciplinary doctoral program, not housed in a traditional academic department but operated by a new center. Some senate members were suspicious of a program so strongly connected to ORNL, while others supported this. One senate member spent a great deal of time reading and studying the proposal and subsequently wrote a long analysis and critique, demanding a number of changes that pertained to, among other issues, who would serve as faculty and how they would be selected. The critical senate member insisted that only senior UT faculty review and approve applications from researchers at UT and ORNL to be part of the CIRE faculty. However, this did not seem appropriate as this was meant to be a *joint* doctoral program with faculty appointed from both institutions and, thus, respected senior researchers at UT *and* ORNL should make these decisions. A compromise was reached such that the chancellor and laboratory director would each appoint four senior researchers to make these faculty appointments.

Arguments continued on four other points of contention relating to more technical issues of the proposed doctoral program, and five amendments to greatly alter the program were to be presented at the imminent Faculty Senate meeting on October 18, when the full proposal would be voted upon.

At this stage, Riedinger resorted to a technique he had learned from Senator Howard Baker "to count the votes"—in effect, getting a read on how many members of a voting body would support a proposed piece of legislation—before a crucial tally is taken. In Riedinger's year (1983–84) working as Baker's science advisor, several contentious issues were voted on in the U.S. Senate, one being the hotly contested proposal to create a national holiday honoring the Reverend Martin Luther King Jr. As an effective majority leader, Baker was careful to talk personally to every senator who was opposed or on the fence, to lay out his arguments and lobby for support. When he had counted the votes and was sure of a successful outcome, he brought the legislation to the floor of the Senate, and approval was granted. Ira Shapiro discusses Baker's consequential run as Senate majority leader in the book *The Last Great Senate: Courage and Statesmanship in Times of Crisis* (PublicAffairs, 2012).

Now, as Riedinger faced a make-or-break determination by the Faculty Senate, it was time for him to count the votes. He talked to and/or visited over several days all members of the Faculty Senate to lay out the plan for the energy science and engineering PhD program, address the objections, and ask for their support. In the end, this strategy worked, as the Faculty Senate voted strongly for the doctoral proposal on October 18. Chancellor Jimmy Cheek's presence at this meeting provided a strong signal of his support. "Counting the votes" had worked.

The next step in this process was a presentation to the Academic Affairs and Student Success Committee of the UT Board of Trustees on October 21, 2010. The committee approved the new program, and the full board agreed the next day. This then allowed the submission of the proposal to the Tennessee Higher Education Commission (THEC). Officials from THEC came to Knoxville on November 16 along with two consultants hired to review the proposal and write a critique and present recommendations to THEC. Both consultants (Marilyn Brown from Georgia Tech—previously an ORNL employee—and Phil Parrish from the University of Virginia) liked what they read and heard; they then wrote a report of strong support with good suggestions on how to improve the plan.[15]

THEC gave approval to the doctoral program on January 12, 2011, opening a new critical UT-ORNL connection built around a joint interdisciplinary center and PhD program. The full review and approval of this novel program were accomplished in the fall semester, at the speed of light, compared with the normal academic process that would have taken several years. It is amazing how rapid an approval process can result from a good idea, new money, and strong support from the governor to the chancellor. This expedited approval process enabled the formation of the initial faculty just before Christmas 2010, even before THEC's formal approval of the program. Having a faculty also led to initiation of an aggressive advertising program to recruit applicants for this new doctoral program from students across the country. The retiring Governor Bredesen was happy about all of this. The inauguration of his successor, Bill Haslam, occurred on January 11, 2011.

A FORMER GOVERNOR LENDS HIS NAME TO A NEW CENTER

Recruiting for this new doctoral program happened in a fast and focused way through the spring of 2011. The first class of graduate students enrolled in the energy science and engineering program numbered eighteen, and school started in mid-August 2011. The recruited class was drawn from across the country and represented academically strong students who had earned their undergraduate degrees from respected schools. Three of the eighteen came with highly competitive graduate fellowships from NSF. Of the original students, all but four would go on to earn their PhDs—an impressive success rate.

The program was built to be interdisciplinary, whereby students could take graduate courses in different departments (in a typical PhD program, a required number of credits must come only from one). The program had a technology-development emphasis, and students could take courses related to entrepreneurship and even explore how to transfer intellectual property (IP) or start companies, drawing on the wealth of relevant IP resources at ORNL.

The program was inextricably linked to ORNL, encouraging students to conduct their dissertation research—in various disciplines related to energy generation and use—at the laboratory, with their ORNL advisors serving as official UT faculty overseeing their work. This was especially attractive since, by this time, ORNL had evolved to be one of the top energy laboratories in the country, pursuing research into batteries and fuel cells, biofuels, climate change, nuclear power, and other energy-related areas. The program also had a policy component, which allowed students to work on the many policy implications of energy technologies and their uses (or abuses).

Chancellor Cheek wanted to keep Bredesen involved in the program that he had inspired and funded. Cheek invited the former governor to visit UT, learn about the status of the program, and meet the graduate students, but the real purpose of Bredesen's invited visit would be to convince him to attach his name to the newly established center. Bredesen's long-time colleague and advisor Jim Murphy warned Cheek that Bredesen would not allow his name used for the center, as it was not his style to have his name attached to state entities or monuments. Nevertheless, Bredesen's and Murphy's visit was arranged for October 7, 2011.

A few days before the visit, Chancellor Cheek asked Riedinger to bring some of the first-year graduate students to the meeting to impress Bredesen with the success of the rapid national recruiting effort. Riedinger chose four CIRE students: Callie Goetz (from Middlebury College), Kyle Sander (Oregon State), Kemper Talley (Clemson), and Stephen Wood (Florida International). On the evening of October 6, Riedinger met with the four students to explain who Bredesen was and the purpose and importance of the meeting the following day. In the middle of Riedinger's explanation, Goetz interrupted to say that she already knew the governor. Mystified, Riedinger asked where and how they had met. Goetz explained that her father was Dave Goetz, who had served as Bredesen's commissioner of finance for eight years in state government. Bredesen had, in fact, known Goetz since she was a child.

PARTICIPANTS IN OCTOBER 7, 2011, MEETING. FRONT ROW, FROM LEFT: STUDENTS STEPHEN WOOD, KEMPER TALLEY, CALLIE GOETZ, AND KYLE SANDER. BACK ROW, FROM LEFT: RIEDINGER, JIM ROBERTO (ORNL), PHIL BREDESEN, JIMMY CHEEK, AND JIM MURPHY. COURTESY OF BETSEY B. CREEKMORE SPECIAL COLLECTIONS AND UNIVERSITY ARCHIVES, UNIVERSITY OF TENNESSEE, KNOXVILLE.

The next afternoon, Chancellor Cheek was nervous about the meeting after Murphy's warning against expecting success on renaming the center for Bredesen. However, when Bredesen entered the conference room and saw Callie Goetz, he lit up, relaxed, and enjoyed learning about the program and talking to the four students about what they were working on. At the end of the meeting, Bredesen assented to the chancellor's request that the new center bear his name. The presence of Callie Goetz clearly broke the ice with the governor and was a key factor in keeping Bredesen engaged in a center to be named after him. Sometimes, it is better to be lucky than good in the selection of graduate students. A formal letter requesting the name change to the Bredesen Center for Interdisciplinary Research and Graduate Education was sent to UT President Joe DiPietro on October 19.[16]

The UT Board of Trustees had to approve the name change, and, to achieve this, Chancellor Cheek had to resort to his own version of "counting the votes." Bredesen would go along with using his name on the center, as long as there would be no controversy or even negative votes when the board convened to consider the issue. So, Cheek went to talk to every board member to ask for support for this change and to point out that Bredesen's name had not been (nor would be) used on any other state entity. Cheek's strategy worked, and the board voted unanimously to approve the change.[17]

In April 2012, Bredesen made the first of what would become annual springtime visits to "his" center to take part in the final segment of Riedinger's two-semester energy technology course. This segment was focused on energy policy and followed an entire academic year's study of the details of various energy technologies. Each year, the graduate students formed into groups and spent a month preparing a policy report on an energy technology of their choosing. Then, at the end of the semester, each group would have ten minutes in a fanciful "Oval Office" to present their national energy policy proposal to stand-ins for the U.S. president, vice president, and secretary of energy. Each year, Riedinger appointed Bredesen to serve as the president, the ORNL director as the vice president, and the UT, Knoxville dean of the engineering college as the energy secretary. With an undergraduate degree in physics and a late-career focus on energy issues, Bredesen was well prepared technically to play this role, and, of course, he understood well the politics and social issues relevant to every energy discussion. He enjoyed playing this role and interacting with the graduate students concerning their policy proposals and continues in this annual process at the writing of this book.

A Notable Policy Panel Stand-in

The ORNL director was not able to participate in the student energy policy presentations in 2013, but Bill Richardson was visiting UT to give a talk at the Baker Center on the morning of April 25.[18] Richardson's credentials were impeccable: he had served President Bill Clinton as the ambassador to the United Nations (1997–98) and then as secretary of energy (1998–2001). In fact, during Richardson's time heading DOE, UT-Battelle LLC won the contract to manage ORNL starting in 2000. He was later elected to two terms as governor of New

FORMER GOVERNORS BILL RICHARDSON AND PHIL BREDESEN AT UT FOR BREDESEN CENTER CLASS—APRIL 25, 2013. COURTESY OF BETSEY B. CREEKMORE SPECIAL COLLECTIONS AND UNIVERSITY ARCHIVES, UNIVERSITY OF TENNESSEE, KNOXVILLE.

Mexico (2003–11). Riedinger attended Richardson's presentation and invited him to be part of the "Oval Office" panel that afternoon.

The student presentations and panel discussion on the afternoon of April 25, 2013, were magical. Bredesen and Richardson greatly enjoyed sitting next to each other and participating, in part because they knew each other well, having served the same eight years as Democratic governors. As usual, Riedinger appointed Bredesen to serve as the U.S. president.[19] He asked Richardson to serve as the vice president, and, although Richardson initially feigned indignation at being outranked by Bredesen, he assented to the assigned role.

The session lasted for two hours as the five groups of students made their policy presentations and were then quizzed by the panel.[20] Bredesen and Richardson have different personalities, the former quiet and deliberative in his thinking and engagement, whereas the latter is outgoing and quick to comment and joke. After one student made a presentation about how to address disposal of spent fuel rods from nuclear reactors, Richardson slammed his hand on the table and said, "That is the best set of ideas on nuclear waste that I have ever heard!" Everyone present knew that Richardson, as secretary of energy, had been responsible for the disposal of these troublesome wastes, so Richardson's enthusiastic response delighted the graduate students.

Jimmy Cheek

In October 2008, the UT Board of Trustees selected Jimmy Cheek as the next chancellor of the Knoxville campus, and he assumed the position on February 1, 2009.[21] Cheek spent most of his career at the University of Florida, rising to the post of senior vice president for agriculture and natural resources. With Cheek's arrival, the campus' faculty and administrators hoped for a return to stability following the abrupt departure of Loren Crabtree in 2008.

Cheek indeed brought stability and achieved remarkable progress on many fronts through his eight years in the position. Perhaps his two biggest accomplishments as chancellor pertained to a revitalized campus and a further strengthening of the partnership with ORNL, especially through the Bredesen Center and its two interdisciplinary doctoral programs—in energy science and engineering and data science and engineering[22]—administered jointly by the university and the national laboratory. Cheek's tenure saw more than $1 billion in campus infrastructure investments[23] and growth in university fundraising. This major renovation of the UT campus occurred in parallel with a huge $1.8-billion modernization of the ORNL facilities.[24]

Cheek met with outgoing interim UT Chancellor Jan Simek on February 1, 2009, Cheek's first day on the job. Simek told Cheek that current UT President

JIMMY CHEEK. COURTESY OF BETSEY B. CREEKMORE SPECIAL COLLECTIONS AND UNIVERSITY ARCHIVES, UNIVERSITY OF TENNESSEE, KNOXVILLE.

Petersen's contract for another five-year term would not be renewed and that Simek would likely be appointed interim president for two years, thus making him Cheek's boss. Simek's prediction did transpire, and the transition in presidents played an important role in Cheek's early successes.

The Bredesen Center's innovative new programs would not have happened or succeeded without Cheek's constant support and the behind-the-scenes backing by Simek. President Simek, like Cheek, strongly agreed that the Bredesen Center and the energy science and engineering PhD program should be administered through a partnership between ORNL and the UT, Knoxville campus, not the UT System, as some had proposed. The center and the doctoral programs broke new ground at the university and spawned a new form of partnership between the university and the laboratory. Cheek enjoyed an excellent working relationship with ORNL Director Thom Mason, which also helped the Bredesen Center to flourish.

Cheek stepped down as chancellor in 2017 and became a distinguished professor in the Department of Educational Leadership and Policy Studies.[25] [26] His contribution to the campus has continued through graduate instruction, service on a host of committees and boards, and his ability and willingness to advise new chancellors and presidents.

THE BREDESEN CENTER BRINGS A CADRE OF TOP-NOTCH SCHOLAR-RESEARCHERS TO EAST TENNESSEE

One hope in establishing the Bredesen Center and an interdisciplinary doctoral program in energy was that they would lead to increased recruitment of top-notch graduate students from across the country. In fact, this worked, and successes came by virtue of this unique academic/research partnership between UT and ORNL. The first graduate students enrolled in August 2011, and, within six years, a total of thirty-two doctorates were awarded through the program, with twenty-five of these students conducting their PhD research at ORNL and seven on the UT campus. After finishing their doctorates, sixteen went to work in industry (including three in local start-up companies and one in the Apple research laboratory in Cupertino, California), nine took positions at ORNL (either short-term postdoctoral positions or regular research staff jobs), and seven went on to postdoctoral research positions at other universities. The interdisciplinary training produced positive impacts on these graduate students' early careers.

The Bredesen Center leadership stressed the importance of personal interviews in Oak Ridge and Knoxville with student applicants for the program. It was (and remains) important to identify students who really wanted to enroll in an interdisciplinary doctoral program and separate them from those who are attracted primarily by the generous fellowship stipends of up to $30,000 per year, plus tuition—awards that were noticeably higher than the national average when the program started in 2011. Further, assessing the candidate's preparation for and interest in research related to energy topics is best done in person. Mallory Ladd, a highly rated chemistry major at the University of Toledo and a star volleyball player, came with fifty other prospective graduate students for interviews on February 25 and 26, 2013.

Two Special Graduate Students

Mallory Ladd was the recipient of one the Bredesen Center fellowship offers sent out to thirty-five students a week after the interviews in February 2013. Over the following month, the center received many acceptances, forming what was shaping up to be an excellent class. However, one of the top recruits, Ladd, was on the fence about coming to the UT-ORNL program or accepting a competing offer from the University of California, Berkeley. The Bredesen Center offered to pay for a second trip so that she could talk further with potential research mentors at ORNL.

MALLORY LADD. COURTESY OF OAK RIDGE NATIONAL LABORATORY.

She was interested in working in the area of climate change, and Berkeley's program in that field was, and remains, quite prestigious. She drove from Toledo, spent all day on Friday, April 12, 2013, at ORNL and was ready to drive home on Saturday, still undecided. Riedinger talked with her on Friday evening and convinced her to come to his home on Saturday morning to have pancakes before driving north. The pancakes Riedinger made were apparently good, as was the discussion of future directions for the graduate program, and, on the drive home, Ladd sent Berkeley an email declining its fellowship offer. The class was now complete.

Ladd excelled as a graduate student and perfectly exemplified the caliber of person whom the university likely could not have recruited without this special doctoral program. The following year, she won a prestigious NSF Graduate Research Fellowship. In 2016, she was invited to be among one hundred graduate students attending the Lindau Nobel Laureate Meeting held in Lindau, Bavaria, Germany. This annual gathering brings together many Nobel laureates (thirty or more) to spend time with bright students invited from around the world. Ladd attended and ended up being the only student on a panel discussion on climate change with two Nobel laureates.[27]

After finishing her PhD at the Bredesen Center, Ladd took a position as a researcher at the Center for Naval Analyses (CNA), a nonprofit research and analysis organization located in Arlington, Virginia. Each Bredesen Center student is required to spend some time on policy studies or technology-transfer activities, in addition to their science and engineering courses and dissertation research. Ladd chose the policy track, which led to employment at CNA, where she focused on military operations and logistics, Arctic science, and climate security.

Ladd's significant other (now husband), Tony Bova, was part of the same Bredesen Center class, and his doctoral research at ORNL focused on the chemistry of lignin. In the production of biofuels from trees and woody plants, lignin, generally regarded as a waste byproduct, fills the spaces in the cell wall between cellulose, hemicellulose, and pectin components—the sugars that can be fermented into combustible fuels—and lends rigidity to the plant.

As a graduate student, Bova started Grow Bioplastics LLC (now called Mobius),[28] a company that produces high-performance plastic and rubber materials based on lignin, and is located not far from Oak Ridge and Knoxville. Bova was not unique in terms of his entrepreneurial bent. In fact, in the first six years of the Bredesen Center program, students started six high-tech companies based on their dissertation research. One (Andrew Lepore) won $50,000 in 2014 in the Next Big Idea competition at ORNL to support his start-up idea of a new way to convert biomass into bio-oil.[29]

The Bredesen Center graduate students are a diverse group of talented people. In the first six years of the program, 173 graduate students came from twenty-three countries and a third of the students were female. They performed research in areas such as nanoscale and macroscopic exploration of the properties of the ceramics and alloys important for the manufacture of more-efficient and higher-temperature automobile engines. Over the first six years, twenty-four of these students conducted research at ORNL's Center for Nanophase Materials Sciences, and many of them published their research findings in respected scientific journals. Other students conducted research on biofuels, batteries and fuel cells, environmental science, and nuclear energy from both fission and fusion.

The success of the Bredesen Center over its first decade resulted in part from the unique offering of joint PhD programs between the university and national laboratory, from excellent research opportunities at the two institutions, and from the dedication of a small staff to the welfare of the graduate students.[30 31] Early on, Riedinger brought into the center two crucial people, Wanda Davis as the UT-based business manager and Mike Simpson as the ORNL-based assistant center director. Simpson knew both institutions well as a joint faculty member and helped new graduate students find research mentors at ORNL. Davis not only took the lead on financial aspects of the Bredesen Center but also served as the unofficial mother figure to the graduate students. She knew all about their families and backgrounds, often brought them to her home for dinner, and helped foster a family atmosphere in the center. Every successful academic program needs such a unifying force and the resulting close-knit environment that results. Two annual gatherings for all Bredesen Center graduate students at Riedinger's home contributed to this family philosophy.

THE BREDESEN CENTER LAUNCHES A SECOND INTERDISCIPLINARY PHD PROGRAM

The success of the energy science and engineering doctoral program led to creation of a second Bredesen Center interdisciplinary doctorate between UT and ORNL. A joint task force formed in July 2015 studied the possibilities and eventually decided that data science and engineering should be the topic for this second doctoral program. The new PhD program's structure would be similar to that of the energy science program, with policy and entrepreneurship components incorporated into the curriculum, along with the core courses grounded in data science. This would be a field of study in which the university could fully leverage ORNL's world-class supercomputing assets.

Riedinger and Russell Zaretzki, a data analytics professor in the Haslam College of Business, took the lead on writing the proposal for this new doctoral program and guiding it through the same steps of the approval process that was required for the energy science and engineering PhD in fall of 2010. Following the Faculty Senate's affirmative vote, the UT Board of Trustees and THEC likewise voted to approve the proposal. Soon after, recruitment of promising doctoral students for the program began.

The only remaining question involved resources. A request for $6 million in state funds—the same amount that was requested for the energy science program—was made to kick-start this second Bredesen Center doctoral offering. However, by then, the political landscape had changed. Bill Haslam succeeded Bredesen as governor in January 2011. The Haslam family is based in Knoxville and has historically and generously supported UT. Nevertheless, Governor Haslam, a Republican, questioned whether he and some members of the legislature should support the special allocation of $6 million for a second doctoral program offered by a center named for his predecessor, a Democrat. Two key circumstances ultimately led to Haslam's approval of the funding.

First, in the spring of 2016, Chancellor Cheek and UT President Joe DiPietro placed funding for the new data science PhD program at the top of their request for new state funds for the coming fiscal year, over funding for new buildings, new programs, or other candidate projects. It came as no surprise that Cheek would place this program at the top of his list, as he had helped to create and subsequently nurture the Bredesen Center. However, it was less expected that DiPietro, the president of UT's statewide system, would agree to prioritize this funding request over the needs and requests of the other campuses.

To help persuade DiPietro and extend the benefits of the new program across the system, Cheek asked Riedinger to invite other UT campuses to participate in the data science doctoral program with ORNL. Subsequently, UT, Chattanooga, asked to join the program, as did the Health Sciences Center in Memphis. This broader inclusion in the program made it easier for DiPietro to prioritize the $6-million request of the state. This marked a major vote of confidence by UT leadership for the concept of building strength through increased partnership with the federal entities in Oak Ridge.

The other circumstance that ensured that this appropriation would be inserted into the governor's budget request was the close personal relationship between Haslam and Bredesen. In fact, Bredesen personally asked Haslam to support the $6 million line item in his next budget. Haslam complied, and the state legislature approved his budget request for the fiscal year beginning July 1, 2017. The UT Board of Trustees approved the plan for the new doctoral program on March 29,[32] the Tennessee Higher Education Commission approved the plan in May,[33] and the inaugural class of data science PhD students started in August 2017. The program was met with enthusiasm across Tennessee.[34]

JOINT UT-ORNL SCIENTIFIC ENDEAVORS ADVANCE IN THE DECADE

The UT-ORNL partnership advanced markedly in this decade, not only through UT-Battelle LLC's capable management and new academic programs offered through the Bredesen Center, but also through gains in a number of strategic research areas, including computing, advanced manufacturing, nuclear physics, and national security.

Computing

In 2010, the ORNL Jaguar, a supercomputer built by Cray Inc. and capable of 1.4 petaflops (a quadrillion floating-point operations per second) of computing speed, was ranked number one in the world. Jaguar was succeeded

ORNL SUMMIT SUPERCOMPUTER. COURTESY OF OAK RIDGE NATIONAL LABORATORY.

by Titan,[35] which utilized a hybrid Cray system composed of both central processing units (CPUs) and graphics processing units (GPUs), resulting in a theoretical peak performance exceeding twenty-seven thousand trillion calculations per second (27 petaflops). Then came the IBM-developed Summit, capable of 200 petaflops (each a thousand trillion floating point operations per second), making it the fastest supercomputer in the world from November 2018 to June 2020. Summit was the first supercomputer to reach exaflop speed (a quintillion operations per second), achieving 1.88 exaflops during a genomic analysis in 2018.

ORNL computational biologist Dan Jacobson, a key Bredesen Center faculty member and mentor to a number of graduate students in the center's two doctoral programs, led this particularly powerful calculation.[36] The large-scale genome sequencing projects of the last twenty years, including the Human Genome Project, have made a wealth of genetic data available to scientists. The challenge is to interpret and mine these mountains of genome data to reach new conclusions on gene function and protein production.

Jacobson's research team capitalized on the computing power of Summit to analyze plant genomic data and discover key regulatory genes of plant cell walls that can be manipulated to enhance production of biofuels and other bioproducts, an effort led by Jacobson's Bredesen Center student Debbie Weighill.[37] Supercomputers like Summit are poised to turbocharge researchers' understanding of genomes at a population scale, Jacobson maintains, enabling a whole new range of science that was simply not possible before the powerful tools' arrival.[38]

Advanced Manufacturing

During this decade, manufacturing science emerged as a major strength of UT and ORNL. One element of that strength is the NTRC, opened in 2000, which includes 65,000 square feet of laboratory space and sits on land halfway between the university and the national laboratory. It became DOE's only designated user facility focused on performing early-stage R&D on transportation technologies. The success of this collaboration was a factor in the next-door construction of the Manufacturing Demonstration Facility (MDF), a 110,000-square-foot facility housing tools that support the development of new materials, software, and systems for advanced manufacturing. MDF opened in 2012.

The world of manufacturing has changed significantly as new techniques of additive manufacturing, also known as 3D printing, have been developed. Using computer-aided design, additive manufacturing allows for the creation of objects with precise geometric shapes, built by depositing materials layer by layer, in contrast to traditional methods which remove bulk pieces of material by subtractive manufacturing of large hunks of that material. Among the advantages of additive manufacturing is the near absence of waste material. MDF tools have been developed to employ additive manufacturing to create objects small and large, the latter using the Big Area Additive Manufacturing (BAAM) system. As a demonstration project, MDF researchers and industry partners printed a replica of a Shelby Cobra, and President Barack Obama viewed the vehicle during a visit to East Tennessee on January 9, 2015.[39] [40] Though Obama was eager to take the vehicle out for a spin, the Secret Service refused to allow the president or his vice president, Joe Biden, to test drive the vehicle.

MDF is not a production facility but a research laboratory, as many issues must be investigated and solved before industry can widely adopt additive manufacturing. Research is taking place to develop, for example, new lower-cost, high-temperature materials for boilers, turbines, and other components of combustion systems that produce electricity. And, new bio-based high-temperature polymers and other composite materials will lead to creation of lightweight, energy-efficient vehicles that will also be highly recyclable.

SHELBY COBRA REPLICA MADE BY 3D PRINTING AT ORNL. COURTESY OF OAK RIDGE NATIONAL LABORATORY.

MDF quickly became a strength of both the university and the national laboratory, in part through the hiring of several UT-ORNL Governor's Chairs to help lead this research laboratory and recruit promising graduate students. Among the Governor's Chair hires are Suresh Babu, in 2013,[41] and Uday Vaidya, in 2015.[42]

The primary purpose of the Obama-Biden visit on January 9, 2015, was to announce funding for a national manufacturing hub. This new entity, IACMI, was established via a national competition and supported by a $70-million commitment from DOE's Advanced Manufacturing Office, plus more than $180 million committed from IACMI's partners. Support from IACMI partners comes from six states—Colorado, Indiana, Kentucky, Ohio, Michigan, and Tennessee—and a 122-member consortium that includes leaders from industry, universities (including UT), and ORNL and other national laboratories. The establishment of this center brought much national attention to UT and ORNL, as did the president's visit, and bolstered an important critical connection between UT, ORNL, and a host of institutions with expertise in additive manufacturing.

Taylor Eighmy, whom Jimmy Cheek hired as UT, Knoxville vice chancellor for research in 2012,[43] succeeding Lee Riedinger (an interim appointment for a year while also the Bredesen Center director), led the effort to draft the winning proposal for the new institute. Eighmy positively impacted the UT-ORNL partnership during his five-year stay at the university, before becoming the president of the University of Texas, San Antonio in 2017. He has maintained his connection to manufacturing and to UT and ORNL in this position, helping to organize the $111-million Cybersecurity Manufacturing Innovation Institute (CyManII). This public-private partnership, established in 2020, engages ORNL, Idaho National Laboratory, and Sandia National Laboratory in protecting and securing vital U.S. industries' supply chains and automated production facilities.

Nuclear Physics

The discovery of new elements has been an intriguing area of research for hundreds of years. In 1945, researchers at Clinton Laboratories, working at the Graphite Reactor, discovered element 61—a previous "hole" in the periodic chart—and later named it promethium (Pm). Some wanted to name this new element clintonium after the laboratory of its discovery, but the chance to name an element after a place or laboratory in Tennessee was lost at that time (see chapter 2). Success in terms of naming rights came seventy-one years later when element 117 was named tennessine in 2016.

The heaviest element found in nature is uranium (element 92), and elements beyond uranium on the periodic table, many of which have very short half-lives, have to be produced in the laboratory to be studied and used. Large accelerators and long experiments (many months) are required. A collaborative effort among researchers from the United States and Russia made the discovery of element 117 possible. ORNL provided the radioisotope berkelium-249 (^{249}Bk), which was bombarded with a beam of calcium-48 (^{48}Ca) at Russia's Joint Institute for Nuclear Research in Dubna, until a few atoms of element 117 were detected and measured.[44] ORNL's key contribution to

LEE RIEDINGER, YURI OGANESSIAN, AND KRZYSZTOF RYKACZEWSKI AT RIEDINGER'S HOME IN JULY 2016. IN POSSESSION OF THE LEAD AUTHOR.

the project was the production, radiochemical separation, and delivery of 22 milligrams of ^{249}Bk, a byproduct of the production of californium-252 (^{252}Cf) in the laboratory's HFIR. ^{249}Bk has a 330-day half-life, but the radioisotope had to decay three months after emerging from the reactor in 2008 to allow intense shorter-lived activities to decay before it could be worked on and separated from other highly radioactive actinide elements, atomic numbers 89 to 103.

The experiment in Dubna ran for seventy days of continuous bombardment and data acquisition, yielding eleven detected atoms of new element 117. Follow-up experiments confirmed the finding. The paper announcing the discovery of the new element was published in 2010,[45] and, in 2016, the International Union of Pure and Applied Chemistry (IUPAC) approved the name tennessine (Ts) after the state of Tennessee, in view of the critical contributions the state's people and facilities had made.[46] On the periodic chart, element 117 is chemically a halogen, and thus it features the same "ine" suffix of other halogens: fluorine, chlorine, bromine, iodine, and astatine.

A specialized reactor and accelerator were required for the discovery of this new element, but this experiment would not have succeeded without the people who conceived, led, and pushed for resources to enable its execution. Physics professor Joseph Hamilton of Vanderbilt University had maintained a long-duration scientific relationship with Yuri Oganessian, director of the accelerator laboratory in Dubna, and, together, they generated the idea for this experiment.

Hamilton brought the idea to Jim Roberto, then ORNL deputy director for science and technology, since HFIR was the only reactor in the world in which the target material (berkelium) could be produced and separated. Producing this material in the reactor through a long-term irradiation and then a difficult radiochemical separation was expensive. However, Hamilton would not let go of this idea and kept lobbying ORNL for permission to undertake the project, and Roberto eventually found the resources and the scheduling to get it done.

Sensitive equipment and electronics were necessary to enable detection of few tennessine atoms, with short half-lives—45 milliseconds. ORNL's Krzysztof Rykaczewski and UT's Robert Grzywacz were experts on the digital electronics necessary for this single-atom detection. The experiments in Dubna produced not only tennessine but also a small number of atoms of new element 118, which was named oganesson (Og), for Yuri Oganessian, by IUPAC in 2016. It is extremely rare to meet a person after whom an element is named, but that was Riedinger's high honor on the occasion of a nuclear physics conference in Knoxville in July of 2016.

National Security

As the new contractor for ORNL, UT-Battelle LLC created a National Security Directorate in support of the National Nuclear Security Administration (NNSA), U.S. DOD, and other agencies. When terrorists attacked the United States on September 11, 2001, ORNL was well positioned to respond to new calls for the science and technology needed to protect the U.S. homeland and national security interests. In June 2002, President George W. Bush proposed the creation of a Cabinet-level Department of Homeland Security (DHS) and chartered a Transition Planning Office to launch this new department, which became operational in January 2003. The staff of the Transition Planning Office included ORNL's deputy director for operations, Jeff Smith, and Battelle's Jeff Wadsworth, who would serve as director of ORNL from 2003 to 2007.

The ORNL National Security Directorate developed a portfolio that spanned nonproliferation and threat reduction, arms and export control, homeland security, and counterterrorism technologies for several federal agencies. Starting in 2007, much of this work was conducted in the $43-million Multiprogram Research Facility, constructed on the ORNL campus to provide state-of-the-art secure space for classified activities. This building was constructed in the same private-sector partnership that resulted in the large addition of new research space five years earlier, providing facilities for many research areas including supercomputing.

History of National Security Work in Oak Ridge

All three federal facilities in Oak Ridge started with an urgent national security charge—to help build an atomic bomb to end World War II. The national security needs in the United States and the contributions from Oak Ridge and from UT have changed greatly in the decades since the Manhattan Project.

Over the ensuing four decades, Y-12 played a vital role in expanding and maintaining the nation's nuclear weapons stockpile. It also enhanced national security by working with DOD on programs such as the Strategic Defense Initiative[47] (SDI, nicknamed "Star Wars") and the Navy's Seawolf submarine. Today, the Y-12 National Security Complex is the nation's only source of enriched uranium components for nuclear weapons; it also provides enriched uranium to fuel the U.S. Navy's nuclear-powered vessels.

Almost all of Y-12's $2.2-billion budget comes from NNSA, which is responsible for enhancing national security through the military application of nuclear science. Y-12 also supports efforts to reduce the risk of nuclear proliferation and performs work for other government agencies that need its specialized resources in materials science and precision manufacturing.

In 1988, Martin Marietta Energy Systems Inc., the contractor for DOE's Oak Ridge facilities, created a Space and Defense Technology Program to capitalize on the unique research, development, and prototype fabrication capabilities distributed across Y-12, ORNL, and K-25. This program consolidated work in support of SDI and other DOD projects, bringing together multidisciplinary teams to solve problems germane to national defense and space exploration.

SDI officially ended in 1993, when the Clinton Administration redirected the efforts toward development of theatre ballistic missiles[48] and renamed the agency the Ballistic Missile Defense Organization. The Oak Ridge Centers for Manufacturing Technology succeeded Martin Marietta's Space and Defense Technology Program. This industrial resource, created to help maintain national security capabilities while bolstering U.S. economic competitiveness, would become a Y-12 program after the transition to separate contracts for ORNL, Y-12, and K-25 in 2000.

Today, about 15 percent of ORNL's research and development portfolio directly supports the national security missions of DOE, NNSA, DOD, DHS, and other agencies. The National Security Directorate has become the National Security Sciences Directorate (NSSD), which focuses on cybersecurity and cyber-physical resilience, data analytics, geospatial science and technology, nuclear nonproliferation, and high-performance computing for sensitive national security missions.

NSSD works closely with researchers in other ORNL directorates to address national security challenges using capabilities in areas such as nuclear and chemical sciences and engineering, applied materials, advanced manufacturing, biosecurity, transportation, and computing. For example, a team at ORNL's Manufacturing Demonstration Facility partnered with the Navy's Disruptive Technology Laboratory in 2017 to create the military's first 3D-printed submersible hull.[49] Using 3D printing reduced hull production costs by 90 percent and shortened production time from months to days.

Relationship building is an important part of ORNL's national security programs. From 2002 to 2020, ORNL hosted fifty-eight military and civilian defense personnel for temporary one-year assignments as part of its military education and outreach program. The Training with Industry program exposes military officers to national laboratory capabilities and R&D processes that can benefit the country and the military as these leaders move on to future assignments.[50]

Peter Tsai and the Now-ubiquitous N95 Mask

In 2006, the UT Research Foundation presented the B. Otto and Kathleen Wheeley Award for Excellence in Technology Transfer to Peter Tsai, a professor in the Department of Materials Science and Engineering. Originally from Taiwan, Tsai received a PhD from Kansas State University in 1981 and then joined the UT faculty to spend a research career working on nonwoven fabrics engineered as sheets of various materials bonded together by entangling fibers mechanically, thermally, or chemically. Nonwoven fabrics are used in a host of protective applications in the construction and medical fields (e.g., surgical gowns and masks, shoe covers, gloves, etc.) and in many forms of filtration.

In 1992, Tsai and his research team developed an electrostatic charging technology to filter out unwanted particles passing through a nonwoven fabric barrier. The fabric's filter contains both positive and negative charges formed by this patented technique. The filter can attract neutral particles, like bacteria and viruses, and polarize these particles, trapping them before they can pass through a protective mask and reach the respiratory tract of the wearer. This invention resulted in a 1995 patent for Tsai, one of twelve he received in his UT career before retiring in 2018. This charging process was used to make the material for the N95 mask, whose name derives from its ability to remove at least 95 percent of submicron particles present in the air.[51]

Tsai's personal-protection invention was soon in use around the world. Then, the COVID-19 pandemic hit. The World Health Organization already recommended the N95 mask to protect against severe acute respiratory syndrome (SARS), bird and swine flu, and other airborne diseases, but the arrival and rapid spread of the COVID-19 pandemic necessitated the manufacture of many millions of N95 masks—a demand well beyond the production capability of industry at that time.

In the absence of adequate supplies of new masks, researchers began to explore ways to clean and reuse the N95 mask without destroying the imbedded positive and negative charges that make the mask so efficient. In response to this challenge, the retired Tsai went to work in his garage in Knoxville, working day and night to experiment with different ways to clean the masks.[52] He tried boiling, steaming, or baking them or exposing them to sunlight for extended periods. By April of 2020, Tsai determined that the best sterilization method was heating the N95 mask to 158°F for sixty minutes.

Meanwhile, as the pandemic continued to surge and the demand for N95 masks skyrocketed, researchers with ORNL's Carbon Fiber Technology Facility asked Tsai to help them find quicker and cheaper ways to manufacture N95 material and treat it with the electrostatic charging system that gives the fabric its high-filtration efficiency. With Tsai's help and guidance, the facility devised ways to build and scale up the electrostatic charging system and produce material for nine thousand masks in an hour. ORNL partnerships with industry led to rapid transfer of this scaled-up technology to the private sector. Soon thereafter, Cummins, a corporation that manufactures engines and filtration products, was producing one million masks per day. In recognition of Tsai's contributions to fighting the COVID-19 pandemic, the UT Research Foundation presented him the inaugural Innov865 Impact Award, presented by Verizon, in October 2020.

ADVANCING THE UT-Y-12 PARTNERSHIP

Early in the term of Jimmy Cheek as UT, Knoxville, chancellor, there was a concerted effort to enhance the partnership between the university and the Y-12 National Security Complex. At UT, this was led by Bill Dunne, associate dean of the College of Engineering, and by Vice Chancellor for Research Wes Hines (followed by Lee Riedinger, who assumed that position in 2012).

Over the decades since the Manhattan Project, collaborations between UT and Y-12 were not nearly as numerous as those with ORNL, since the latter is primarily a research facility while the former is a secure technology development facility and the country's uranium center of excellence. Nevertheless, leaders of UT and Y-12 wanted to enhance the partnership for the benefit of both institutions. This started in January 2011 as Y-12 appointed Debbie Reed as the liaison with UT, and a series of monthly meetings began. Associate deans of research in most of the UT colleges participated, as did various Y-12 program managers, all focused on areas in which the two institutions could engage more deeply. This led to an April 6, 2011, Partners in Technology forum at the Y-12 New Hope Center, during which Chancellor Cheek and Y-12 General Manager Darrel Kohlhorst signed an MOU to guide future joint programs. After the signing, parallel technical sessions were devoted to manufacturing, advanced materials, and nonproliferation.

Progress in collaborative programs followed from the MOU signing and the monthly meetings of the joint working group. Nine of eleven UT, Knoxville, colleges became involved in some form of partnership with Y-12. A Joint Assignment Agreement (JAA) was signed to allow the formal exchange of personnel—the first of its kind for a NNSA production site. An Engineering Management Graduate Program was initiated and taught in Oak Ridge on Fridays, initially for thirty or more Y-12 employees.

A graduate research assistant program was established, with twenty-five to thirty UT students working at Y-12 on specific projects. Approximately

$1 million of Y-12 funds were allocated in 2012 to nine UT faculty working jointly with Y-12 staff on various programs. In the summer of 2012, long-time UT chemistry professor George Schweitzer taught a radiochemistry course for the benefit of Y-12 employees, leading to a radiochemistry certificate for these employees. Collaborations were started that year in eight areas, including nuclear forensics, testing of radioactive shipping containers, archiving Manhattan Project documents and records, strengthening nuclear-science education in the K-12 curricula, and launching Y-12's Total Health Risk Intervention and Education (THRIVE) initiative, a comprehensive disease-management program.

Ten years later, in 2022, the announcement of a $9.5-million agreement between UT and Consolidated Nuclear Security, the manager of Y-12, gave the partnership between UT and Y-12 a boost.[53] Gene Sievers, the Y-12 site manager, and Deborah Crawford, the UT, Knoxville, vice chancellor for research, championed this agreement. In the previous decade, the two institutions collaborated on more than thirty R&D programs spanning materials research, nuclear nonproliferation, cyber security, advanced manufacturing, and supply-chain management. The goal of the newly announced agreement (extending to 2026) is for UT to support Y-12 in driving the implementation of new technologies, while Y-12 serves as an incubator for innovations applied to key national security mission areas. These collaborations sometimes lead to the hiring of graduates at Y-12, as now roughly 20 percent of the Y-12 workforce are UT alumni. This is a partnership born in the Manhattan Project that continues to expand eighty years later.

UT-Battelle's Contributions to the Region

Oak Ridge and the East Tennessee region benefited from the generosity of UT-Battelle LLC over the twenty years of its management and operation of ORNL. UT-Battelle each year donates part of its award fee from DOE to local charities and civic organizations. The average annual contribution to local nonprofits has been around $550,000. In 2014, UT-Battelle donated $100,000 to the capital campaign for the expansion of East Tennessee Children's Hospital in Knoxville.[54] In 2019, UT-Battelle donated $175,000 to restore the Oak Ridge Playhouse, which dates from the 1940s. Substantial contributions were made to the renovation of Oak Ridge High School, and ORNL director Thom Mason served as one of the community leaders of this project.

In 2020, these UT-Battelle funds supported 175 East Tennessee nonprofit organizations. A new disaster-relief program received $10,000 in matching funds, and more than ten thousand people benefited from STEM outreach programs. The UT-Battelle Scholarship Award goes annually to a child of an

ORNL employee, providing a renewable scholarship to UT, Knoxville, aimed at encouraging careers in science.

In 2021, the city of Oak Ridge received a $500,000 contribution from UT-Battelle to assist with the planned airport project at the East Tennessee Technology Park, located near the former site of K-25 in west Oak Ridge.[55] This small airport would spur future economic growth of the region and become an important feature for potential business development.

In addition, private-sector funding of research buildings at ORNL requires the UT-Battelle Development Corporation to pay City of Oak Ridge property tax on land leased from DOE for the facilities' construction. In 2021, City of Oak Ridge tax records indicated that this amounted to $875,202.

SEVERAL OF THE PARTNERSHIP'S LEADING PROPONENTS TRANSITION TO RETIREMENT

For a partnership that has evolved over eighty years, there have been many transitions that affected its development—transitions in politicians, programs, leadership, and people at all levels. Major transitions occurred on multiple fronts toward the end of the 2010s, on the heels of a successful decade.

A long line of governors and senators who have supported the development and continued growth of the UT-ORNL partnership has blessed Tennessee. However, this partnership actually got off to a rocky start in the early 1940s, when Governor Prentice Cooper felt blindsided by the federal purchase of land and the top-secret start of the Manhattan Project in East Tennessee, without consulting or informing him. This rift was quickly remedied, and the relationships between state government and the federal research facilities has been harmonious ever since.

Indeed, over the past forty years, those relationships have been significantly strengthened, to the benefit of both parties. Five consecutive governors—Alexander, McWherter, Sundquist, Bredesen, and Haslam—placed increased focus on the UT-ORNL partnership and invested the state resources necessary to grow it. A new governor, Bill Lee, came into office in January 2019. Lee spent his first year learning how to be governor (his first political office) and the rapid spread of COVID-19 across Tennessee consumed much of his second. The signs are positive that he will continue his predecessors' legacy of support for ORNL and its university partner, as he has pledged state funds for the new Oak Ridge Innovation Institute at UT.

The same story applies to the roles of U.S. Senators from Tennessee. Howard Baker was elected to the Senate in 1967 and served three terms before he declined to run for a fourth; Albert Gore Jr. succeeded him in 1993. Lamar Alexander was elected to the Senate in 2002 and served three terms before retiring at the end of 2020. All three of these senators were well aware of the important research taking place at the national laboratory and the university and worked to enhance the partnership. With Alexander's retirement and the installation of Tennessee Senators Marsha Blackburn in 2019 and Bill Hagerty in 2021, ongoing support for the UT-ORNL partnership has not yet stood the test of time.

Succession in leadership at national laboratories is generally well planned and usually proceeds smoothly. This has certainly been the case at ORNL. In the UT-Battelle era of ORNL management, the laboratory directors have been Bill Madia, Jeff Wadsworth, Thom Mason, and Thomas Zacharia. The decade of the 2010s ends with Zacharia in charge and the laboratory strong and well positioned for the future. Stephen Streiffer became ORNL director in 2023.

This type of seamless transition is not always the case at universities, including UT. Since 2000, there have been seven UT System presidents, including two appointed on an interim basis after the contracts of three sitting presidents ended early. Six of these seven supported a strong relationship with the national laboratory, and, as a result, the partnership grew despite the changes in leadership. The decade ends with local businessman Randy Boyd at the helm of the UT system. Boyd had no background in higher education prior to being named to the post but is widely viewed as being a highly effective president. Creation of the new UT-Oak Ridge Innovation Institute was announced after Boyd took office, and he has since supported the UT-Oak Ridge partnership and the region's capabilities in innovation.

There have been six UT, Knoxville chancellors in the UT-Battelle era, including two interims. As the decade ends, Donde Plowman, a former professor of business, helms the Knoxville campus and builds on the work of Jimmy Cheek, who did much to support the partnership over his eight years as chancellor. Plowman became chancellor of UT, Knoxville on July 1, 2019, and, in her first week on the job, spoke highly of the emerging UT-Oak Ridge Innovation Institute as the next step in the long partnership between UT and ORNL.[56]

Beyond the elected politicians and the institution leaders, numerous other dedicated individuals helped to build, sustain, and strengthen this partnership. Among them is Homer Fisher, a UT senior vice president, who, with Bill Madia of Battelle, formed UT-Battelle LLC, the managing contractor of ORNL since 2000. Fisher is now fully retired, as are Jim Roberto of ORNL and Wayne Davis of UT, who co-chaired the task force to create the Bredesen Center and the interdisciplinary energy PhD program. Davis spent his career in UT's College of Engineering and, while serving as the college's dean, facilitated the hiring of a number of Governor's Chairs. Davis finished his career by serving a stint as interim chancellor prior to Plowman's appointment.

Three members of the original UT-Battelle leadership team came from UT, and they all departed ORNL in this decade: Frank Harris (biology), Lee Riedinger (physics), and Billy Stair (executive assistant to the UT president). Likewise, the members of the Bredesen Center's leadership team—UT's Riedinger as director, ORNL's Mike Simpson as assistant director, and UT's Wanda Davis as business manager—all stepped down from their administrative roles in the final years of the 2010s.

Transitions in leadership have, of course, been commonplace over the eighty years of this partnership. William Pollard of the UT physics faculty was the first to work extensively to build a relationship between UT (and other universities) and Clinton Laboratories, starting in 1943. His work was at least partially responsible for keeping Clinton Laboratories alive at the end of the Manhattan Project, which paved the way for the later transition to ORNL. A host of others who kept the innovation going and the partnership expanding succeeded Pollard. William Bugg (one of this book's authors, who retired in 2002) headed the UT physics department for twenty-six years, until Riedinger succeeded him in 1996, and did much to expand the ranks of joint UT-ORNL hires and to recruit department faculty who could leverage the considerable assets of the national laboratory.

Based in part on the long tenure of the UT-Oak Ridge partnership and the way it has been woven into so many aspects of the institutions, new leadership naturally develops to replace those who no longer serve. In February 2021, Joan Bienvenue was hired as the first executive director of the UT-Oak Ridge Innovation Institute and brought excellent credentials to lead this newest development in the eighty-year partnership.[57] She was succeeded by ORNL's David Sholl in 2023.[58]

One purpose of this book is to provide a history of the UT-Oak Ridge partnership—a playbook of sorts—to a new class of leaders, charting administrative and scientific successes, along with the chains of decisions, that bolstered the partnership over its long existence. All four of the book's authors were direct participants—and, in some cases, prime actors—in this extensive history, and, collectively, they represent 180 years of studying or working professionally at one or both of the partnering institutions. In that sense, all are primary sources for much of the information contained in this volume.

The hope is that new leaders will help build a partnership future that leverages well the experience and accomplishments of the past.

EPILOGUE

The Key to Predicting the Future Trajectory of the UT-ORNL Partnership May Lie in Its Past

THIS STORY BEGAN in 1940, when Oak Ridge did not exist and UT offered no doctoral programs and its professors conducted little in the way of meaningful scientific research. In many ways, UT *needed* the formation of Oak Ridge and the Manhattan Project to spur progress in the sciences and engineering. The university had suffered from uninspired leadership through the 1930s and languished under a state government that did not see much value in higher education and, thus, did little to support it.

As a devastating world war raged, the nation's scientists looked to a new and powerful weapon to bring the conflict to a close, and the federal government selected—quickly and secretly—a large amount of farmland west of Knoxville and south of Clinton as the place to prepare crucial material for the weapon, under the auspices of the Manhattan Project. Soon after, contractors began construction on project facilities and housing to accommodate the thousands of workers and their families soon to arrive at a top-secret community christened Oak Ridge.

With Japan's surrender in August of 1945, news of the Manhattan Project and its pivotal role in winning the war went public,[1] and, soon after, the federal facilities in Oak Ridge and the state's flagship university began a long and mutually beneficial partnership that has since spanned eighty years, starting in 1943. This book has chronicled the step-by-step development of that partnership, along with the successes and missteps that occurred along the way. Many critical connections between UT and Oak Ridge contributed to the growth in this partnership, and twenty-three of them have been highlighted over the course of this book and the eight decades it covers.[2]

However, where does the story go from here? Indeed, how might the partnership evolve over the next two decades as it approaches its centennial year? Predicting the future is fraught with risk, and there is a rich history of bold predictions that have proven glaringly wrong. In 1943, for instance, IBM chairman and CEO Thomas Watson speculated that the future global market for computers likely would not exceed five machines.

In 1952, AEC Chair Lewis Strauss, in addressing the future promise of nuclear power, said, "It is not too much to expect that our children will enjoy in their homes electricity too cheap to meter."[3] In 1981, Microsoft co-founder Bill Gates may have said that "640K of memory ought to be enough for anybody."[4] Around that time, Ken Olsen, president of Digital Equipment Company, offered that "there is no reason anyone would want a computer in their home."[5]

Sobered by these bold but incorrect prognostications, we nevertheless are willing to risk a look forward, but informing that projection will be a look back in time. Indeed, an eighty-year retrospective study of the development of the UT-Oak Ridge partnership brings the possibilities of the future into sharper perspective.

1940 TO 1960

The first twenty years saw the construction and development of the federal facilities in Oak Ridge, the creation of UT's first PhD programs (in 1946), and the first partnership programs that became building blocks for the future. The Manhattan Project sponsored the construction of three facilities in Oak Ridge serving three different purposes. Y-12 bore responsibility for developing methods for separating isotopes of uranium and producing uranium enriched in the fissionable isotope ^{235}U. This represented a massive effort fraught with technical difficulties, but it succeeded, producing tens of kilograms of highly enriched uranium that were sent to Los Alamos for incorporation into the atomic bomb dropped on Hiroshima on August 6, 1945.

The switch from Y-12 calutrons to K-25 gaseous diffusion for enrichment of uranium became complete in December 1946, as electromagnetic separation proved too energy intensive and expensive. Y-12 began to convert ^{235}U powder to metal in 1948, initiating the nuclear weapons mission that has remained the primary direction for Y-12 to date.[6] This was a key strategic element in fighting the Cold War and ultimately helped break the Soviet Union's economy, as it attempted to match Y-12's considerable output, leading to the nation's collapse in 1991. K-25 developed gaseous diffusion as a more cost-effective way to enrich uranium and served as the centerpiece of the country's enrichment industry until 1983.

X-10 was the site of the world's first continuously operating nuclear reactor. Scientists and engineers learned how to use the Graphite Pile (reactor) at Oak Ridge to create a fissionable isotope of ^{239}Pu and radiochemically separate it from the uranium fuel rods. The successful separation of the plutonium isotope quickly prompted construction of larger reactors at Hanford, Washington, to produce the kilogram quantities of ^{239}Pu used in the atomic bomb dropped on Nagasaki, bringing World War II to an end on September 2, 1945, when the surrender documents were signed.

The Oak Ridge X-10 site, also called Clinton Laboratories during World War II, first became Clinton National Laboratory and then, in March 1948, Oak Ridge National Laboratory. In September 1945, with the ink on the recently signed terms of Japan's surrender barely dry, Clinton leaders approached UT about offering physics courses at the Oak Ridge site to entice Manhattan Project scientists to remain in East Tennessee. Many of them were pulled onto the project and forced to abandon, at least for the duration of the war, graduate studies at universities across the country. This was the first of the critical connections, and, by spring of 1946, this modest initial educational offering expanded to include a host of science and engineering courses offered in Oak Ridge and administered by UT faculty. The first PhD programs in chemistry and physics allowed former Manhattan Project researchers to continue working while completing their doctoral studies. These developments mark the beginning of a collaborative educational relationship that continues to thrive between the national laboratory and the university.

UT physics professor William Pollard saw and seized an opportunity to build an official presence for UT and other universities in Oak Ridge through creation of ORINS, which is still flourishing in 2022 as ORAU. This innovative new connection was critical for UT's developing role in Oak Ridge programs. ORINS/ORAU has worked with the Oak Ridge facilities and the partnering universities (which now number over a hundred) on a host of activities since 1946.

ORNL developed into a prominent nuclear reactor R&D facility throughout the 1950s. Scores of UT faculty worked as consultants at the Oak Ridge nuclear plant starting around 1950.[7] This reactor emphasis and expertise in Oak Ridge led to the 1957 formation of the UT Department of Nuclear Engineering, now ranked as one of the top such departments in the United States. Through the 1950s, many Oak Ridge scientists and engineers received their doctorates from UT, and many UT graduate students performed their dissertation research at ORNL. The UT students' presence at ORNL has increased substantially in subsequent decades, to the benefit of both institutions.

1960 TO 1980

Momentous changes in the United States, including the birth of the environmental movement in the 1960s, prompted a shift in the federal research emphases and funding patterns, which forced a change in the nature of ORNL's research enterprise. After being chiefly a nuclear energy R&D laboratory in the 1940s and 1950s, ORNL became a more balanced laboratory in the 1960s, which benefited UT and the partnership in substantial ways. For example, ORNL and then UT developed ecology programs, building on wide national interest spurred in part by Rachel Carson's book *Silent Spring*.

In the 1970s, ORNL physicist-turned-energy researcher John Gibbons came to UT to create the Environment Center, which conducted research on U.S. energy-use patterns and opportunities for conservation following the disruptive OPEC oil embargo of late 1973. Before Gibbons left for Washington to direct the Congressional Office of Technology Assessment in 1979, he renamed the UT organization the Energy, Environment, and Resources Center, which merged with other environmental research units and renamed the Institute for a Secure and Sustainable Environment in 2005.

Alvin Weinberg, who served as ORNL director from 1955 to 1972, was a reactor expert who nevertheless steered ORNL toward a greater emphasis on environmental programs and nuclear safety issues. Weinberg had a unique vision for national laboratories (especially *his* national laboratory) to play a new and prominent role in educating the next generation of scientists and engineers.[8] He suggested that ORNL partner with UT in offering PhD degrees, with laboratory researchers serving as university faculty and mentors to graduate students through a joint-institute arrangement. Weinberg's concept, perhaps a bit too novel and ambitious in the 1960s, achieved fruition with creation of the Joint Institute for Heavy Ion Research in the early 1980s, the first of several established over subsequent years between UT and ORNL.

Created in 1967, the UT-Oak Ridge Graduate School of Biomedical Sciences—another product of Weinberg's vision—received funding for many years from the NIH. This school officially was part of UT, but almost all of the faculty members were researchers from ORNL's Biology Division; they taught the courses in Oak Ridge and mentored graduate students conducting their research there. This innovative school operated for a quarter century, before NIH funding ceased and opposition escalated from UT's campus-based biology faculty, who saw the Oak Ridge doctoral program as an unwelcome competitor for top students.

Accelerators constructed at ORNL in these twenty years led to a dramatic increase in participation by nuclear physicists from UT, Vanderbilt, and other universities in experiments and research programs in Oak Ridge. The first user facility at ORNL (and perhaps at any national laboratory) was created in 1970 with the installation of UNISOR on a beam line of ORIC. In the ensuing decades, this "user" concept would lead to the creation of many other such facilities at ORNL and at other national laboratories and universities.

1980–2000

These two decades saw the initiation of formal collaborative programs that brought the partnership to a new level of intimacy and productivity, and newly elected Tennessee Governor Lamar Alexander played a critical role. Alexander served as Tennessee governor from 1979 to 1987 before beginning his three-year term as UT president in 1988. Alexander was the first of six consecutive Tennessee governors (succeeding Alexander were Ned McWherter, Don Sundquist, Phil Bredesen, Bill Haslam, and Bill Lee) to see the obvious value to the state of bolstering the connection between UT and Oak Ridge and to promote and invest in the partnership.

As governor, Alexander led the successful push to boost the state sales tax specifically for increased investment in education. The higher-education portion of that bill focused on formation of centers of excellence at state universities, and the first such competitive award supported the creation of the UT-ORNL Science Alliance, which recruited high-level researchers and engineers (termed Distinguished Scientists) into positions shared equally by the two institutions. The visit of President George H. W. Bush to UT in 1990 brought increased focus on the Science Alliance and the partnership it represented. UT-ORNL joint programs flourished in these twenty years, especially in research into properties of materials, ranging from the nanoscale (nanoscience) to the macro-scale (creation of new alloys and composites).

This new level of institutional partnership emboldened UT to consider competing for the contract to manage ORNL, when the thirty-seven-year tenure of Union Carbide ended in 1984. After considerable work by a committee to study all options for moving forward, university leadership decided that the time was not ripe for assuming such a sweeping and complex role in managing, not just the national laboratory, but the Y-12 and K-25 facilities as well. Martin Marietta won the management contract, but, fourteen years later, UT would overcome its initial reticence and compete for management responsibilities at the national laboratory.

In 1998, DOE again opened competition for the operating contracts of the Oak Ridge facilities, but, in this instance, the department decided to award separate management contracts for the three facilities (ORNL, Y-12, and K-25), positioning UT to bid only on the contract for the national laboratory. This time, the university entered the fray and aggressively vied for the contract, with the strong support of UT President Joe Johnson. In developing the proposal to manage ORNL, UT teamed with the Battelle Memorial Institute in Columbus, Ohio and formed UT-Battelle LLC, which succeeded in winning the contract and began managing the laboratory in April 2000. A key component of the winning proposal was the commitment by Governor Sundquist to fund construction of three new state-supported joint institutes at ORNL: JICS, JIBS, and JINS. Also, JIAM (now called the Institute for Advanced Materials and Manufacturing)[9] was constructed at the UT Research Park at Cherokee Farm.

2000–2020

Using a combination of state, federal, and private dollars, UT-Battelle undertook a sweeping renovation to modernize the ORNL campus and to make it more welcoming to visiting scientists and staff. With the new management contract in place, the number of UT-ORNL joint faculty increased significantly, and construction of several expensive new user facilities was completed on budget and on time, including CNMS (2005) and SNS (2006).

Meanwhile, in coming years, the laboratory would become a global supercomputing stalwart, in part because the ORNL campus renovation included construction of a 40,000-square-foot computer facility—equipped with abundant power and cooling capacity—even though, at the time of construction, the supercomputer had yet to be funded. However, with the requisite infrastructure in place, the laboratory was in a position to compete aggressively for the people, programs, and funds necessary to spur its eventual emergence as a supercomputing powerhouse. In 2007, with funding from

DOE, the laboratory acquired the Cray XT4 Jaguar supercomputer, rated at a top processing speed of 119 teraflops (a teraflop equals 1 trillion floating-point operations per second) followed by the Cray XT5 rated at 1.4 petaflops (or 1,400 teraflops).

UT was able to exploit these new ORNL assets to expand its own research capabilities and introduce new programs. And, in 2007, the university secured its own supercomputing funding from the National Science Foundation in part because it planned to house its powerful new machine—Kraken, rated at 1-petaflop (1,000 teraflops)—at the ORNL facility that contained the laboratory's Jaguar. With the university's acquisition of Kraken, faculty strength in supercomputing, neutron science, and nanoscience increased dramatically. JICS provided a framework for laboratory and university researchers to leverage these two supercomputers for new research in various areas.

Governor Phil Bredesen came into office in 2003 and provided another major boost to the UT-ORNL partnership. A new program (with attached funding) facilitated the hiring of the next cadre of top scientists and engineers—the Governor's Chairs—into joint university-national laboratory positions. The UT College of Engineering leveraged Governor's Chair funding to boost new research capabilities, which led to the construction of new joint research facilities located on a new campus situated halfway between Knoxville and Oak Ridge along Pellissippi Parkway. The National Transportation Research Center opened on this new campus in 2000, and the Manufacturing Research Facility—focused on additive manufacturing—was added in 2012. President Obama visited the latter facility in 2015 to announce a major federal award for the creation of the IACMI.

In 2010, under Bredesen's leadership, an interdisciplinary center (now called the Bredesen Center) was formed at UT and linked to ORNL. The center would go on to create two interdisciplinary UT-ORNL PhD programs, the first in energy science and engineering (2011) and the second in data science and engineering (2017).

ESSENTIAL INGREDIENTS FOR THE PARTNERSHIP'S PAST SUCCESS

The ingredients for the successful development of the UT-ORNL partnership grew from a void (no Oak Ridge and no UT PhD programs in 1940) and progressed to the level of perhaps the finest university-national laboratory collaborative relationship in the country. Among the critical ingredients were the following.

Dedicated people at all levels. Institutions advance through foresight, initiative, and dogged pursuit of excellence by people dedicated to making change. In 1946, UT physics professors William Pollard and Kenneth Hertel saw the need and opportunity for the first partnership organization in Oak Ridge (ORINS). Pollard organized the formation, growth, and development of this important new entity. This brought UT and other universities into an official role in Oak Ridge, and this university engagement only increased in succeeding decades. Further, the formation of ORINS helped justify the continuation of Clinton Laboratories after the end of the Manhattan Project. Another example is ORNL's Thomas Zacharia who took the lead on making ORNL and UT leaders in the world of supercomputing in the decade of the 2000s.

Strong and consistent leadership at the top. Institutions grow only if there is consistency in imaginative and respected top-level leadership. This has certainly been the case at ORNL, with its succession of capable leaders who served as laboratory directors: Weinberg, Postma, Trivelpiece, Madia, Wadsworth, Mason, and Zacharia. In different but substantial ways, each of these laboratory directors has supported the partnership with UT, and each contributed substantially to the university's growth. The impact of UT's presidents on the partnership has ranged from the uninspired (Hoskins in 1940) to the energetic and transformative (Holt, Boling, Alexander, and Johnson). Then, of course, there were the UT presidential tenures clouded by controversy (Gilley, Shumaker, and Petersen), but even during these difficult times, the leadership at the chancellor level remained exceptional (Reese, Crabtree, and Cheek).

Support from governors. Tennessee is not a financially rich state and is ranked close to the bottom of all states in support of K-12 education. In a 2021 survey, Tennessee ranked eighth from the bottom in average expenditure of state funds per student.[10] On the other hand, Tennessee ranks twelfth among states in expenditure per student for higher education.[11] The UT-Oak Ridge partnership has prospered over the decades from generally strong and consistent support by the state's governors. Six consecutive governors took the lead on investing directly in the partnership through funds for specific projects. There may be no other state or federal laboratory that can boast of this level of support for programs in partnership with a land-grant university.

Sharing of people, resources, and facilities. A university and a national laboratory are different institutions with different driving principles and agendas. However, strength comes when there is the sharing of ideas and resources in education, research, and/or development areas of shared interest. ORNL strength in nuclear reactor R&D led to the formation of a nationally strong nuclear engineering department at UT. Research strength in materials science R&D has been built in a consistent and joint way at both the university and national laboratory. Joint institutes funded by the state, owned by UT, and located at ORNL have become bustling centers of collaboration.

The sharing of people through the appointment of joint faculty, Distinguished Scientists, and Governor's Chairs has enabled growth at both institutions. This strong partnership has had, and will continue to have, a big impact on attracting the best and brightest to the UT faculty and to the ORNL staff. In fact, in 2020, Clarivate Analytics included nine UT and ORNL staff on its list of Highly Cited Researchers, people in the top 1 percent of the world's researchers in terms of publication citations over the past decade, according to the Web of Science. Six of these nine are jointly engaged at the two institutions.[12] [13]

In addition, a study conducted by Stanford University named 156 UT faculty members among the top 2 percent of scientists in the world for research citations. The database includes about 156,000 scientists worldwide whose work has been most-frequently cited by peers. The discipline with the most UT faculty members on the list, twelve, is applied physics, followed by seven each in electrical and electronics engineering, energy, and materials, all areas of joint strength with ORNL.[14]

Focus on interdisciplinary topics. While universities are historically disciplinary in scope, national laboratories are inherently interdisciplinary, since they usually focus on addressing our nation's R&D needs, which almost always engage multiple disciplines. The university is certainly more interdisciplinary in focus today as a result of the partnership with Oak Ridge. In 1973, ORNL's John Gibbons became the first director of what was later called UT's Energy, Environment, and Resources Center, clearly a multidisciplinary entity ranging across several departments and colleges.

UT's Bredesen Center initiated two interdisciplinary doctoral programs jointly with ORNL, engaging a multitude of faculty across the university's departments. UT's social scientists made important contributions to various ORNL initiatives relating to logistics and transportation studies for the military. Supercomputing, advanced manufacturing, genome science, bioenergy, and neutron science are all interdisciplinary research areas that have prospered in this joint environment.

Continuing success in all of the above areas is necessary for the partnership to remain strong and grow even more. Over the next twenty years, it will be important for both UT and ORNL leadership to commit to a joint vision, to facilitate joint hiring in areas of common strength, to share facilities for joint research, and to recruit jointly graduate students who can enrich both the university and the national laboratory. Ongoing commitment to a joint agenda will support forefront research and development and the necessary financial support from sponsors. Continuing joint focus on growing startup companies based on the joint research agenda will provide return to the state of Tennessee for its long-term investment in the university and its partnership with Oak Ridge.

THE UT-OAK RIDGE INNOVATION INSTITUTE: THE PARTNERSHIP'S NEXT EVOLUTIONARY STEP

Creation of the UT-Oak Ridge Innovation Institute, a newly formed component of the UT system that received approval by the Board of Trustees in June 2019, will certainly influence the UT-ORNL partnership's future direction. The report of the steering committee formed to draft plans for this new entity describes its overarching rationale and purpose: "Because of its close relationship to Oak Ridge National Laboratory, the University of Tennessee is uniquely positioned to meet the opportunity, and indeed, its responsibility, to increase the availability of the Nation's highest quality talent ready to compete in today's rapidly changing world."[15] In this instance, "talent" refers both to individual research standouts as well as to future administrators and research leaders.

Senator Lamar Alexander's reputation and position played a critical role in the DOE's decision to award $20 million to the new institute to support workforce development in emerging energy fields.[16] [17] [18] In fact, in an appropriations bill, Alexander outlined a competitive process for securing the federal dollars for workforce training, and, among other eligibility stipulations, the training must be conducted by a national laboratory that had existing connections to a land-grant research university. Though impossible to divine Alexander's intent in drafting the bill, it is reasonable to assume that he might have had the UT-ORNL partnership in mind when he established the criteria. Energy Secretary Dan Brouillette was quoted at the time of the award as remarking, "Oak Ridge is going to lead us into the technological future."[19]

Governor Bill Lee has continued the long history of state support of the UT-Oak Ridge partnership by pledging $8 million of annual funding for the UT-Oak Ridge Innovation Institute, over ten years.[20] This bodes well for the continuing development of this partnership, which is so important to UT and Oak Ridge institutions.

In the last forty years, a variety of successful partnership programs has developed, but their linkages have not always been fully exploited. Consider, for instance, that currently there are around three hundred UT-ORNL joint faculty, including seventeen Governor's Chairs, and four joint institutes on the ORNL campus. Meanwhile, each year, about two hundred UT graduate students perform thesis or dissertation research at ORNL.

One of the UT-Oak Ridge Innovation Institute's primary goals is to bring greater coordination to existing partnership programs by unifying them under one organizational roof. Another goal ultimately is to more than double the population of UT graduate students working at the laboratory, up to a level of around five hundred. This will not only benefit UT in its recruitment of top talent for its graduate programs but will also serve ORNL's interests by bringing a larger cohort of bright young minds to the laboratory's research divisions.[21]

Further, the institute will facilitate a much higher level of joint strategic planning between the two institutions and bring funding for large interdisciplinary projects and programs. New funding awards will, in turn, provide for the hiring of perhaps sixty new UT faculty, matched by a like number of new research staff at ORNL. Bringing further harmony to the UT-ORNL collaboration, the institute's director reports to both the provost of UT, Knoxville, and the ORNL director (and also will serve as part of the laboratory's leadership team). In February 2021, Joan Bienvenue was hired as the first executive director of the UT-Oak Ridge Innovation Institute.[22]

The institute's plans also call for the hire of a "capture manager," a person to take the lead on pursuing and winning new major funding awards. While national laboratories have traditionally appointed these fund-raising champions, universities historically have tended to allow individual PIs (principal investigators) to secure funding for their projects. The institute's appointment of a capture manager will likely lead to increased awards for large, interdisciplinary projects and allow UT's faculty to focus more intently on their research.

ASSESSING THE PARTNERSHIP

Continuing development of the UT-Oak Ridge partnership will depend on the people involved—leaders of the institutions and contributors at all levels. The institutional intimacy started in part by the vision of UT physics professor William Pollard and post-war Oak Ridge leaders who needed the university to start providing science courses in Oak Ridge. The partnership developed step-by-step over eighty years due to the work of many people and the initiation of many programs of cooperation.

People come and go and programs rise and fall. Arrangements for both institutions to share researchers officially began around 1950 and got a big boost when the UT-ORNL Distinguished Scientist program was initiated in 1982. Through this program, a dozen such top scholars were hired into fifty-fifty joint positions. However, over twenty years, they all eventually gravitated back to UT as they failed to generate new ORNL programs and funding. As a result, the laboratory ultimately backed out of its funding commitment to each shared hire.

The track record of success at both institutions so far has been better on the similar Governor's Chair positions initiated in 2006, partly because the two institutions had learned from the shortcomings of the initial program initiated twenty years earlier. The number of joint faculty (shared researchers at all academic levels) started small (a few in the early 1990s), rose to around three hundred by 2020, but has since decreased somewhat due to real or perceived structural DOE problems at ORNL, e.g., concern over how a staff member splits their work time.

Joint institutes were a vehicle of cooperation first proposed by Alvin Weinberg in 1962, but did not start to exist for another twenty years. Five joint institutes were initiated, four at ORNL and one at UT, each built around a key area of research strength. But these institutes have undergone significant change—and not necessarily for the better—in part due to the shifting focus of research. The one at UT (Institute for Advanced Materials and Manufacturing) is no longer called a joint institute, as it now fully focuses on UT faculty research programs. Likewise, the first one (1984) at ORNL (JIHIR) is no longer a joint institute, as DOE has closed the on-site experimental facilities in nuclear physics (as a new radioactive-beam facility was constructed at Michigan State University), the lease of land for this university-owned building has expired, and the state-constructed building will henceforth be used for janitorial services at ORNL.

Other developments appear more promising. New joint research centers continue to arise, among them the Manufacturing Demonstration Facility located halfway between Knoxville and Oak Ridge. Meanwhile, jointly conducted research by scientists at UT and ORNL continues to flourish, though not necessarily through joint faculty appointments. A renewed partnership between UT and Y-12 was announced in 2022. And, the UT-Battelle partnership has flourished since 2000, even though some feel that UT has not consistently achieved as much of an impact as it co-manager, Battelle. DOE has renewed the five-year contract to manage ORNL four times, and many await the decision to renew or open the contract to competition in 2025.

After a failed UT-Battelle search for a successor to retiring ORNL director Thomas Zacharia, a search firm was hired in early 2023 to find a new laboratory director[23]—a selection critical to the 2025 contract-renewal question. In the interim, former deputy director for operations Jeff Smith was brought out of retirement to serve as interim laboratory director, certainly a positive sign for the partnership with UT. He and Riedinger started together as the two UT-Battelle deputy directors in 2000, and Smith stayed in that position for twenty years. Stephen Streiffer, appointed ORNL director in 2023, will be important for setting the tone for the continuing partnership.

From 1943 onward, the partnership between UT and Oak Ridge has helped the university enormously, as it increased in programs, faculty expertise, and stature. The university still is far from the desired ranking in the top twenty-five publicly supported universities in the United States, a goal voiced by former chancellor Loren Crabtree, even though some programs and departments (e.g., nuclear engineering) have ascended to those heights. Still, it is clear that partnerships with ORNL and Y-12 are essential for UT to move in this direction.

As UT needed the partnership with Oak Ridge to advance as a nationally respected university, so also the young city of Oak Ridge and its federal facilities needed the involvement of universities and their infusion of faculty and graduate student talent to help establish prominence on the national stage. UT recognized that need and opportunity early in the history of Oak Ridge and worked to bring courses, degree opportunities, graduate students, and academic possibilities for people working at the federal facilities. This has proved a mutually beneficial partnership and will continue to be as people and programs continue to progress.

CONSULTING THE TEA LEAVES: UT-ORNL FUTURE RESEARCH AREAS

The research agendas of UT and the federal facilities in Oak Ridge evolved and changed greatly over the history of the partnership, and the next twenty years will likely see an acceleration of that evolution as new national priorities emerge and as scientists and engineers discover and develop new technologies. Based on the partnership's past accomplishment and current capabilities, the following potential future growth areas show particular promise.

Supercomputing. For the foreseeable future, it is clear that ORNL will continue to host one of the more powerful supercomputers in the country, if not the world, for scientific research. ORNL's next supercomputer (Frontier) became operational in 2021 and is capable of more than 1.5 billion billion calculations per second (1.5 exaflops). In 2022, it was declared the world's fastest supercomputer.[24] Funded by DOE via a contract with manufacturer Cray Inc., Frontier builds on the series of world-leading computers housed at ORNL since 2005: Jaguar, Titan, and Summit. Many research areas in science and engineering will benefit from these ever-more-powerful supercomputers, including materials science, astrophysics, fusion reactor dynamics, and drug design.

Quantum science. Supercomputers have continued to become more powerful as central processing units are equipped with increasing numbers of silicon semiconductors containing larger and larger numbers of integrated circuits. However, developers are approaching physical limits in terms of the number of chips and circuits that can be incorporated into a processor's core. As a result, future increases in computing power and speed, based on conventional chip technology, will not match the rate of advance that has persisted over past decades, when Moore's law, which predicts that the number of transistors on a chip would double every year, has more or less held true since 1965.

This apparent roadblock will prompt scientists to explore novel ways to process information, including a new generation of computers based on quantum mechanics. Greatly simplified, quantum mechanics describes the properties of matter at the smallest scale—molecules, atoms, and the electrons, neutrons, and protons that atoms comprise—that are distinct from the properties of matter as understood via classical physics and on a larger scale. For instance, atomic particles can exist as both particles and waves. Machines devised to exploit these properties could lead to breakthroughs in speed, performance, and encryption as dramatic as those that gave rise to current digital age.

To spur development of this new technology, the U.S. Congress passed the National Quantum Initiative Act, which became law in December 2018, and DOE is establishing research centers in this field, including the Quantum Science Center led by ORNL.[25] UT created the Appalachian Quantum Initiative with financial support from NSF and leadership from the physics department. Partners include ORNL, Microsoft, Tennessee Tech University, and Volkswagen of America.[26] Faculty recently were hired in various departments in this broad area of quantum science.

One direction of quantum science is to develop quantum computers that could be a major direction for high-performance computing in the future.

This involves not only building and selling quantum computers (companies such as IBM are making early progress) but also developing quantum algorithms that allow a quantum computer to perform certain types of calculations quickly and efficiently.

Electric vehicles. The world is gradually moving away from fossil-fuel-powered vehicles and toward battery-powered cars, trucks, and buses. Research programs will continue at ORNL and UT on a host of topics related to electric vehicles, including development of more robust batteries and the use of composites to achieve lighter vehicle components. In 2020, Volkswagen of America announced creation of its first innovation hub in North America at the UT Research Park at Cherokee Farm, in a partnership with UT and ORNL.[27] [28] This center will research lightweight car components and develop electric vehicles, to complement the efforts of Volkswagen's new electric vehicle production facility near Chattanooga.[29]

Many other battery-related enterprises continue to spring up in East Tennessee. One example is Safire, a start-up company that licensed ORNL technology to produce a drop-in additive for lithium-ion batteries that prevents explosions and fire resulting from the impact sustained during automobile collisions.[30] The Safire Technology Group recently established a research and development laboratory at UT in the Spark Innovation Center.[31] The center, located at the UT Research Park, focuses on entrepreneurship development and commercialization of regionally developed technologies. In a related development, DOE announced a new grant to the UT/ORNL IACMI for research and development of advanced composite materials used in clean energy technology relating to electric vehicles.[32]

Neutron science. This research area should continue to prosper at the two institutions. DOE plans to expand greatly the capabilities of SNS, which began operating in 2006. A second target station will be added, including another suite of twenty-two instruments for performing neutron-scattering measurements of various types.[33] Also, the linear proton accelerator that leads to production of pulsed neutrons received a power upgrade (from 1 to 1.55 MW)[34] so that the beam can be split between the two target halls and two suites of detectors.

The upgraded accelerator will still produce sixty proton pulses per second, with three fourths of these being sent to the first target station and one fourth going to the (new) second target hall. Because there will be a longer time interval between produced neutron pulses at the second target station, researchers will be able to utilize longer neutron wavelengths (so-called "colder" neutrons) and therefore study larger molecules by neutron scattering. These experimental capabilities will enable studies of important biological compounds, including proteins important to disease and drug development.[35]

SNS will remain the top neutron-scattering facility in the world, which is appropriate since Shull and Wollan started the field of neutron scattering at ORNL's Graphite Reactor in the late 1940s. There are also preliminary plans to upgrade and renew HFIR, opened in 1965. A new pressure vessel, which encloses the reactor, may be installed to replace the current one that is more than fifty years old. Of course, all of these upgrades and additions likely will require strong political support to secure timely funding.

Next-generation nuclear fission reactors. The nuclear power industry has declined for the past forty years. Indeed, few new reactors were added to the U.S. electricity grid over recent decades, and quite a few reactors were closed, due in part to the availability of cheaper electricity produced by burning natural gas. Additionally, the attrition problem will get worse as more than half of U.S. reactors (currently generating 19 percent of the country's electricity) are retired over the next twenty years, when their operating licenses expire.

An increased focus on reducing the U.S. carbon footprint could prompt renewed interest in power-producing nuclear reactors, which emit little or no carbon dioxide (CO_2). However, the new reactor fleet likely will not include the large 1,000-MW plants of the past but rather reflect a new breed of smaller and inherently safer reactors that are incapable of experiencing meltdown in the event of an accident.

A small test reactor of this type, Hermes, is slated to be built on the Oak Ridge K-33 site[36] and will be followed by larger units that could play a role in the nation's future power-production portfolio. Hermes is the first step in start-up company Kairos Power's effort to build and commercialize pebble-bed reactors, in which the enriched uranium oxide fuel is safely encased in small triple-coated pellets, only a half millimeter in size, to ensure that the radioactive fission products are confined to the pellets as the reactors operate.[37] Farrington Daniels conceived the pebble-bed reactor during the Manhattan Project. Pioneering studies led by Daniels were conducted at ORNL in the 1940s. In addition, Ultra Safe Nuclear Corporation announced in 2022 the siting of its Pilot Fuel Manufacturing operation in on the former K-25 site.[38] The purpose is to manufacture pellets of enriched uranium fuel encased in three coatings of impervious materials for advanced reactor projects.

Kairos is constructing this first-step reactor in Oak Ridge because of the proximate reactor expertise at ORNL and UT. Hermes and other pebble-bed reactors, with their improved safety profile, high-efficiency operation, and lower carbon footprint, could replace the more conventional reactors slated to be shut down in coming decades and, in the process, deliver electricity at a lower cost than natural gas. Senator Lamar Alexander, a sub-committee chairman of the important Appropriations Committee, played a critical role in spurring this new direction for nuclear power by supporting funding for DOE's Advanced Reactor Development Program. Kairos received significant funding from this program to design the Hermes prototype.

Meanwhile, ORNL is exploring another novel reactor design through its Transformational Challenge Reactor Demonstration Program.[39] In this instance, the new design will exploit UT and ORNL strength in additive manufacturing and nuclear engineering to create a 3D-printed nuclear reactor core using new materials able to withstand a reactor's extreme environment. The device will include numerous sensors allowing for near-autonomous operation and providing a constant flow of data on the core's operation. A team of engineers from three national labs (ORNL, Argonne, and Idaho) and UT is working on the design of the reactor core.[40]

TVA is considering creation of an advanced nuclear reactor technology park at the Clinch River Nuclear Site in Roane County, Tennessee. The park would contain one or more advanced nuclear reactors with a cumulative electrical output not to exceed 800 MW. This is the site where the Clinch River Breeder Reactor was to be built before being canceled in 1983. The new reactor scheme being contemplated is a Small Modular Reactor, now under development by several U.S. companies. These advanced reactors will vary in size from tens of megawatts up to hundreds of megawatts, have inherent safety features that render a core meltdown all but impossible, can be sited in locations not possible for larger nuclear plants and allow for incremental power additions. This reactor scheme could be the centerpiece of a rebuild of the nation's nuclear power industry.

This direction became more likely when the Nuclear Regulatory Commission (NRC) certified the design for what will be the United States' first small modular nuclear reactor.[41] Utilities could now pick the design for a 50-MW, advanced light-water small modular reactor by Oregon-based NuScale Power and apply to the NRC for a license.

Oak Ridge started with a primary focus on developing nuclear fission for a weapon and then for power reactors. While mostly a nuclear reactor town in the 1950s, Oak Ridge and East Tennessee are returning in part to a focus on reactor and isotope development. A 2023 survey by the East Tennessee Economic Council shows 229 nuclear companies in the State of Tennessee, including 154 in Oak Ridge-Knoxville area.[42] To further position Tennessee as a national leader for nuclear energy innovation, Governor Bill Lee appointed twenty members to the newly formed Tennessee Nuclear Energy Advisory

Council, and partnered with the Tennessee General Assembly to create a $50 million Nuclear Fund to provide assistance to growing or relocating nuclear power-related businesses in the state. The emphasis on nuclear energy in Oak Ridge, UT, and the state will continue to increase in future years.[43]

The continuing march toward nuclear fusion. The science and engineering assets of six countries plus the European Union are combining to build ITER, the world's forefront fusion experiment, in southern France. ORNL currently leads the U.S. contribution to the project, whose ultimate goal is to build a reactor able to sustain generation of electricity through the nuclear fusion of isotopes of hydrogen. Scientists understand well the physics of the fusion process, which creates the thermal energy of burning stars, but replicating that process in a human-built fusion reactor has remained a daunting challenge ever since Herman Postma and others started investigating this technology at ORNL in the 1950s.

ITER is scheduled to be operational around 2025, but, even if the test device operates as planned, the construction of commercial-scale fusion reactors capable of producing more energy than they consume—and in abundant quantities—will not come until mid-century. Meanwhile, DOE has provided funding to ORNL to build the Materials Plasma Exposure Experiment facility, which will study materials capable of withstanding the extremely harsh conditions (high temperature and intense neutron flux) inside fusion reactors.[44] [45] The next big step for the United States could be to build a smaller and cheaper fusion reactor.[46] In fact, a 2021 report by the U.S. National Academy of Sciences[47] calls for the construction, in the 2035 to 2045 timeframe, of a 50-MW pilot plant to demonstrate fusion power production, and ORNL is well positioned to compete for this project.

Although fusion reactors are farther in the future than advanced fission reactors, private companies are involved in fusion. Type One Energy is initiating a small facility in Oak Ridge to leverage local expertise in their use of advanced manufacturing and high-performance computing to build superconducting magnets for stellarators they envision to be the best confinement strategy for the eventual fusion reactor.[48]

Direct capture of CO2 from the atmosphere. The global temperature would continue to rise through this century due to the massive amount of carbon dioxide humans have pumped into the atmosphere over the past century—even if the United States and the rest of the world stopped burning fossil fuels; even if every country transitioned quickly to electric vehicles; even if wind, solar, and nuclear generation of electricity became universal in the next twenty years. In fact, the only plausible ways to lower the CO_2 content of the atmosphere are to double the number of carbon-sequestering forests on Earth (unlikely, in view of the ever-increasing global population and the infrastructure necessary to support it) or to remove the CO_2 from the atmosphere.

The latter approach is far beyond the chemistry and economics of existing processes (the current industrial CO_2 scrubber treats gases released from industrial plants using potassium hydroxide as the sorbent). A promising new CO_2-scrubbing technology under investigation at ORNL uses amino acids loaded with a bicarbonate and a synthesized organic compound containing guanidines (chemical groups common in proteins).[49] A chemical reaction of CO_2 with this solution produces an insoluble carbonate salt (representing the captured carbon), which precipitates out and regenerates the amino acid sorbent. The sorbent could then be recycled.[50] While this technology is still a long way from scaling up and solving the Earth's carbon problem, it does hold some promise, and one can perhaps envision future large plants that would slowly remove CO_2 from the atmosphere.

Isotope Separation—Back to the Future. Oak Ridge began during the Manhattan Project with enrichment of the uranium fissionable isotope as a primary driver. A fleet of calutrons using electromagnetic separation of isotopes was rapidly developed and built at Y-12, producing the ^{235}U serving as the core of one of the atomic bombs detonated during World War II. After the war ended, these calutrons were used for fifty years to separate and ship more than 230 enriched stable isotopes to customers around the country and the world, for a host of uses. These isotopes benefited researchers at many institutions, among them the UT and Vanderbilt nuclear physicists doing experiments at ORNL. After a thirty-year hiatus, DOE and Oak Ridge are returning to the stable isotope enrichment business.

On October 24, 2022, DOE Secretary Jennifer Granholm visited ORNL to break ground for the U.S. Stable Isotope Production and Research Center.[51] The facility is slated to receive $75 million in construction funding and will produce stable isotopes to meet the nation's increasing demand for enriched isotopes in medicine, industry, science, and national security.[52] This renewed capability should greatly benefit nearby university researchers, much as the calutrons did for decades.

TOWARD UT'S EXPANDED MANAGEMENT ASPIRATIONS AND CREATION OF A TECHNOLOGY CORRIDOR

Through eighty years of developing its partnership with Oak Ridge, UT greatly increased in reputation, expertise, and maturity, and, after twenty years of successfully managing ORNL, the university considered expanding its management portfolio to include the Y-12 National Security Complex in Oak Ridge and a sister nuclear weapons plant (Pantex) in West Texas.[53] [54] As the reader may recall, in 1982, UT administrators explored competing for the contract to manage the three federal facilities in Oak Ridge (plus one in Paducah, Kentucky). At the time, the university ultimately decided that it was not quite ready to manage Y-12, part of DOE's weapons complex.

However, times have changed, and the university explored an opportunity to team with Texas A&M University to manage Y-12 and the Pantex Plant near Amarillo, both of which fall under the purview of DOE's NNSA. Providing UT and Texas A&M with an opening, NNSA decided that it would not extend the contract of Consolidated Nuclear Security, LLC, then manager of the two facilities. NNSA announced[55] the award of a $2.8 billion contract to Nuclear Production One LLC to manage and operate the Oak Ridge facility and Pantex Plant, beginning in December 2021.

This was a short-lived arrangement, as NNSA terminated[56] this contract in May 2022 after the firms Bechtel and BWXT filed complaints with the Government Accountability Office, citing conflicts of interest in the contracting process.[57] On October 3, 2022, NNSA awarded a contract extension to Consolidated Nuclear Security for the management of Y-12 and Pantex.[58] The extension period is up to five years for Y-12 and up to three years for the Pantex Plant. Parallel to these management-contract challenges and opportunities, UT and Y-12 continue to build new partnerships programs. For example, a $9.5 million contract will spur more collaboration in areas ranging from engineering and materials science to business analytics and professional development and training.[59]

In the 1980s, Governor Lamar Alexander spoke frequently about developing the Oak Ridge Corridor, a virtual technology highway stretching between Oak Ridge and Knoxville and featuring a continuum of companies exploiting discoveries spun off from research at UT and ORNL or serving as subcontractors in support of the two institutions' ongoing projects. In addition to the anchor tenants, NTRC, the Manufacturing Research Facility, and IACMI, some enterprising start-ups have opened shop along the imagined high-tech highway situated midway between Knoxville and Oak Ridge along Pellissippi Parkway.

While the Oak Ridge Corridor has not reached the level of high-tech entrepreneurship found in North Carolina's Research Triangle or California's renowned Silicon Valley, success of the new UT-Oak Ridge Innovation Institute will help bridge the physical distance between UT and ORNL and inspire a new generation of entrepreneurs to capitalize on technologies emerging from the partnering institutions and turn them into profitable businesses.

However, building high-tech clusters—whether corridors, triangles, or valleys—is not an easy task. In fact, more than 90 percent of the nation's innovation-sector growth between 2005 and 2017 occurred in only five metro areas: Boston, San Francisco, San Jose, Seattle, and San Diego, according to a recent Brookings Institution report.[60] Another survey shows that Tennessee is not among the top-ten states receiving the most venture capital funding.[61] Breaking into that select company will be challenging for the East Tennessee region, but the prerequisites—chiefly the combined expertise and technical assets of UT and ORNL in supercomputing, neutron sciences, nanotechnology, and additive manufacturing, among other specialized fields—are already in place. As Lamar Alexander retired from the U.S. Senate in 2021, after serving three terms, he still held out hope that the corridor would, one day, reach its full potential.

ORNL is doing its part to encourage and support aspiring entrepreneurs—and perhaps prompt establishment of start-up businesses along the Oak Ridge Corridor—through its Innovation Crossroads program.[62] The program, started in 2017 and supported by DOE's Advanced Manufacturing Office and TVA, connects innovators with ORNL expert mentors and provides access to the laboratory's suite of technological tools and equipment over the two years of each innovator's stay at ORNL.

Four of the thirty-three participants in the first six cohorts[63] in this program were PhDs from the UT/ORNL Bredesen Center. Another participant, Anna Douglas from Vanderbilt University, was in the first cohort and has started a small company, SkyNano,[64] to make carbon nanotubes from captured pollutants at a TVA natural-gas plant. These nanotubes are a valuable material for making tires, ultralight bicycles, computer chips, and advanced batteries.[65] SkyNano is part of UT's Spark Cleantech Accelerator focused on developing decarbonization solutions and advancing supporting startup companies in East Tennessee.[66]

Techstars, an American seed-money accelerator founded in 2006 in Boulder, Colorado, performed in 2020 an assessment of the Knoxville-area innovation and entrepreneurial ecosystem.[67] The report identified this market's challenges, including attracting early-stage funding, establishing support for growth-stage companies, and opening doors to local research institutions and resources. A key recommendation of the report was to provide entrepreneurs with more early-stage funding through a national accelerator.

In response to this, ORNL, UT, and TVA pledged $9 million to launch a startup accelerator that will foster the growth of thirty technology companies in three years.[68] In 2022, the Techstars Industries of the Future accelerator in Knoxville sponsored ten startup companies, mentoring them to scale up their business plans while leveraging technical contacts with ORNL and UT and searching for investor funding.[69] Two of these startups plan to establish permanent operations in the area: liquid computer cooling developer FLUIX Inc. (from Florida) and bio-based materials developer Silvis Materials (from Colorado). The work of the former relates to cooling of supercomputer chips at ORNL, and the latter will set up a laboratory to work closely with Oak Ridge and UT. Techstars will sponsor ten more startups in 2023 and in 2024. Work is ongoing to develop more sources of startup funding for these new companies.[70] All of these activities will help move the Knoxville–Oak Ridge corridor along to eventually becoming a national innovation hub like the Research Triangle and Silicon Valley.[71]

THE FUTURE OF INNOVATION

Many examples of transformative innovations emerged in the world and in the United States over the years. In the past century, two American examples stand out. One resulted from the Manhattan Project, where a national crisis (World War II) prompted the federal government to assemble talented people and invest a great deal of money to see if the newfound nuclear fission process could be harnessed to build an atomic bomb. The fundamental science was not yet understood, and the way fission might be used to build a bomb posed an enormous challenge, even if the basic science could be worked out.

At the time, it was absolutely imperative that the United States succeed in building an atomic weapon before Nazi Germany's scientists could beat them to the punch. This came to pass, and the two atomic bombs dropped on Japan ended the war without need for land invasion of Japan, which would have cost many lives.

A second example of innovation is the successful development of telecommunications technology by AT&T Bell Laboratories Inc., better known by the abbreviated name Bell Labs, over a period of sixty years. This private industrial laboratory received generous funding from revenue generated by the then-monopolistic Bell Telephone Company, not through grants from the federal government. As with the Manhattan Project, the best and brightest were brought together at Bell Labs, but in the absence of an over-riding national emergency. Rather, Bell Labs provided ample funding and a creative environment to scientists, who were given the freedom to explore new fields that could perhaps contribute to improved telecommunications. Bell Labs also had a stable of fine engineers who could turn the results of discoveries in fundamental science into products that transformed the industry. As detailed in *The Idea Factory,* a book published a decade ago,[72] the transistor, solar panels, communications satellites, and lasers, are just a few of the innovative technologies that were developed by Bell Lab scientists, eight of whom received Nobel Prizes.

How does innovation occur in the 21st century, and what model will work best in driving it forward and sustaining it? Many of the technical challenges now facing our nation and the world relate to the interconnected topics of 1) energy generation and use, 2) restoration and preservation of the environment, and 3) climate change. Neither the Manhattan Project nor Bell Labs provides a useful model for addressing these challenges. Indeed, the single focus of the Manhattan Project would be far too limited, and clustering people and resources in a few related areas (e.g., materials science and electronics, as at Bell Labs) would not cast a wide enough net to begin to solve the amazingly complex challenges of energy, environment, and climate change.

To solve these difficult challenges, institutions must bring together experts from many fields to create a truly interdisciplinary environment, and a diversity of institutional participants will be required: national laboratories, universities, industries, and policy centers. Enlightened social policies encouraging a diversity of views and social activities are also paramount, to attract and retain the "best and brightest." Development of the forefront instruments needed for conducting exploratory R&D will require resources on a large scale, necessitating the direct involvement of the federal government.

While the hyper-focused Manhattan Project did not meet all these criteria of wide and open collaborations, the UT-Oak Ridge partnership that developed from it does and represents just the kind of interdisciplinary vehicle needed to address the pressing challenges of today. And, it is a collaboration model that will be increasingly emulated in future decades.

EIGHTY YEARS ON: AN ENLIGHTENED PERSPECTIVE

It is difficult to imagine a more daunting challenge than the one that led to the creation of the town we now know as Oak Ridge: to build a powerful weapon and end a devastating and deadly war—and to do it within a greatly compressed timeframe. And, it is equally difficult to imagine that a national laboratory will ever achieve, in a single stroke, a technological breakthrough or scientific discovery that will surpass, or even rival, its first.

However, the timeframe is no longer compressed—indeed, it now traces eighty years into the past but also stretches unencumbered into the future—and the tools and capabilities of ORNL and its cadre of scientists, along with those of its longstanding partner, the University of Tennessee, advanced in astounding ways that would undoubtedly have left those of early Manhattan Project days incredulous and amazed.

The impact of those early efforts to enrich uranium into a fissionable isotope continues to ripple through time and inform and influence scientific fields too numerous to count. Near the beginning, there was the Graphite Reactor, whose core sparked to life on November 4, 1943, creating fissionable plutonium but also—and perhaps more importantly—laying the groundwork for all the marvelous discoveries that were to follow at the laboratory and the university, engaging untold numbers of scientists, engineers, students, and educators in pursuits that would come to define their lives, shape their legacies, and change the world.

The people that led these many developments are gone, but their legacies continue as new leaders will emerge in the next twenty years. This is how partnerships form and continue over such a long time, just as the breakthroughs in science and technology stand the test of time. The future will be built on the past.

APPENDIX 1

Acronyms

AEC	Atomic Energy Commission
AEC-ORO	Atomic Energy Commission-Oak Ridge Operations
ANP	Aircraft Nuclear Propulsion
ANS	American Nuclear Society
ARE	Aircraft Reactor Experiment
ATLC	Atomic Trades and Labor Council
AVIDAC	Argonne Version of the Institute's Digital Automatic Computer
BAAM	Big Area Additive Manufacturing
BESC	Bioenergy Sciences Center
BS	Bachelor of Science
BSR	Bulk Shielding Reactor
CEW	Clinton Engineer Works
CIRE	Center for Interdisciplinary Research and Graduate Education
CO_2	Carbon dioxide
CPU	central processing unit
CRBR	Clinch River Breeder Reactor
DARPA	Defense Advanced Research Projects Agency
DHS	U.S. Department of Homeland Security
DOD	U.S. Department of Defense
DOE	U.S. Department of Energy
DOPE	Doctor of Pile Engineering
EERC	Energy, Environment, and Resources Center
EGCR	Experimental Gas-Cooled Reactor
EPA	Environmental Protection Agency
ERDA	Energy Research and Development Administration
ETEC	East Tennessee Economic Council
ETSU	East Tennessee State University
Exaflop	quintillion floating-point operations per second
FBI	Federal Bureau of Investigation

GeV	billions of electron volts
Gigaflop	billion floating point operations per second
GIS	Geographic information systems
GOCO	government-owned, contractor-operated
GPU	graphics processing unit
HBCU	Historically Black Colleges and Universities
HEW	U.S. Department of Health, Education, and Welfare
HFIR	High Flux Isotope Reactor
HRE	Homogeneous Reactor Experiment
HUAC	House Un-American Activities Committee
I&C	Instrumentation and Controls
IACMI	Institute for Advanced Composites Manufacturing Innovation
IEA	Institute for Energy Analysis
IP	Intellectual property
IUPAC	International Union of Pure and Applied Chemistry
JCAE	Joint Committee on Atomic Energy
JFAST	Joint Flow and Analysis System for Transportation
JIBS	Joint Institute for Biological Sciences
JICS	Joint Institute for Computational Sciences
JIHIR	Joint Institute for Heavy Ion Research
JINS	Joint Institute for Neutron Sciences
KCDC	Knox County Development Corporation
kW	kilowatts (1000 watts)
kWh	kilowatt hour
LITR	Low Intensity Test Reactor
LSU	Louisiana State University
M&O	management and operation
MAPS	Mobility Analysis and Planning System
MDF	Manufacturing Demonstration Facility
Megaflop	million floating point operations per second
MeV	million electron volts
MOU	Memorandum of Understanding
MS	Master of Science
MTR	Materials Test Reactor
MV	million volts
MW	megawatts (one million watts)
NAA	Neutron activation analysis
NASA	National Aeronautics and Space Administration
NEPA	National Environmental Policy Act
NIH	National Institutes of Health
NNSA	National Nuclear Security Administration
NRC	Nuclear Regulatory Commission

NSF	National Science Foundation
NTRC	National Transportation Research Center
ORACLE	Oak Ridge Automatic Computer and Logical Engine
ORAU	Oak Ridge Associated Universities
ORGDP	Oak Ridge Gaseous Diffusion Plant
ORIC	Oak Ridge Isochronous Cyclotron
ORINS	Oak Ridge Institute of Nuclear Studies
ORNL	Oak Ridge National Laboratory
ORR	Oak Ridge Research Reactor
ORSORT	Oak Ridge School of Reactor Technology
ORTEC	Oak Ridge Technical Enterprises Corporation
Petaflop	quadrillion floating point operations per second
PNNL	Pacific Northwest National Laboratory
Ppm	parts per million
R&D	research and development
RAM	Random Access Memory
REDC	Radiochemical Engineering Development Center
SDI	Strategic Defense Initiative
SNS	Spallation Neutron Source
SSC	Superconducting Super Collider
STEM	Science, technology, engineering, and mathematics
SURA	Southeastern Universities Research Association
TAT	Training and Technology Program
TCRD	Tennessee Center for Research and Development
TEC	Technology for Energy Corporation
Teraflop	trillion floating point operations per second
TeV	trillion electron volts
THEC	Tennessee Higher Education Commission
THRIVE	Total Health Risk Intervention and Education program
TRIUMF	TRI University Meson Facility
TSF	Tower Shielding Facility
TSR	Tower Shielding Reactor
TSSAA	Tennessee Secondary School Athletic Association
TVA	Tennessee Valley Authority
UNISOR	University Isotope Separator at Oak Ridge
URA	University Research Association
UT	University of Tennessee
WPA	Works Progress Administration
WWII	World War II

APPENDIX 2

Critical Connections in Building the UT-Oak Ridge Partnership

1945 October: first UT graduate physics courses taught on site in Oak Ridge; start of Oak Ridge Resident Graduate Program of many UT courses taught in Oak Ridge

1946 October: formation of Oak Ridge Institute of Nuclear Studies (ORINS), led by William Pollard

1948 Formation of UT-AEC Agricultural Research Laboratory (the name changed to the Comparative Animal Research Laboratory in the 1970s)

1950 Thirty UT faculty, representing ten departments, serving as consultants in three plants in Oak Ridge

1954 December: UT President Brehm notified that the Presidential executive order barring racial discrimination by private contractors receiving government funds would apply to UT when it renewed its contracts with AEC

1957 UT Department of Nuclear Engineering formed as a result of collaborations in reactor R&D

1963 Award of Ford Foundation grant to bring ORNL researchers to UT departments for teaching and research at UT on 20-percent appointments

1967 Formation of UT-Oak Ridge Graduate School of Biomedical Sciences located in Oak Ridge

1969 Formation of UT graduate ecology program with strong ties to ORNL

1970 Formation of University Isotope Separator at Oak Ridge (UNISOR) by 12 universities led by UT and Vanderbilt; first user facility at ORNL

1972 Formation of UT Environment Center by ORNL's John Gibbons, renamed in 1979 to Energy, Environment, and Resources Center

1983 February: signing of memorandum of understanding to establish the Distinguished Scientist Program, by Chancellor Jack Reese and ORNL director Herman Postma

1983	Award of state center of excellence grant to form the Science Alliance to promote greater cooperation between UT and ORNL
1984	October: opening of the Joint Institute for Heavy Ion Research, operated by UT, Vanderbilt University, and ORNL; first joint institute
1991	Initiation of collaborating scientist program of joint hires, later renamed joint faculty program
1998	November: formation of UT-Battelle LLC
2000	April: start of UT-Battelle managing ORNL
2005	Formation of Governor's Chair program of joint hires
2006	Completion of Spallation Neutron Source, a new user facility of importance to UT
2007	National Science Foundation award of Kraken supercomputer to UT, housed at ORNL
2010	Formation of Bredesen Center and first of two UT/ORNL PhD programs
2015	Award of grant to form the Institute for Advanced Composites Manufacturing Innovation at UT to bolster research in additive manufacturing
2020	Formation of UT-Oak Ridge Innovation Institute

APPENDIX 3

UT-Oak Ridge Innovation Institute

Bredesen Center: formed in 2010, two joint PhD programs

Governor's Chairs: formed in 2005; twenty have been hired; predecessor program (Distinguished Scientists) began in 1984

Joint Faculty Program: predecessor program (Collaborating Scientists) began in 1991; current incarnation as Joint Faculty Program began in 2001

Joint Institute for Advanced Materials: formed in 2005; construction of its research facility at the UT Research Park at Cherokee Farm was completed in 2016

Joint Institute for Biological Sciences: formed in 1997; located in state building at ORNL

Joint Institute for Computational Science: formed in 1991; located at ORNL

Joint Institute of Nuclear Physics and Applications: predecessor institution (Joint Institute for Heavy Ion Research) formed in 1983; renamed in 2013; formerly located at ORNL in building constructed with state, private, and federal funds

Science Alliance: formed in 1984

Shull Wollan Center: a Joint Institute for Neutron Sciences; formed in 1998; located in state building at ORNL

Graduate School of Genome Science and Technology: formed in 1998; PhD program

APPENDIX 4

Directors of Oak Ridge National Laboratory (and Predecessor Organizations)

Martin Whitaker, 1943–1945
James Lum, 1945–1947: executive director, operations
Eugene Wigner, 1946–1947: director for research and development
Prescott Sandidge, 1947–1948: executive director, operations
Alvin Weinberg, 1948–1955: director for research and development
Nelson Rucker, 1948–1950: executive director, operations
Clarence Larson, 1950–1955
Alvin Weinberg, 1955–1973
Floyd Culler, interim,1973–1974
Herman Postma, 1974–1988
Alex Zucker, interim, 1988–1989
Al Trivelpiece, 1989–2000
Bill Madia, 2000–2003
Jeff Wadsworth, 2003–2007
Thom Mason, 2007–2017
Thomas Zacharia, 2017–2022
Jeff Smith, interim, 2023
Stephen Streiffer, 2023–

Notes

Introduction

1. Appendix 2 gives a listing of these twenty-three critical connections.

1. The Early 1940s

1. Milton M. Klein, *Volunteer Moments: Vignettes of the History of the University of Tennessee, 1794–1994*, 2nd ed. (Knoxville: Office of the University Historian, the University of Tennessee, 1996). For the early history of UT, see James Riley Montgomery, "Threshold of a New Day, the University of Tennessee 1919–1946," *University of Tennessee Record* 74, no. 6 (November 1971), https://academic.oup.com/jah/article-abstract/59/2/458/820211.

2. T. R. C. Hutton, *Bearing the Torch: The University of Tennessee, 1794–2010* (Knoxville: University of Tennessee Press, 2022), 9.

3. Montgomery, 132.

4. Ibid., 129.

5. William Andrew Gault, "Investigations of Some Rhenium Compounds and the Structure of the Mesoperrhenate Ion" (PhD diss., Iowa State University, 1969), https://lib.dr.iastate.edu/rtd/3577.

6. W. Noddack and I. Noddack, "Die Herstellung von einem Gram Rhenium," *Zeitschrift für Anorganische und Allgemeine Chemie* 183, no. 1 (1929): 353–75, doi:10.1002/zaac.19291830126.

7. Klein, *Volunteer Moments*, 63–64.

8. "A. D. Melavan-Rhenium Scholarships," University of Tennessee Department of Chemistry, https://chem.utk.edu/undergraduate-programs/a-d-melaven-rhenium-scholarships/.

9. "History of the Department," University of Tennessee Department of Chemical and Biomolecular Engineering, http://cbe.utk.edu/history/.

10. For a history of the discovery of the neutron in England in 1932 up through the beginning of the Manhattan Project, see Richard Rhoades, *The Making of the Atomic Bomb* (New York: Simon and Schuster Paperbacks, 1986), 162–167 for discovery of the neutron; 256–275 for discovery of nuclear fission; and 279–442 for events leading to Manhattan Project.

11. Many books discuss the history of the Manhattan Project. In addition to the ones referenced herein, see Richard G. Hewlett and Oscar E. Anderson, *The New World, 1939/1946: A history of the United States Atomic Energy Commission* (University Park: Pennsylvania State University Press, 1962), https://www.energy.gov/sites/default/files/2013/08/f2/HewlettandAndersonNewWorldNoBookmarks.pdf.

12. Arthur Holly Compton, *Atomic Quest: A Personal Narrative* (New York: Oxford University Press, 1956), 6–7.

13. The Army Corps of Engineers led the initial activity; their units were typically named after the headquarters city of the engineer group leading the project, so it was called the Manhattan Engineer District and then the Manhattan Project.

14. Talks delivered by Senator Howard Baker in Oak Ridge venues and witnessed by Ray Smith and Lee Riedinger.

15. "Atom Secret Only One He Ever Kept, McKellar Says," *Oak Ridge Journal*, August 22, 1946.

16. Kenneth D. Nichols, *The Road to Trinity: A Personal Account of How America's Nuclear Policies Were Made* (New York: William Morrow, 1987), 120.

17. "Atom Secret Only One He Ever Kept, McKellar Says," 1.

18. Leslie R. Groves, *Now It Can Be Told: The Story of the Manhattan Project* (1962; repr., Cambridge, MA: Da Copo Press, 1983), 13–14.

19. "Oral History of Lester Fox," interview by Keith McDaniel, October 19, 2004, Oak Ridge Public Library Digital Collections, available at https://cdm16107.contentdm.oclc.org/digital/collection/p15388coll1/search.

20. "Oral History of Jerry Shattuck," interview by Keith McDaniel, February 2, 2012, Oak Ridge Public Library Digital Collections, available at https://cdm16107.contentdm.oclc.org/digital/collection/p15388coll1/search.

21. For an account of the early development of Oak Ridge, see George O. Robinson Jr., *The Oak Ridge Story: The Saga of a People Who Share in History*, 5th ed. (Oak Ridge Heritage and Preservation Association, 2019). The original edition of the book appeared in 1950.

22. Dewaine A. Speaks and Ray Clift, *East Tennessee in World War II (Military)* (Charleston, SC: The History Press, 2016).

23. *Manhattan District History*, bk. 1, vol. 8, *Appendix A, Chart 1: Manhattan District Employment Chart for Contractors, August 1942* (December 1948), available at https://www.osti.gov/includes/opennet/includes/MED_scans/Book%201%20-%20General%20-%20Volume%208%20-%20Personnel.pdf.

24. "New Fares, Same as Knoxville, Begin on Buses Here August 20," *Oak Ridge Journal*, August 17, 1944.

25. "Statistics Tell Women Workers Part of Project," *Oak Ridge Journal*, November 23, 1945.

26. Ruth Carey, "Change Comes to Knoxville," in *These Are the Voices: The Story of Oak Ridge 1942–1970*, ed. James Overholt (Oak Ridge, TN: Children's Museum of Oak Ridge, 1987), 214–18.

27. Dixon Johnson, "Brass Hat Who Doubles—Lt. Col. Donald C. Williams, Responsible for Safety and Fire Protection in the Reservation at Oak Ridge Serves as Piano Soloist with Its Symphony Orchestra," *Tennessean*, March 16, 1947. For a description of many of the early relationships, see a book by two UT history professors: Charles W. Johnson and Charles O. Jackson, *City Behind a Fence: Oak Ridge, Tennessee, 1942–1946*, (Knoxville: University of Tennessee Press, 1981). Their

second chapter, "Good Fences Make Bad Neighbors," includes points of view of those displaced, those newly arrived, the Army's "benevolent dictatorship," and frequent interactions among members of the various communities.

28. Courtney H. Hodges, "Military Area No. 1 of the State of Tennessee," *Federal Register* (June 1943): 8924–8925, https://www.govinfo.gov/content/pkg/FR-1943-06-30/pdf/FR-1943-06-30.pdf.

29. Nichols, *The Road to Trinity*, 100–120.

30. "Governor to Confer on Clinton Project," *Tennessean*, October 29, 1943.

31. Lenore Fine and Jesse A. Remington, *The Corps of Engineers: Construction in the United States* (Washington, D.C.: Center of Military History, U.S. Army, 1989), 594, 686.

32. Cameron Reed, "Kilowatts to Kilotons: Wartime Electricity Use at Oak Ridge," *American Physical Society Forum on the History of Physics*, 12, no. 6, 5 (2015), https://www.aps.org/units/fhp/newsletters/spring2015/oak-ridge.cfm.

33. A calutron is a type of mass spectrometer in which a sample is ionized and then accelerated by electric fields and deflected by magnetic fields. Since the ions of the different isotopes of an element have the same electric charge but different masses, the magnetic field deflects more the heavier isotopes, causing the beam of particles to separate into several beams by mass, striking the collection plate at different locations. This provided the separation of isotopes of uranium needed for the Manhattan Project, utilizing many separate calutron units to achieve the needed enrichment in fissionable ^{235}U.

34. For a description of the electromagnetic separation process, see: H. D. Smyth, "Electromagnetic Separation of the Uranium Isotopes," in *Atomic Energy For Military Purposes* (York, PA: Maple Press, 1945); Vincent C. Jones, "The Electromagnetic Process," in *Manhattan: The Army and the Atomic Bomb* (Washington, D. C.: Center of Military History, U.S. Army, 1985).

35. Clarence E. Larson, "The Role of Chemistry in the Oak Ridge Electromagnetic Project," *Bulletin of the History of Chemistry* 28, no. 2 (2003): 101–109; Gregg Herken, *Brotherhood of the Bomb* (New York: Henry Holt & Co., 2002), 115.

36. Smyth, "Electromagnetic Separation of the Uranium Isotopes."

37. Nichols, 42.

38. There are 14.58 troy ounces per pound and 2,000 pounds per ton.

39. Nichols, 100–120.

40. For a description of the gaseous diffusion process, see: Smyth, "Separation of the Uranium Isotopes by Gaseous Diffusion"; Jones, "The Gaseous Diffusion Process"; William J. Wilcox Jr., *K-25: A Brief History of the Manhattan Project's 'Biggest' Secret* (TN: Oak Ridge Public Library, 2008), http://docshare01.docshare.tips/files/7716/77162930.pdf; and, for more popular explanations of gaseous diffusion, *K-25 Virtual Museum*, http://www.k-25virtualmuseum.org/site-tour/index.html.

41. Donald B. Trauger, *Horse Power to Nuclear Power: Memoir of an Energy Pioneer* (Hillsboro, KS: Hillsboro Press, 2002), 111.

42. Each year, the Southeastern Section of the American Physical Society gives

the Francis G. Slack Award for service to this society; Lee Riedinger won this award in 2005.

43. Alex Barshai, "Nickel Powder: Manhattan's secret ingredient," *emew Blog*, May 12, 2021, https://blog.emew.com/why-nickel-powder-is-the-secret-ingredient.

44. Trauger, *Horse Power to Nuclear Power*, 126.

45. William G. Pollard and R. D. Present, "On Gaseous Self-Diffusion in Long Capillary Tubes," *American Physical Society* 73 (April 1948): 762.

46. Trauger, 127.

47. *Voices of the Manhattan Project* (Atomic Heritage Foundation), "James C. Stowers's Interview" (n.d.), https://www.manhattanprojectvoices.org/oral-histories/james-c-stowerss-interview; Stephane Groueff, *Manhattan Project: The Untold Story of the Making of the Atomic Bomb* (Boston, MA: Little Brown, 1967), 277. The first source in this note is an interview of James C. Stowers and was conducted by Stephane Groueff on March 3, 1965; the Atomic Heritage Foundation was granted exclusive rights for its use on their website.

48. Groves, *Now It Can Be Told*, 95, 111.

49. R. P. Prince and A. Milton Stanley, *Journal of East Tennessee History* 72 (2000): 82.

50. For a description of the plutonium production process, see: Smyth, "The Plutonium Production Problem as of February 1943" and "The Plutonium Problem, January 1943 to June 1945"; Jones, "The Pile Process."

51. The pile contained 400 tons of graphite, six tons of uranium metal, and 50 tons of uranium oxide, according to: Smyth, *A General Account of the Development of Methods of Using Atomic Energy for Military Purposes under the Auspices of the United States Government, 1940–1945* (Washington, D.C.: U.S. Government Printing Office, 1945).

52. U.S. Department of Energy, *The First Reactor*, Report DOE/NE-0046 (December 1982, https://www.osti.gov/includes/opennet/includes/Understanding%20the%20Atom/The%20First%20Reactor%20V.2.pdf; "How First Controlled Reaction Looked on Paper—Birth Certificate for a New Era," *Oak Ridge National Laboratory The News* (November 1951): 1.

53. A major X-10 contribution was developing the radiochemical techniques to separate minute amounts of created ^{239}Pu from the uranium fuel rods irradiated in the X-10 Pile.

54. D. Ray Smith, "Historically Speaking: Sam Beall—Director of Two ORNL Divisions," *Oak Ridger*, October 1, 2013, https://www.oakridger.com/story/news/local/2013/10/01/historically-speaking-sam-beall-8212/42797742007/.

55. Georgiana Vines, "Where Are They Now: Rotary Scholarship Launched Sam Beall's Engineering Career in Nuclear Industry," *Knoxville News Sentinel*, January 2, 2016; http://archive.knoxnews.com/entertainment/life/where-are-they-now-rotary-scholarship-launched-sam-bealls-engineering-career-in-nuclear-industry-280–363854391.html.

56. Marion Alexander, "The Schools of Oak Ridge," in *These Are the Voices*, ed. James Overholt, 135–42.

57. D. Ray Smith, "An Adventure in Democratic Administration in Oak Ridge Schools," *Oak Ridger*, November 28, 2006, 1.

58. D. Ray Smith, *Historical Sketches of Oak Ridge Schools* (self-pub., Oak Ridge, TN, 2007); Smith's self-published book also appeared in weekly newspaper columns of *Historically Speaking* in 2007, as well as the annual edition of *2007 Historically Speaking* book.

59. Ibid.

60. Charles W. Johnson and Charles O. Jackson, *City Behind a Fence: Oak Ridge Tennessee 1942–1946* (Knoxville: University of Tennessee Press, 1981), 47.

61. Klein, *Volunteer Moments*, 62.

62. "University of Tennessee Will Offer Free Courses in Safety Engineering," *Oak Ridge Journal*, September 7, 1944.

63. Harry S. Truman, "Statement on the Atomic Bomb, August 8, 1945," *Public Papers of the Presidents: Harry S. Truman* 1 (1945), https://web.mit.edu/21h.102/www/Primary%20source%20collections/World%20War%20II/Truman,%20Atomic%20Bomb.htm.

64. Richard G. Hewlett and Oscar E. Anderson, *The New World, 1939/1946: A History of the United States Atomic Energy Commission* (University Park: Pennsylvania State University Press, 1962), 71, https://www.energy.gov/sites/default/files/2013/08/f2/HewlettandAndersonNewWorldNoBookmarks.pdf.

65. Groves, *Now It Can Be Told*, 138–48.

66. Composed of 4,900 civilian guards (each facility had its own guard force), 740 military police, 400 civilian police, plus a Women's Army Corps detachment of 275.

67. "Oral History of Joseph Dooley," interview by Jennifer Thonhoff, May 18, 2005, Oak Ridge Public Library.

68. "Oral History of Reba Holmberg," interview by Anne Garland, April 30, 2002, Oak Ridge Public Library.

69. J. Parnell Thomas and S. V. Jones, "Reds in Our Atom-Bomb Plants," *Liberty*, June 21, 1947, 15–17.

70. D. Ray Smith, "Why a Soviet spy in Oak Ridge was never arrested or recognized," *Oak Ridger*, June 17, 2022.

71. Ann Hagedorn, *Sleeper Agent: The Atomic Spy in America Who Got Away* (New York: Simon & Schuster, 2021), 42.

72. Ibid., 63.

73. Ibid., 77.

74. Michael Walsh, "George Koval: Atomic Spy Unmasked," *Smithsonian Magazine*, May 2009, https://www.smithsonianmag.com/history/george-koval-atomic-spy-unmasked-125046223/.

75. Polonium is a highly radioactive element discovered in 1898 by Marie

Curie, and its isotope polonium-210 (^{210}Po), which is highly toxic to humans, was suspected in the poisoning death of former Russian spy Alexander Litvinenko in London in 2006. This isotope emits an alpha particle which can then produce a neutron when a beryllium foil encloses ^{210}Po, making this an initiator of a chain reaction in an atomic bomb.

76. Hagedorn, 201–2.

77. D. Ray Smith, "Spy Worked at Y-12, Tennessee Eastman Co," *Oak Ridger*, December 29, 2014, https://www.oakridger.com/article/20141229/NEWS/141229904.

78. "CIA Official Dispatch, Subject: Espionage—Russell Alton McNutt," August 25, 1950, https://www.cia.gov/library/readingroom/docs/DOC_0000014625.pdf; John Earl Haynes, Harvey Klehr, and Alexander Vassiliev, *Spies: The Rise and Fall of the KGB in America* (New Haven, CT: Yale University Press, 2010), 34–35.

79. Hagedorn, 73.

80. *Voices of the Manhattan Project* (Atomic Heritage Foundation), "James Earl Haynes's Interview" (n.d.), https://www.manhattanprojectvoices.org/oral-histories/john-earl-hayness-interview/; Bob Considine, "'The Great A-Bomb Robbery': Most Daring Theft in History," *Tennessean*, December 9, 1951. James Earl Haynes's interview was conducted by Cindy Kelly and the Atomic Heritage Foundation has full rights.

81. Hagedorn, 88.

82. Testimony of James Sterling Murray and Edward Tiers Manning Regarding Clarence Hiskey and Arthur Adams. Hearings, Eighty-first Congress, First Session. August 14 and October 5, 1949 (2015), 877-899, by James Sterling Murray and Edward Tiers Manning, available at Amazon; "Teachers Cited by Former Red," *Tennessean,* September 1, 1949, 2.

83. "Clarence Hiskey," Wikipedia, last modified on August 14, 2023, https://en.wikipedia.org/wiki/Clarence_Hiskey.

2. *The Late 1940s*

1. *ORNL Review* 25, nos. 3–4 (1992): 30.

2. As the Clinton Pile started operating, the X-10 scientific emphasis expanded to include research in physics, chemistry, biology, and materials and became known as Clinton Laboratories.

3. Montgomery, "Threshold of a New Day, the University of Tennessee 1919–46," 181–82.

4. Connie L. Lester, "Kenneth L. Hertel," in *Tennessee Encyclopedia* (Knoxville: University of Tennessee Press, 2017), https://tennesseeencyclopedia.net/entries/kenneth-l-hertel/.

5. "UTRF History," UT Research Foundation, accessed August 15, 2023, https://utrf.tennessee.edu/about/utrf-history/.

6. Alvin H. Nielsen, "The University of Tennessee Physics Department: The

Early Years" (talk, Sigma Pi Sigma [University of Tennessee], Knoxville, TN, May 17, 1979).

7. Henry DeWolf Smyth, *Atomic Energy for Military Purposes: A General Account of the Development of Methods of Using Atomic Energy for Military Purposes* (NJ: Princeton University Press, 1945).

8. William G. Pollard, *ORAU: From the Beginning* (Oak Ridge, TN: Oak Ridge Associated Universities, 1980), 2–3.

9. K. L. Hertel, Letter to Dean L. R. Hesler submitting the Annual Report for 1945–46, June 26, 1946, 2, 3b.

10. Pollard, *ORAU*, 3.

11. Hilton A. Smith, "The Oak Ridge Resident Graduate Program of the University of Tennessee" in Pollard, *ORAU*, 57–59.

12. K. L. Hertel, Letter to Dean L. R. Hesler, submitting the Annual Report for 1945–46, June 26, 1946, 9–11, courtesy of Betsey B. Creekmore Special Collections and University Archives, University of Tennessee, Knoxville.

13. Jagdish Mehra and Helmut Rechenberg, *The Historical Development of Quantum Theory* (New York: Springer Publishing, 1982), 990–91.

14. Katharine Way, "The Liquid-Drop Model and Nuclear Moments," American *Physical Society* 55, no. 10 (May 1939): 963–65.

15. John Archibald Wheeler, *Some Men and Moments in the History of Nuclear Physics: The Interplay of Colleagues and Motivations* (Minneapolis: University of Minnesota Press, 1979), 266.

16. D. Ray Smith and Carolyn Krause, "Katherine Way and Her Influence on Oak Ridge," Historically Speaking column, *Oak Ridger*, December 6, 2016.

17. Thomas F. X. McCarthy, "My! How ORAU Has Grown," *These Are Our Voices*, ed. James Overholt, 374–81.

18. Pollard, 6–7.

19. Ibid.

20. Ibid., 9–10.

21. *Manhattan District History*, bk. 1, vol. 4, *Auxiliary Activities, Chapter 10, The Oak Ridge Institute of Nuclear Studies* (December 1948). Available at https://www.osti.gov/includes/opennet/includes/MED_scans/Book%20I%20-%20General%20-%20Vol.%204-Chapters%209-10.pdf

22. Richard G. Hewlett and Francis Duncan, *Atomic Shield, 1947–1952*, vol. 2, *A History of the United States Atomic Energy Commission* (Springfield, VA: National Technical Information Service, 1972), 635–36.

23. *ORNL Review* 9, no. 4 (1976): 64–65.

24. Hewlett and Duncan, *Atomic Shield, 1947–1952*, vol. 2, *A History of the United States Atomic Energy Commission*, 70.

25. The sphere of plutonium (Pu) would not become critical and support a chain reaction in normal geometry since the emission of neutrons from plutonium fission would not be sufficient to cause enough other Pu nuclei to fission and start a

chain reaction. A rapid and symmetrical compression of the sphere of plutonium would drive the material to a critical stage, as more neutrons would be absorbed and more fission events would occur.

26. The curie is a quantity of radioactive material, named after the famous scientist Marie Curie.

27. J. A. Marinsky, L. E. Glendenin, and C. D. Coryell, "The Chemical Identification of Radioisotopes of Neodymium and of Element 61," *Journal of American Chemical Society* 69, no. 11 (1947): 2781–85.

28. *ORNL Review* 9, no. 4 (1976): 50.

29. Julie E. Coryell, ed., *A Chemist's Role in the Birth of Atomic Energy: Interviews with Charles Coryell*, (Menlo Park, CA: Promethium Press, 2012).

30. The Executive Committee, Oak Ridge Institute of Nuclear Studies, "A Nuclear Research Institute at Oak Ridge," *American Association for the Advancement of Science* 103, no. 2685 (1946): 705–6, https://www.science.org/doi/10.1126/science.103.2685.705.

31. Pollard, 133.

32. The Atomic Energy Commission was formed August 1, 1946, taking over operation of the Manhattan Project facilities.

33. Pollard, 24–25.

34. Hilton A. Smith, in Pollard, *ORAU*, 58.

35. "1100 Persons Enrolled in Education Courses," *Oak Ridge Journal*, December 5, 1946.

36. "U-T's Top Instructors to Conduct Graduate Training Program Here," *Oak Ridge Journal*, September 26, 1946.

37. "Psychology, Botany Training Planned," *Oak Ridge Journal*, March 11, 1948.

38. Hilton A. Smith, in Pollard, *ORAU*, 58.

39. "Igor Gouzenko, Unmasked Spies," *The New York Times*, June 30, 1982, https://www.cia.gov/library/readingroom/docs/CIA-RDP90-00552R000202330005-8.pdf.

40. "ORES Celebrates First Anniversary," *Oak Ridge Journal*, September 26, 1946.

41. "Scientists to Tour State in Atomic Energy Forums," *Oak Ridge Journal*, August 15, 1946.

42. "Seitz to Direct School at Clinton Laboratories," *Oak Ridge Journal*, September 29, 1946.

43. "Clinton Lab School to Open Sept. 15–Post Doctorate Course to Draw Leading Scientists and Engineers for Year's Study," *Oak Ridge Journal*, September 6, 1946.

44. "Naval Engineers Here to Study Atomic Energy," *Oak Ridge Journal*, November 7, 1946.

45. Leland Johnson and Daniel Schaffer, "High-Flux Training," *ORNL Review* 25, nos. 3–4 (1992): 43–46.

46. "MIT Will Establish Engineering School at Carbide Plants Here," *Oak Ridge Journal*, April 15, 1948.

47. Theodore Rockwell, *Creating the New World: Stories & Images from the Dawn of the Atomic Age*, 2nd ed. (Bloomington, IN: AuthorHouse, 2003), 88–89.

48. *ORNL Review* 25, nos. 3–4 (1992): 91–92.

49. Built and operated by the H. K. Ferguson Company late in the Manhattan Project, the S-50 plant shut down in September 1945.

50. William J. Wilcox Jr., *An Overview of the History of Y-12, 1942–1992* (Oak Ridge, TN: American Museum of Science and Energy, 2001), 19–22.

51. Steve Polston, "What is gaseous diffusion?" *The Paducah Sun*, October 14, 2022, https://www.paducahsun.com/news/what-is-gaseous-diffusion/article_69e2005c-0a7d-55c2-baeb-4903f8752453.html.

52. *ORNL Review* 25, nos. 3–4 (1992): 23.

53. Ibid., 36.

54. "Fifty News, Radio Men Tour Clinton Lab–First Radioisotopes Shipped from Area," *Oak Ridge Journal*, August 8, 1946.

55. Alvin M. Weinberg, "1948-Survival and Purpose," *ORNL Review* 9, no. 4 (1976): 57–60.

56. Hewlett and Duncan, 38–39.

57. Louis Slotin was on duty at the Clinton Pile on November 4, 1943, when uranium slugs were being loaded one by one to approach criticality in the reactor (enabling a chain reaction). In the middle of the night, Slotin realized that criticality would be reached sooner than expected. He drove into Oak Ridge and, at 4 a.m., he knocked on the door of the houses of M. D. Whitaker and R. L. Doan (Clinton Laboratories directors) to wake them and bring them to the Clinton Pile. (A large mural depicting this historic wakening hangs on a wall at the Graphite Reactor building.) After their arrival, a few more slugs were inserted, a critical mass of uranium was achieved, and the reactor went critical for the first time.

58. Arthur H. Snell and Alvin M. Weinberg, "History and Accomplishments of the Oak Ridge Graphite Reactor," *Physics Today* 17, no. 8 (1964): 32.

59. "Radiation Burns Fatal to Former CEW Scientist," *Oak Ridge Journal*, June 6, 1946.

60. Louis Henry Hempelman, Clarence C. Lushbaugh, and George L. Voelz, "What Has Happened to the Survivors of the Early Los Alamos Nuclear Accidents?" (report [LA-UR-79-2802] presented at the *Conference for Radiation Accident Preparedness*, Oak Ridge, TN, October 1979.

61. Hewlett and Duncan, 66–67.

62. Ibid.

63. Trauger, 133.

64. Lucie Levine, "Going Nuclear: The Manhattan Project in Manhattan," https://www.6sqft.com/going-nuclear-the-manhattan-project-in-manhattan/.

65. D. Ray Smith, *A Look Back at Union Carbide's* [first] *20 Years in Nuclear Energy [The Gaseous Diffusion Plants]*, accessed June 8, 2020, https://www.y12.doe.gov/sites/default/files/assets/document/2012-12-28.pdf.

66. Trauger, 164.

67. Ibid., 181.

68. William J. Wilcox, Jr., *A Brief History of K-25* (Oak Ridge, TN: The Secret City Store, 2007), 105, http://docshare01.docshare.tips/files/7716/77162930.pdf.

69. Hewlett and Duncan, 66–68.

70. Pollard, 11.

71. *ORNL Review* 9, no. 4 (1976): 30.

72. "National Lab Established at Oak Ridge; University to Take over from Monsanto Chemical Company, Industrial Groups, Southern Universities Will Participate in Research Work Here," *Oak Ridge Journal*, September 25, 1947.

73. "Clinton Research Group Directors Issue Statement," *Oak Ridge Journal*, September 25, 1947.

74. Hewlett and Duncan, 123–24.

75. "U. of C. Withdraws From X-10 Bid," *Oak Ridge Journal*, January 2, 1948.

76. Pollard, 26–28.

77. "Atom Research Unit Shifted to Chicago," *Tennessean*, January 1, 1948.

78. "Work to Continue Vigorously at Clinton Lab, States AEC's Dr. Bacher in Speech to Rotary," *Oak Ridge Journal*, January 22, 1948.

79. H. B. Teeter, "Tennessee May Keep Oak Ridge Laboratory under New AEC Plan," *Tennessean*, February 1, 1948.

80. "Clinton National Lab to Be Renamed Soon, Radio Talk Here Reveals: Oak Ridge National Laboratory to Be New Title; Status Unchanged," *Oak Ridge Journal*, February 5, 1948.

81. "Ridge Seen as Giant Center of Research," *Oak Ridge Journal*, March 11, 1948.

82. "U-T Now Using Radioisotopes for Research," *Oak Ridge Journal*, September 11, 1947.

83. "Radiations Laboratory Is Established by U-T," *Oak Ridge Journal*, November 20, 1947.

84. "Isotope Research Is Planned at U-T," *Oak Ridge Journal*, May 8, 1947.

85. David E. Lilienthal, Robert F. Bacher, Sumner T. Pike, and Lewis L. Strauss, *Fifth Semiannual Report to Congress: Atomic Energy Development* (Washington, D.C.: U.S. Government Printing Office, 1949), 92–93, https://www.osti.gov/biblio/1362100.

86. Edward R. Ricciuti, "Atomic Age Animals: Granny," in *Animals in Atomic Research* (Atomic Energy Commission, 1967), 8–9, https://www.osti.gov/includes/opennet/includes/Understanding%20the%20Atom/Animals%20in%20Atomic%20Research.pdf, see https://files.eric.ed.gov/fulltext/ED042652.pdf for full text and bibliographic information.

87. D. Ray Smith, "A Brief History of the UT-AEC Agricultural Research Laboratory: Part Three," Historically Speaking column, *Oak Ridger*, November 3, 2009, http://smithdray1.net/historicallyspeaking/2009/11-3-09%20A%20brief%20history%20of%20the%20UT-AEC%20Agricultural%20Research%20Laboratory%20-%20part%20three.pdf.

88. “White Cows of Los Alamos Becoming White Elephants,” *Oak Ridge Journal*, February 13, 1947.

89. “No Irradiated Milk from Still Grey Cows,” *Oak Ridge Journal*, April 24, 1946.

90. Smith, “A Brief History of the UT-AEC Agricultural Research Laboratory,” Historically Speaking column, *Oak Ridger*, October 20, 2009, https://www.oakridger.com/article/20091020/NEWS/310209998.

91. “UT, AEC Widen Field of Radiation Study: Experiments on Exposed Cattle Now Cover Plant and Animal Life,” *Tennessean*, September 18, 1949.

92. Daniel N. Tapper, Robert H. Wasserman, and Edgar L. Gasteiger, “Cyril L. Comar: March 28, 1914–June 11, 1979” (Cornell University Library, 1979), https://ecommons.cornell.edu/handle/1813/17965.

93. Lilienthal, Bacher, Pike, and Strauss, *Sixth Semiannual Report to Congress: Atomic Energy and the Life Sciences* (Washington, D.C.: U.S. Government Printing Office, 1949), 101–2, https://www.osti.gov/biblio/1362103.

94. Twentieth Anniversary Biology Division of Oak Ridge National Laboratory, (Atomic Energy Commission, 1966) 6, http://dx.doi.org/10.2172/1603086.

95. Marshall H. Brucer, “Early Clinical Research” in Pollard, 67–75.

96. Ibid.

97. Pollard, 31–32, 99-100.

98. Lilienthal, Bacher, Pike, and Strauss, *Fifth Semiannual Report to Congress*, 97.

99. Lilienthal, Bacher, Pike, and Strauss, *Sixth Semiannual Report to Congress*, 9, 77, 110–16.

100. “Atomic Puff Opens Gates at Oak Ridge: 40,000 Visitors Crowd into War City for Festivities; Notables Are Speakers,” *Tennessean*, March 20, 1949.

101. Richard Smyser, *Oak Ridge 1942–1992, A Commemorative Portrait* (TN: Oak Ridge Community Foundation, 1992), 24–28.

102. Pollard, 33–34.

103. Alvin M. Weinberg, “Oak Ridge National Laboratory,” *Science* 109, no. 2828 (1949): 245–48.

3. *The Decade of the 1950s*

1. Louis T. Iglehart, “Impact of ORNL’s Atomic Energy Research Program Felt Vividly at UT Where ‘Atom Shines In Academic Sun,’” *Oak Ridge National Laboratory—The News* 3, no. 50 (June 1951): 1.

2. Gordon Dean, *Eleventh Semiannual Report of the United States Atomic Commission* (Washington, D.C.: U.S. Government Printing Office, 1952), 14.

3. Manhattan District History, bk. 1, vol. 12, “Clinton Engineer Works Central Facilities” (December 1946), 9. Available at https://www.osti.gov/includes/opennet/includes/MED_scans/Book%20I%20-%20General%20-%20Volume%2012%20-%20Clinton%20Engineer%20Work%20-%20Centr.pdf.

4. *Eighth Annual Report, U. S. Federal Works Agency* (Washington, D.C.: U.S. Government Printing Office, 1947), 46.

5. *Emergency Educational Aid for Government Reservations, Hearings before the Committee on Education and Labor*, House of Representatives, vol. 1, May 5 and 6, 1947 (Washington, D.C.: U.S. Government Printing Office, 1947), 84–96.

6. Dean, *Eleventh Semiannual Report of the United States Atomic Commission*, 14.

7. Fred W. Ford and Fred Peitzsch, *A City is Born: The History of Oak Ridge, Tennessee*, Oak Ridge Operations, USAEC, June 1961, 64; reprinted 2009 by the Oak Ridge Heritage and Preservation Association, available from them and from the Oak Ridge History Museum.

8. Atomic Energy Commission, "Major Activities in Atomic Energy Programs, January-June 1952" (Washington, D. C.: U.S. Government Printing Office, 1952), 36.

9. *Emergency Educational Aid for Government Reservations, Hearings before the Committee on Education and Labor*, 84–96.

10. Ford and Peitzsch, *A City is Born*, 83.

11. Smyser, *Oak Ridge 1942–1992, A Commemorative Portrait*, 104–105.

12. William J. Wilcox Jr., "The Transformation and Second Birth of Oak Ridge: Independence Day, June 4, 1960," talk, Oak Ridge Heritage & Preservation Association, May 13, 2010. Available at: https://www.oakridgeheritage.com/wp-content/uploads/2015/12/Bill-Wilcox-The_Transformation_and_Second_Birth_of_Oak_Ridge.pdf, page 8.

13. Ibid., 10.

14. Ray Smith, "The Oak Ridge Boys Tie to Oak Ridge," Historically Speaking column, *Oak Ridger*, February 13, 2007, available at http://smithdray1.net/historicallyspeaking/2007/2-13-07%20The%20Oak%20Ridge%20Boys%20ties%20to%20Oak%20Ridge.pdf; Smith, "My Day with the Oak Ridge Boys," Historically Speaking column, *Oak Ridger*, June 19, 2007, http://smithdray1.net/historicallyspeaking/2007/6-19-07%20My%20day%20with%20The%20Oak%20Ridge%20Boys.pdf.

15. Smyser, 93.

16. Angela N. H. Creager, *Life Atomic: A History of Radioisotopes in Science and Medicine* (IL: University of Chicago Press, 2013), 195.

17. "New Helicopter Service from Oak Ridge to Knoxville Airport to Start in August," *Oak Ridge National Laboratory—The News*, July 16, 1954, 1.

18. Ford and Peitzsch, *City is Born*, 290.

19. Bill Wilcox, "Reflections on the College of Oak Ridge," *Oak Ridger*, December 21, 2011, https://www.oakridger.com/story/news/local/2011/12/22/reflections-on-college-oak-ridge/63357430007/.

20. Roane State Community College occupied other facilities in Oak Ridge beginning in 1988, before moving to the new Coffey-McNally Building in August 1999.

21. Quitclaim Deed between the United States of America acting by and between the U.S. Department of Health, Education, and Welfare and the University of Tennessee, Knoxville. Property identification 02 025CC 025CC70300 00 of 2259.642 acres, book B8, page 367, dated August 21, 1961, received for filing Sep-

tember 14, 1961, available from Anderson County Register of Deeds, Anderson County, Tennessee.

22. Matthew Alan Cate, "The City of Oak Ridge and the Quest for a General Aviation Airport" (*Chancellor's Honors Program Projects,* 1997), https://trace.tennessee.edu/utk_chanhonoproj/204.

23. In 2021, there is renewed hope for building a general aviation airport near the site of the former K-25 plant; "Oak Ridge Airport Frequently Asked Questions" https://www.oakridgetn.gov/oak-ridge-airport updated May 2023.

24. Tennessee Municipal League Secretary Herbert Bingham quoted in Joe Hatcher, "Politics," *Tennessean*, November 6, 1956.

25. "5 Firms To Pay State Back Tax: Oak Ridge Group Agrees on Compromise of $413,00 Levy," *Tennessean*, June 28, 1957.

26. "Payments in Lieu of Taxes; Revisions to DOE Order Could Provide Better Assurance That Payments Meet Goals," *United States Government Accountability Office*, GAO-20–122, October, 2019, 46.

27. Ford and Peitzsch, *A City is Born,* 227–228.

28. Klein, *Volunteer Moments*, 80.

29. All UT students wore uniforms in the late nineteenth century, lived under military discipline, and were called cadets.

30. Klein, 66.

31. Ibid., 65.

32. Lucas Johnson, "Legislative committee says TSU could receive more than $540 million in unmet land-grant agreement," *Tennessee State University Newsroom*, April 5, 2021, https://tnstatenewsroom.com/archives/27636.

33. Brown v. Board of Education of Topeka, 347 U.S. 483 (1954).

34. Dwight D. Eisenhower, "Executive Order 10557—Approving the Revised Provision in Government Contracts Relating to Nondiscrimination in Employment," Federal Register (September 1954): 5655.

35. *Non-Discrimination Clause in Research Contracts, to C. E. Brehm, December 31, 1954*, President's Papers, AR.0006, University of Tennessee, Knoxville, Special Collections.

36. Cynthia Griggs Fleming, "Theotis Robinson Jr.," in *Tennessee Encyclopedia,* October 8, 2017, last updated March 1, 2018, http://tennesseeencyclopedia.net/entries/theotis-robinson-jr/.

37. "Theotis Robinson Jr.," Trailblazer Series, University of Tennessee, 2014, https://trailblazer.utk.edu/2014–2015/theotis-robinson/.

38. Klein, *Volunteer Moments,* 70–71.

39. "J. Ernest Wilkins Jr.," Atomic Heritage Foundation, https://www.atomicheritage.org/profile/j-ernest-wilkins-jr.

40. Smyser, 132.

41. Ken Morrell, "The Equation at Oak Ridge," *Southern Education Report,* March 1968, 15–18, https://files.eric.ed.gov/fulltext/ED020986.pdf; *African Americans*

at Los Alamos and Oak Ridge: A Historic Context Study, U.S. Department of the Interior, National Park Service, prepared for Manhattan Project National Historical Park, September 2019, https://www.nps.gov/mapr/learn/historyculture/upload/Study-African-Americans-at-Los-Alamos-and-Oak-Ridge.pdf.

42. D. Ray Smith, "Education in Oak Ridge–Pre-Oak Ridge and Early-Oak Ridge Schools, Part 2," Historically Speaking column, *Oak Ridger*, November 21, 2006, http://www.smithdray1.net/historicallyspeaking/2006/11-21-06%20Education%20in%20Oak%20Ridge%20-%20pre%20Oak%20Ridge%20part%202.pdf.

43. D. Ray Smith, "A 1950's letter and the integration of area schools," Historically Speaking column, *Oak Ridger*, January 21, 2011, http://smithdray1.net/historicallyspeaking/2011/1–21–11%20A%20look%20at%20the%201950%20integration%20story.pdf.

44. "Oak Ridge Uses First Negro Player," *Tennessean*, February 28, 1957.

45. Smyser, 135.

46. D. Ray Smith, "Brief History of Integration in Oak Ridge, TN," September 8, 2020, video, 19:08, https://www.youtube.com/watch?v=82yI2YHIb2c.

47. All quotations come from Ray Smith's YouTube presentation cited in previous note.

48. "U. S. Accused of Run-Around on Clinton Aid," *Tennessean*, October 10, 1958; "They Rush to School," *Tennessean*, October 10, 1958.

49. June Adamson, "Bobby Lynn Cain (1939–)," in *Tennessee Encyclopedia*, October 8, 2017, last updated June 26, 2019, https://tennesseeencyclopedia.net/entries/bobby-cain/.

50. Jack M. Holl, *Argonne National Laboratory, 1946–96* (Urbana: University of Illinois Press, 1997), 122–125.

51. Leland Johnson and Daniel Schaffer, "Information Acceleration," *ORNL Review* 25, nos. 3–4 (1992): 86.

52. Holl, *Argonne National Laboratory, 1946–96*, 62–68.

53. The cladding is a thin tube that encloses the nuclear fuel rod and prevents corrosion of the fuel by the coolant and the release of fission products into the coolant.

54. D. Ray Smith, "John Googin: The scientist of Y-12," *Oak Ridger*, April 24, 2009, https://www.oakridger.com/story/opinion/columns/2009/04/24/john-googin-scientist-y-12/63372897007/.

55. Lithium deuteride is the primary fuel of a hydrogen nuclear fusion bomb; a uranium fission explosion produces neutrons that bombard the lithium deuteride to form a deuteron and a triton (two isotopes of hydrogen), which then fuse to power the hydrogen bomb, which is more powerful than a fission bomb.

56. WBIR Staff, "ORNL Helps Build New Facility to Make Isotopes Used in Medical Treatments and Many Other Fields," October 24, 2022, https://www.wbir.com/article/tech/science/stable-isotope-new-facility-ornl-helped-create-for-doe/51-b90200ed-607c-44a1-bed2-c752d4198fe9.

57. Lewis L. Strauss and John A. McCone, *Twenty-Fourth Semiannual Report of*

the United States Atomic Commission (Washington, D.C.: U.S. Government Printing Office, 1958), 409–10.

58. Gould A. Andrews, Beecher W. Sitterson, Arthur L. Kretchmar, and Marshall Brucer, "Criticality Accident at the Y-12 Plant," *Diagnosis and Treatment of Acute Radiation Injury* (World Health Organization, Geneva, 1961), 27–48.

59. Alvin M. Weinberg, *The First Nuclear Era: The Life and Times of a Technological Fixer* (Woodbury, NY: AIP Press, 1994), 2–3.

60. Murray W. Rosenthal, *An Account of Oak Ridge National Laboratory's Thirteen Nuclear Reactors,* Report ORNL/TM-2009/181, Oak Ridge National Laboratory, August 2009, revised March 2010.

61. Leland Johnson and Daniel Schaffer, *Oak Ridge National Laboratory: The First Fifty Years* (Knoxville: University of Tennessee Press, 1994), 80–98.

62. Rosenthal, *An Account of Oak Ridge National Laboratory's Thirteen Nuclear Reactors,* 42.

63. Ibid., 44.

64. Ibid., 43.

65. Johnson and Schaffer, *Oak Ridge National Laboratory,* 80–81.

66. Rosenthal, 14.

67. Johnson and Schaffer, 58–59.

68. Rosenthal, 48–51.

69. Several other studies were performed at the Tower Shielding Facility, including drops of shipping casks of radioactive fuel to test their structural integrity. Another test involved a roaring Pratt & Whitney engine and sought to determine damage to the engine after fifty hours of high-radiation exposure. TSR-2 was shut down in 1992.

70. Rosenthal, 24–28.

71. Ibid., 52–54.

72. Nielsen, "The University of Tennessee Physics Department: The Early Years.

73. George K. Schweitzer, *Chemistry at UTK: A History of Chemistry at the University of Tennessee, Knoxville from 1794 through 1987* (Knoxville: University of Tennessee, Dept. of Chemistry, 1988); published by the Department of Chemistry and available from them.

74. The stories of these six people are told from the personal experience of William Bugg, who was a physics graduate student at UT in the 1950s before becoming a faculty member in this department.

75. Schweitzer, *Chemistry at UTK.*

76. Joseph Martinez, Herbert Krause, and Ben Brederson, "Sheldon Datz," *Physics Today* 55, no. 11 (November 2002): 88.

77. Personal recollection of William Bugg.

78. "ORNL takes active part in launching new mobile radiological laboratory," Oak Ridge National Laboratory News, October 12, 1951: 1 and 3.

79. D. Ray Smith and Carolyn Krause, "Insights into Rufus Ritchie, Father of the Surface Plasmon," Historically Speaking column, September 25, 2018, https://

www.oakridger.com/story/opinion/columns/2018/09/25/historically-speaking-insights-into-rufus/10067444007/.

80. Rufus Ritchie, "Plasma Losses by Fast Electrons in Thin Films," *Physical Review* 106 (1957): 874–881.

81. Smith and Krause, "Insights into Rufus Ritchie, Father of the Surface Plasmon."

82. Jake Yoder, "Say Hello to Elo," *Knoxville News Sentinel*, June 19, 2021.

83. "A Tribute to Isabel H. Tipton," UT Alumni News, October 15, 2018, https://giving.utk.edu/s/1341/2/20/adv.aspx?sid=1341&gid=2&pgid=14572.

84. "U-T is studying metal content of bones and tissues," *Oak Ridge National Laboratory, The News*, August 1, 1952: 1.

85. Wilford M. Good, "Remembrances of an Accelerator Past," *ORNL Review* 13, no. 2 (1980): 24–27.

86. J. H. Gibbons and R. L. Macklin, "Neutron Capture and Stellar Synthesis of Heavy Elements," *Science* 156 (May 1967): 3778.

87. Carolyn Krause, "Alex Zucker: A Scientific and Managerial Pillar at ORNL," Historically Speaking column, *Oak Ridger*, March 3, 2014, http://smithdray1.net/historicallyspeaking/2014/3-3-14%20Alex%20Zucker.pdf.

88. "The Oak Ridge Isochronous Cyclotron: Enhancements to ORNL's HRIBF Driver Accelerator" CP600, *Cyclotrons and Their Applications 2001*, Sixteenth International Conference, edited by F. Marti (American Institute of Physics 0-7354-0044-X/01, 2001).

89. C. Mazzocchi, Robert Grzywacz, S. Liddick, Krzysztof Rykaczewski, H. Schatz, J. Batchelder, Carrol Bingham, Carl Gross, J. Hamilton, J. Hwang, S. Ilyushkin, Agnieszka Korgu, W. Krolas, Kang Li, R. Page, D. Simpson, and Jeff Winger, "α Decay of ^{109}I and Its Implications for the Proton Decay of ^{105}Sb and the Astrophysical Rapid Proton-Capture Process," *Physical Review Letters* 98, 212501 (2007); 10.1103.

90. Holl, *Argonne National Laboratory*, 187–89.

91. Atomic Energy Commission, "Curtiss-Wright Corp.–Notice of Proposed Issuance of Construction Permit," *Federal Register* (June 1957): 3902–4; Atomic Energy Commission, "American Radiator and Standard Sanitary Corp.—Notice of Proposed Issuance of Construction Permit," *Federal Register* (October 1957): 8189–8190.

92. "McCord Asks State Academy of Science," *Tennessean*, March 7, 1958.

93. "State Building Work at Hand," *Tennessean*, July 23, 1959.

94. "Atom Materials Accepted by UT," *Tennessean*, October 3, 1959.

95. Oak Ridge Institute for Science and Education, "Nuclear Engineering Enrollments and Degrees Survey, 2107-2018 Data," ORISE report Number 80, April 2019, 1-6, https://orise.orau.gov/stem/reports/ne-brief-80-2017-18-data.pdf.

96. Alfred Nobel Memorial Prize in Economic Sciences (Professor Robert M. Solow), press release, Royal Swedish Academy of Sciences, October 21, 1987, https://www.nobelprize.org/prizes/economic-sciences/1987/press-release/.

97. R. Clinton Fuller, "Forty Years of Microbial Photosynthesis Research: Where It Came from and What It Led to," *Photosynthesis Research* 62 (1999): 1–29.

4. *The Decade of the 1960s*

1. Klein, *Volunteer Moments*, 159.

2. Ibid., 77–83.

3. Albert H, Teich and W. Henry Lambright, "The Redirection of a Large National Laboratory," *Minerva* 14 (Winter 1976–77): 447–74.

4. D. Ray Smith, "A Brief History of the UT-AEC Agricultural Research Laboratory," Historically Speaking column, *Oak Ridger*, October 20, 2009, https://www.oakridger.com/article/20091020/NEWS/310209998.

5. D. Ray Smith, "An Oak Ridge Treasure–the UT Arboretum," Historically Speaking column, *Oak Ridger*, September 22, 2009, http://smithdray1.net/historicallyspeaking/2009/9-22-09%20An%20Oak%20Ridge%20Treasure%20-%20the%20UT%20Arboretum%20final.pdf.

6. Meeting of the Executive and Business Committees, Board of Trustees, the University of Tennessee, September 19, 1960; Exhibit 8, R_BOT_1960_09_19_176.pdf, courtesy of Betsey B. Creekmore Special Collections and University Archives, University of Tennessee, Knoxville.

7. The transfer of deed is available in Anderson County Register of Deeds office in Deed Book "B," Volume 8, 367–82.

8. D. Ray Smith, "UT Arboretum–Oak Ridge Treasure, part 2," Historically Speaking, *Oak Ridger*, Sept. 29, 2009, http://smithdray1.net/historicallyspeaking/2009/9-29-09%20UT%20Arboretum%20part%202.pdf.

9. James R. Montgomery, "The Relationship between the University of Tennessee and the Oak Ridge Installations," March 7, 1961-letter and 22-page report to Hilton Smith, A. R. Nielsen, P. F. Pasqua, George Gleaves, and Gordon Carlson, UT archives AR 361, box 15, folder 39.

10. Ibid.

11. Ibid.

12. Ibid.

13. Ibid.

14. Alvin M. Weinberg, "Federal Laboratories and Science Education," *Science* 136, no. 3510 (April 1962): 27–30.

15. PhD programs in Energy Science and Engineering and Data Science and Engineering both started in the Bredesen Center, the first in 2011 and the second in 2017.

16. "Draft of Report of University Relations Committee to the Council," meeting held in Oak Ridge, June 18, 1962, courtesy of Betsey B. Creekmore Special Collections and University Archives, University of Tennessee, Knoxville.

17. Ibid.

18. Johnson and Schaffer, 124.

19. *Environmental Implications of the New Energy Plan: Hearings Before the Subcommittee on the Environment and the Atmosphere of the House Committee on Science and Technology*, 95th Congress, 1977, 454–55.

20. U.S. Department of Energy, Office of Science: https://science.osti.gov/fermi/Award-Laureates/1980s/weinberg.

21. U.S. Atomic Energy Commission, "Annual Report to Congress of the Atomic Energy Commission for 1964" (Washington, D.C.: U.S. Government Printing Office, January 1965), 44.

22. Wilcox, *K-25*, 99–100; see Chapter 2 #68.

23. "Oral History of Harold Conner, Jr.," interview by Keith McDaniel, Center for Oak Ridge Oral History, July 1, 2018, http://coroh.oakridgetn.gov/corohfiles/Transcripts_and_photos/Conner_Harold/Conner_Final.doc.

24. U. S. Department of Energy, Office of Environmental Management, "Harold Conner: From Oak Ridge Intern to Site Manager and Beyond," February 22, 2022, https://www.energy.gov/em/articles/harold-conner-oak-ridge-intern-site-manager-and-beyond.

25. D. Ray Smith, "A Better Life through Centrifuges," Historically Speaking column, *Oak Ridger*, October 25, 2017, http://smithdray1.net/historicallyspeaking/2017/10-25-17%20A%20better%20life%20through%20centrifuges.pdf.

26. D. Ray Smith, "Beyond Influenza: More Centrifuge-Based Oak Ridge Medical Instrumentation," Historically Speaking column, *Oak Ridger*, February 12, 2018, http://smithdray1.net/historicallyspeaking/2018/2-12-18%20Beyond%20Influenza.pdf.

27. A shortage of copper during WWII led to the use of silver for the electrical coils to power the large magnets in the calutrons. The silver was borrowed from the U.S. Treasury for this wartime need.

28. L. O. Love, "Electromagnetic Separation of Isotopes at Oak Ridge: An Informal Account of History, Techniques, and Accomplishments," *Science* 182, no. 4110 (October 1973): 343–52; "$124 Million (in Silver) Returned to U.S. Treasury," *ORNL News*, March 7, 1969.

29. There are 14.58 troy ounces per pound and 2,000 pounds per ton.

30. U.S. Atomic Energy Commission, "Annual Report of the Atomic Energy Commission for 1969" (U.S. Government Printing Office, Jan. 1970) 321.

31. D. Ray Smith, "Apollo "Moon Boxes" made at Y-12," *From Our Historian, Oak Ridger*, July 17, 2009, https://www.y12.doe.gov/sites/default/files/assets/document/09-07-17.pdf; D. Ray Smith, "Apollo "Moon Boxes" made at Y-12, part 2—Y-12's moon mission, use of Teflon," *From Our Historian, Oak Ridger*, July 24, 2009, https://www.y12.doe.gov/sites/default/files/assets/document/09-07-24.pdf.

32. "In Memoriam: Lawrence 'Larry' Taylor," *UTK News*, September 20, 2017, https://news.utk.edu/2017/09/20/memoriam-lawrence-larry-taylor/.

33. Comptroller General of the U.S., "Review of Manned Aircraft Nuclear Propulsion Program—Atomic Energy Commission and Department of Defense," February 1963, https://fas.org/nuke/space/anp-gao1963.pdf.

34. "X-10 Reactor Made National Landmark: 'Old Soldier' Becomes Monument to Pioneers of Nuclear Energy," *ORNL News*, September 16, 1966.

35. "LITR—First 'Blue Glow' Reactor Retired After 20 Years' Service," *ORNL News*, October 18, 1966.

36. "Termination of the Experimental Gas Cooled Reactor Project," *ORNL News*, January 14, 1966.

37. "Oak Ridge Prepares Experimental Reactor for Deactivation, Demolition," U.S. Department of Energy, Office of Environmental Management, May 25, 2021, https://www.energy.gov/em/articles/oak-ridge-prepares-experimental-reactor-deactivation-demolition.

38. "MSRE Brought to Power with Uranium-233 Fuel," *ORNL News,* October 11, 1968.

39. Rosenthal, *An Account of Oak Ridge National Laboratory's Thirteen Nuclear Reactors*, 29–34.

40. Leland Johnson and Daniel Schaffer, "Balancing Act: Desalting the Waters," *ORNL Review* 25, nos. 3–4 (1992): 128–30.

41. James Conca, "How 1,500 Nuclear-Powered Water Desalination Plants Could Save the World from Desertification," *Forbes*, July 14, 2019, https://www.forbes.com/sites/jamesconca/2019/07/14/megadroughts-and-desalination-another-pressing-need-for-nuclear-power/#46038df67fde.

42. D. Chandler, R. T. Primm, III, G. I. Maldonado, Reactivity Accountability Attributed to Beryllium Reflector Poisons in the High Flux Isotope Reactor, Report ORNL/TM-2009/188, Oak Ridge National Laboratory, December 2009, 12.

43. Rosenthal, 55–59.

44. Barbara Lyon, "New Neutron Spectrometers at HFIR Provide Advanced Research Facilities," *ORNL News*, Nov. 17, 1967, 1 and 3.

45. Leland Johnson and Daniel Schaffer, "Neutrons and JFK," *ORNL Review* 25, nos. 3–4 (1992): 143.

46. *Report of the President's Commission on the Assassination of President Kennedy (Warren Commission Report),* President John F. Kennedy Assassination Records Collection, PR 36.8:K 38/R 29, September 24, 1964.

47. Local lore maintains that Oswald, with a Dallas address and USSR citizenship, visited the American Museum of Atomic Energy in Oak Ridge on July 26, 1963, four months before the Kennedy assassination in Dallas on November 22. During that summer, Oswald was known to be living in New Orleans and engaged in pro-Cuba activities. The museum registration sheet suggests that others from Texas accompanied Oswald. The signature on the visitor register has not been verified, and no one knows if Oswald was in fact in Oak Ridge in July of 1963 or why he would have visited. He had no known connections to Tennessee. See: Frank Munger, "Was Lee Harvey Oswald in Oak Ridge?"; https://jenniferlake.wordpress.com/2015/03/26/atomic-agent-oswald/.

48. Leland Johnson and Daniel Schaffer, "President Zachary Taylor and the Laboratory," *ORNL Review* 25, nos. 3–4 (1992): 254.

49. "Laboratory to Open Libraries, Seminars to Non-Employees," *ORNL News*, December 7, 1962.

50. "R. S. Livingston Describes Successful Operation of ORNL's Electron Cyclotron," *ORNL News*, September 15, 1961.

51. Stephen Stow and Marilyn McLaughlin, "An Interview with Alex Zucker," Department of Energy Oral History Presentation Program, Center for Oak Ridge Oral History, March 4, 2003, http://coroh.oakridgetn.gov/corohfiles/Transcripts_and_photos/ORNL/ORNL_Zucker_Alex_transcript.pdf.

52. Leland Johnson and Daniel Schaffer, "Responding to Social Needs—Accelerators," *ORNL Review* 25, nos. 3–4 (1992): 146, 158.

53. "User Facilities," Oak Ridge National Laboratory, https://www.ornl.gov/content/user-facilities.

54. Leland Johnson and Daniel Schaffer, "Big Biology," *ORNL Review* 25, nos. 3–4 (1992): 130–34.

55. Ibid.

56. Ibid.

57. "Carbide Sparks New U-T Project in Oak Ridge," University of Tennessee *Torchbearer*, University of Tennessee, November 1966, 3.

58. "U-T Trustees Authorize Graduate School of Biomedical Sciences—First Class Opens Next Fall," *ORNL News*, November 12, 1965.

59. "UT Moves Fast, Grows Fast to Meet New Responsibilities," *Tennessean*, September 17, 1967.

60. Clinton Fuller, "The Campus in 9207," *ORNL Review* 4, no. 3 (Spring 1971): 12–17. 9207 was the number of the building located at Y-12 and used in part for the Biomedical School.

61. An E building contained two two-bedroom and two one-bedroom apartments. Rick Reynolds was a graduate student from 1971 to 1975 and lived in an E2 on 123 Vance Street.

62. F. D. Hamilton, "Undergraduate Research Training," in *Minorities in Science: The Challenge for Change in Biomedicine*, eds. V. L. Melnick and F. D. Hamilton (Boston, MA: Springer Publishing, 1977), 225–230; https://doi.org/10.1007/978-1-4757-5851-1_26.

63. Reinhold Mann, "Life Sciences Division Progress Report for the Period CYs 1997–1998[Oak Ridge National Laboratory]," ORNL/TM-1999/121, June 1999, https://www.osti.gov/servlets/purl/771200.

64. Rachel Carson, *Silent Spring* (Boston, MA: Houghton Mifflin Company, 1962).

65. Frank Harris did his graduate work at UT during this period (MS 1966, PhD 1970) and spent thirteen years as a member of the ORNL staff (1970–1983), before moving to the UT faculty and returning to ORNL in 2000 as part of the UT-Battelle leadership team.

66. Louis J. Gross, "A History of the Graduate Program in Ecology at UT Knoxville," June 1987, http://www.nimbios.org/~gross/Ecology/EcologyHistoryUTK.html.

67. William Bugg, "The Nielsen Years 1956–1969" (unpublished manuscript), *History of the UT Physics Department.*

68. Eugene Dietz, "UT A-Program Gets $750,000," *Tennessean*, November 7, 1963.

69. "Seven ORNL Staff Members to Join UT Faculty Part Time," *ORNL News*, December 20, 1963.

70. "Physics Division Annual Progress Report for Period Ending December 31, 1965," ORNL Report 3924, 153; https://www.osti.gov/servlets/purl/4520567/.

71. "ORNL Scientists to Teach," *Physics Today* 17, no. 3 (March 1964): 84.

72. Winfred L. Godwin, "Promising Endeavor in Nuclear Sciences," *Tennessean,* October 12, 1964.

73. "ORNL Scientists at Tennessee," *Physics Today* 17, no. 11 (November 1964): 99.

74. "UT Departments Get $1.45 Million Grant," *Tennessean,* June 15, 1968.

75. "L. L. Riedinger," *ORNL News*, April 11, 1968.

76. "Physics Division Annual Progress Report for Period Ending Dec. 31, 1965," ORNL Report 3924, 154, https://www.osti.gov/servlets/purl/4520567/.

77. "Thirteen AEC-PHS-sponsored Students Begin Eleven-weeks of Study at ORNL," *ORNL News*, July 15, 1966.

78. Lawrence Radiation Laboratory later became Lawrence Berkeley and Lawrence Livermore National Laboratories and the National Reactor Testing Station later became Idaho National Laboratory.

79. "Forty-Two Students Presently Working As Co-ops in 10 Divisions at Laboratory," *ORNL News*, November 23, 1962.

80. "Science and Engineering Students at ORNL for Summer Trainee Program," *ORNL News*, July 17, 1967.

81. "ORNL to Participate in Ecology Program for Undergraduates," *ORNL News*, February 2, 1968.

82. "Forty-eight College, University Teachers Take part in Summer Research Program—Assigned to 15 ORNL Divisions," *ORNL News*, July 22, 1966.

83. "Graduate Resident Program Extended through June 30, 1972," *ORNL News,* August 11, 1967; "OR Resident Graduate Program Lists 21 Spring Quarter Courses," *ORNL News,* March 21, 1969; "100 Percent Reimbursement Available for Study Leading to Graduate Degree," *ORNL News,* June 9, 1967.

84. "Times Points to ORNL's Novel Education Program," *ORNL News,* April 11, 1969.

85. U.S. Atomic Energy Commission, "Annual Report to Congress from the Atomic Energy Commission for 1966" (Washington, D.C.: U.S. Government Printing Office, 1967), 306.

86. E. L. Merrill and Wendell H. Russell, "The Training and Technology Project Experimental Research Program for Vocational-Technical Teachers—Final Report," Oak Ridge Associated Universities and the University of Tennessee, December 1968.

87. U.S. Atomic Energy Commission, "Annual Report to Congress on the Atomic Energy Commission for 1968" (Washington, D.C.: U.S. Government Printing Office, 1969), 270.

88. "A Model for Training the Disadvantaged: TAT at Oak Ridge Tenn.," Manpower Research Monograph No. 29, U.S. Department of Labor, Manpower Administration, 1973.

89. "Oak Ridge Training Means Big Pay Gain," *The New York Times,* December 14, 1969.

90. U.S. Atomic Energy Commission, "Annual Report to Congress on the Atomic Energy Commission for 1969" (Washington, D.C.: U.S. Government Printing Office, 1970), 270–271.

91. D. Ray Smith, "Oak Ridge Technical Enterprises Corporation–ORTEC," Historically Speaking column, *Oak Ridger*, seven episodes from November to December 2011.

92. From the papers of the Eugene Joyce collection held by the East Tennessee Economic Council in Oak Ridge.

93. ORNL funding problems resulted in part from the decrease in nuclear reactor programs and the rise in national challenges due to the energy crisis and the OPEC oil embargo.

94. Oak Ridge and its federal facilities lie in Roane and Anderson counties.

95. Muddy Boot awards recipients Al Trivelpiece (1999), Herman Postma (2001), Alvin Weinberg (2003), Bill Madia (2003), Ron Townsend (2004), Homer Fisher (2005), Billy Stair (2006), Jeff Smith (2008), Joe Johnson (2011), Bonnie Carroll (2012), Ray Smith (2013), Thom Mason (2014), Frank Munger (2016), David Millhorn (2017), Tom Ballard (2017), Harold Connor (2019), Lee Riedinger (2019), and Thomas Zacharia (2020) contributed to the success of the UT-Oak Ridge partnership.

96. Alvin Weinberg, "Science, government, and information: The responsibilities of the technical community and the government in the transfer of information" (Presidents Science Advisory Committee, 1963); can be found at Analysis and Policy Observatory, https://apo.org.au/node/56798.

97. D. Ray Smith, "Bonnie Carroll: Librarian to Successful Entrepreneur," Historically Speaking column, *Oak Ridger,* August 18, 2022, https://www.oakridger.com/story/lifestyle/2022/08/19/bonnie-carroll-librarian-successful-entrepreneur/10332734002/.

98. "AEC Unit Plans State Site View," *Tennessean,* November 10, 1965.

99. "State Loses Atom-Smasher Project," *Tennessean,* May 15, 1966.

100. Michael D. Lemonick, "A Controversial Prize for Texas," *Time,* November 18, 1988, 79.

101. "ORNL Ranks First in AEC Investment among Laboratories," *ORNL News,* November 4, 1966.

102. Alvin M. Weinberg, "Redeployment of the National Laboratories," *Nuclear Applications* 3 (July 1967): 394.

103. Jonathan Spivak, "Achievements at Oak Ridge Prototype for Future Labs," *ORNL News,* January 5, 1968.

104. Ed Willingham, "Baker, Muskie, Evins Unveil Huge Anti-Pollution Plan," *Tennessean*, February 15, 1970.

105. Leland Johnson and Daniel Schaffer, "Oak Ridge National Laboratory: The First 50 Years," *ORNL Review* 25, no. 3 and 4 (1992): 144–145.

106. Ibid.

5. *The Decade of the 1970s*

1. "Plan to Sell AEC Plants Major Threat to Region," *Tennessean*, July 24, 1970.

2. "AEC Hints Sale Still On," *Tennessean*, July 24, 1970.

3. Elaine Shannon, "Do Republicans Plan to Sell Oak Ridge?" *Tennessean*, September 27, 1970.

4. Dwight Lewis, "Escapee from Center Here Listed as Hijacker," *Tennessean*, November 12, 1972.

5. Clelly Johnson, "Race Relations, Terrorism, and Nuclear Obliteration: The Hijacking of Southern Airways Flight 49," *Vulcan Historical Review* 16 (2012,): 12.

6. Johnson, "Race Relations, Terrorism, and Nuclear Obliteration," 13.

7. Ibid, 18–19.

8. Ibid.

9. Wilcox, *A Brief History of K-25*, 71.

10. "Oral History of James Alexander," interview by Keith McDaniel, December 9, 2011, http://coroh.oakridgetn.gov/corohfiles/Transcripts_and_photos/Alexander_James/Alexander_James.docx.

11. Johnson, 18–19.

12. Bob Hutton, "Volunteers for America, 1970 and 2020," UT Department of History Archives; https://history.utk.edu/volunteers-for-america-1970-2020/.

13. John Shearer, "50 Years Ago, Billy Graham Crusade Ignited Knoxville. Were you there?" *Knoxville News Sentinel*, May 18, 2020, https://www.knoxnews.com/story/shopper-news/bearden/2020/05/19/billy-graham-crusade-knoxville-neyland-stadium-1970/3094565001/.

14. Paul Huray had a long career at UT, ORNL and in Washington, D.C., before being appointed vice president for research at the University of South Carolina. He did much to build the UT–ORNL partnership. Gil Nussbaum later transitioned to medical physics and joined the faculties of medicine at Tufts University and, later, Washington University in St. Louis. He was an internationally renowned authority on the use of heat therapy for the treatment of cancer.

15. Johnson and Schaffer, *Oak Ridge National Laboratory*, 124.

16. Ibid, 129.

17. Ibid, 124.

18. Ibid, 124.

19. Louis J. Gross, "A History of the Graduate Program in Ecology at the University of Tennessee, Knoxville," June 1987; private communication.

20. National Environmental Policy Act, Section 102[42 USC § 4332] (2)(c), 1970.

21. Johnson and Schaffer, *Oak Ridge National Laboratory*, 144–45.

22. Chapter 10 (pages 175–200) of Alvin M. Weinberg's book, *The First Nuclear Era: The Life and Times of a Technological Fixer* (AIP Press, New York, 1994), gives a detailed discussion of the evolving problems of nuclear energy.

23. Witness the accidents at: Three Mile Island; Chernobyl, which occurred in 1986 in the former Soviet Union; and Fukushima Daiichi, which occurred in 2011 in Japan. The Three Mile Island accident involved a partial meltdown of a reactor core with little or no release of radioactivity to the atmosphere. Chernobyl was the result of failed reactor tests that led to a core meltdown and a huge release of radioactivity in Ukraine, since this type of Soviet reactor had no containment vessel (unlike reactors in the West). A reactor core at Fukushima melted due to a disabled cooling system caused by a huge tsunami, with only small release of radioactivity.

24. Weinberg, *The First Nuclear Era*, 196.

25. Ibid, 175.

26. Ibid, 176.

27. Ibid, 199.

28. Ibid, 127–131.

29. ^{235}U and ^{233}U emit on average 2.4 neutrons per fission, whereas ^{239}Pu emits 2.9, which is a reason why plutonium is better as a fissionable fuel for a nuclear chain reaction.

30. Technical CRBR references beyond Wikipedia are difficult to find. One source is: L. E. Strawbridge, "Safety Related Criteria and Design Features in the Clinch River Breeder Reactor Plant," ANS Fast Reactor Safety Meeting, Beverly Hills, CA, April 2–4, 1974, https://www.osti.gov/servlets/purl/4311171.

31. Weinberg, 125–127.

32. Ibid., 129–31.

33. Johnson and Schaffer, *Oak Ridge National Laboratory*, 137.

34. Weinberg, 129–131.

35. Gus Manning and Haywood Harris, *Once a Vol, Always a Vol!: The Proud Men of the Volunteer Nation* (Champaign, IL: Sports Publishing, 2006), 1.

36. Cochran received a PhD in physics from Vanderbilt University in 1967, where he studied high-energy physics while Lee Riedinger was also a graduate student. Riedinger admired Cochran's ability to ride a unicycle in the hallways of the Stevenson Science Center at Vanderbilt, at night, of course, when no professors were present.

37. Thomas B. Cochran, *The Liquid Metal Fast Breeder Reactor: An Environmental and Economic Critique* (Baltimore, MD: Resources for the Future, 1974, distributed by John Hopkins University Press).

38. Henry Sokolski, "The Clinch River Folly," *Backgrounder* #231, The Heritage Foundation, December 3, 1982.

39. Ibid.

40. Kurt Anderson, "Clinch River: A Breeder for Baker," *Time*, August 3, 1981, http://content.time.com/time/subscriber/article/0,33009,949264,00.html.

41. Charles Barton, *Milton Shaw and the Decline of the American Nuclear Establishment*, Sept. 2008; https://energyfromthorium.com/2008/09/23/milton-shaw-and-the-decline-of-the-american-nuclear-establishment/.

42. Cheryl Rofer, "Why Did the U.S. Abandon a Lead in Reactor Design?" *Physics Today, Commentary and Reviews,* August 7, 2015, https://doi.org/10.1063/PT.5.2029.

43. Weinberg, 199.

44. Kimberly Amadeo, "OPEC Oil Embargo, Its Causes, and the Effects of the Crisis," *the balance*, April 20, 2022, https://www.thebalance.com/opec-oil-embargo-causes-and-effects-of-the-crisis-3305806?utm_source=emailshare&utm_medium=social&utm_campaign=shareurlbuttons.

45. Greg Myre, "The 1973 Arab Oil Embargo: The Old Rules No Longer Apply," *Parallels, NPR*, October 16, 2013, https://www.npr.org/sections/parallels/2013/10/15/234771573/the-1973-arab-oil-embargo-the-old-rules-no-longer-apply.

46. "Oil Embargo, 1973–1974," in *Milestones in the History of U.S. Foreign Relations: 1969-1976,* Office of the Historian, Foreign Service Institute, United States Department of State, https://history.state.gov/milestones/1969–1976/oil-embargo.

47. Weinberg, 217–221.

48. Ibid., 201–6.

49. William Greenberg, "U.S. Uranium Enriching Pays off: $100 Million Earned by AEC," *Tennessean*, November 1, 1970.

50. Ed Willingham, "Shall U.S. Sell Atomic Plants?" *Tennessean*, April 27, 1969.

51. Elaine Shannon, "Enigma: Money & Uranium: Where's Cash Coming From?" *Tennessean*, June 30, 1974.

52. "TVA Eyes Plant for Uranium Use," *Tennessean*, June 16, 1974.

53. Nat Caldwell, "Exxon Buys Rhea County Land as Possible Site for N-Facility," *Tennessean*, February 28, 1976.

54. "Fusion Energy, Boeing Plants Save Oak Ridge," *Tennessean*, September 13, 1980.

55. "Boeing expansion to create 430 jobs," *Tennessean*, June 27, 1986.

56. Philip Shabecoff, "Ford Would End Nuclear Fuel Monopoly," *Tennessean,* June 19, 1975.

57. Frank Cormier, "Ford Talks Out Decisions, Acts Quickly, Backs Business," *Tennessean,* August 10, 1975.

58. Kirk Loggins, "U.S. Panel Feels TVA Bite," *Tennessean*, February 10, 1976.

59. U.S. Department of Energy, Office of Nuclear Energy, "Building a Uranium Reserve: The First Step in Preserving the U.S. Nuclear Fuel Cycle," May 11, 2020, https://www.energy.gov/strategy-restore-american-nuclear-energy-leadership.

60. Uranium enriched between 5 and 20 percent is required for most U.S. advanced reactors to achieve smaller designs that get more power per unit of volume.

61. Weinberg, 169.

62. On December 5, 2022, scientists at Lawrence Livermore National Laboratory's

National Ignition Facility conducted the first controlled fusion experiment to produce more energy than the fusion process consumed; see: https://www.energy.gov/articles/doe-national-laboratory-makes-history-achieving-fusion-ignition. Still, experts generally believe that the pathway to the eventual fusion reactor for electricity production will involve magnetic confinement in a tokamak as opposed to inertial confinement used in the Livermore experiment.

63. Leland Johnson and Daniel Schaffer, "Responding to Societal Needs: Gold-Plated Fusion," *ORNL Review* 25, nos. 3–4 (1992): 160–62.

64. Johnson and Schaffer, "Responding to Societal Needs," 163.

65. J. Rand McNally Jr., Arthur H. Snell, and Alvin M. Weinberg, "E.D. Shipley," *Physics Today*, February 1982, 92.

66. Bonnie Nestor, "The Origins of Fusion Energy at ORNL," *ORNL Review* (blog), January 20, 2021, https://www.ornl.gov/blog/origins-fusion-energy-ornl.

67. U.S. Department of Energy, Alternative Fuels Data Center, https://afdc.energy.gov/data/.

68. Johnson and Schaffer, *Oak Ridge National Laboratory,* 167.

69. Weinberg, 197.

70. Daniel Schaffer and Jack Barkenbus, *First Score: 20 Years at the Energy, Environment, and Resources Center, 1973–1993* [unpublished manuscript] (University of Tennessee Archives, *Volopedia*).

71. Ibid.

72. "Friedrich "Bio" Schmidt-Bleek," World Resources Forum, https://www.wrforum.org/people/friedrich-bio-schmidt-bleek/.

73. Schaffer and Barkenbus, *First Score.*

74. Ibid.

75. Ibid.

76. Frank Munger, "Frozen Mice: Celebrating a 40-year success story," June 17, 2012, *Knox News,*
https://archive.knoxnews.com/news/local/frozen-mice-celebrating-a-40-year-success-story-ep-360515503-356932181.html/.

77. "Oral history of Peter Mazur," interview by Stephen H. Stow and Marilyn Z. McLaughlin, February 24, 2003. Available at: Oak Ridge Public Library, Oral History collection, http://coroh.oakridgetn.gov/corohfiles/Transcripts_and_photos/ORNL/ORNL_Mazur_peter_transcript.pdf.

78. Leland Johnson and Daniel Schaffer, "The Carter Visit," *ORNL Review* 25, nos. 3–4 (1992): 198.

79. "Oral history of Liane B. Russell," interview by Stephen H. Stow and Marilyn Z. McLaughlin, April 23, 2003; http://coroh.oakridgetn.gov/corohfiles/Transcripts_and_photos/ORNL/ORNL_Russell_Liane_transcript.

80. Johnson and Schaffer, "The Russells," *ORNL Review* 25, nos. 3–4 (1992): 88–89.

81. Johnson and Schaffer, "Of Mice and Mammals," *ORNL Review* 25, nos. 3–4 (1992): 91.

82. Liane B. Russell, "Birth of a Trail and an Organization," in *These Are the Voices*, ed. James Overholt, 456–62.

83. Johnson and Schaffer, "Responding to Social Needs: Accelerators," *ORNL Review* 25, nos. 3–4 (1992): 146, 158.

84. University of Alabama at Birmingham, Georgia Tech, Emory, Furman, Kentucky, Louisiana State, Massachusetts, South Carolina, Tennessee, Tennessee Tech, Vanderbilt, and Virginia Tech.

85. J. H. Hamilton, "University Isotope Separator at Oak Ridge: The UNISOR Consortium," *Science* 185, no. 4154 (September 1974): 819–24.

86. This facility was named after U.S. Congressman Chet Holifield (California), a long-time member of the Joint Committee on Atomic Energy. To honor Holifield upon his retirement in 1974, Congress attached a rider to an appropriations bill to rechristen ORNL as Holifield National Laboratory. This produced an outcry in Oak Ridge and the Tennessee delegation took the lead on convincing the Congress to change the name back to ORNL late in 1975. See: Johnson and Schaffer, *Oak Ridge National Laboratory*, 180.

87. Alvin Weinberg, "Transuranic Elements and the High-flux Isotope Reactor," *Physics Today* 20, no. 6 (1967): 23.

88. To date, californium (element 98) is the only actinide with significant industrial application. For example, it is used as a neutron source to help get the uranium chain reaction going in starting a nuclear reactor.

89. Richard Hahn, "Life at the End of the Periodic Table," *ORNL Review* 16, no. 2 (1983): 26–35.

90. "Brookhaven honors four scientists for Distinguished R&D efforts: Richard Hahn Chemistry," *Brookhaven Bulletin* 52, no. 5 (January 30, 1998): 1 and 4, https://www.bnl.gov/bnlweb/pubaf/bulletin/files/1998/19980130.pdf.

91. C. E. Bemis, Jr., R. J. Silva, D. C. Hensley, O. L. Keller, Jr., J. R. Tarrant, L. D. Hunt, P. F. Dittner, R. L. Hahn, and C. D. Goodman, "X-Ray Identification of Element 104," *Physical Review Letters* 31, no. 10 (September 1973): 647–650.

92. Ronald J. Cohn, "Element 104 identified by characteristic x rays," *Physics Today*, 26(9), September 1973, 17-18.

93. Emission of an alpha particle (two protons and two neutrons) is a type of radioactive decay found commonly for the heavy elements, those beyond uranium.

94. Robert V. Gentry, "Giant Radioactive Halos: Indicators of Unknown Radioactivity?" Science 169 (1970): 670–673; "Radiohalos in a Radiochronological and Cosmological Perspective," *Science* 184 (1974): 62–66.

95. Ronald L. Numbers, *The Creationists: From Scientific Creationism to Intelligent Design* (Cambridge, MA: Harvard University Press, 2006), 280.

96. "Girls' Half-Court Rule Subject of State Suit," *Tennessean,* Aug. 15, 1976.

97. Pat Head Summitt and Sally Jenkins, *Sum It Up: 1,098 Victories, a Couple of Irrelevant Losses, and a Life in Perspective* (New York: Crown Publishing, 2013), 130.

98. Bill Haltom and Amanda Swanson, *Full Court Press: How Pat Summitt, a*

High School Basketball Player, and a Legal Team Changed the Game (Knoxville: University of Tennessee Press, 2018), 99.

99. "Girls Half-Court Game Ruled 'Unconstitutional,'" *Tennessean,* November 25, 1976.

100. Haltom and Swanson, *Full Court Press,* 100-102.

101. Tom Squires, "It'll be Six-Girl Basketball: Appellate Court Reverses Decision," *Tennessean,* October 4, 1977.

102. "5-on-5 Gets Okay from West," *Tennessean,* January 25, 1978.

103. "School Told To Make Girls Equal: Oak Ridge School Gets HEW Order," *Tennessean,* January 25, 1978.

104. F. M. Williams, "Vols' Head Makes Five-on-Five Pitch: pleas to TSSAA on Upcoming Vote," *Tennessean,* March 4, 1979.

105. "Senators Urge TSSAA To Accept Five-on-Five," *Tennessean,* March 15, 1979.

106. Haltom and Swanson, *Full Court Press.*

6. The Decade of the 1980s

1. Doug Hall, "Blanton Frees Humphreys; Clemency for 51 Others," *Tennessean,* January 16, 1979.

2. Larry Daughtrey, "'Inappropriate' Oath Rejected by Alexander," *Tennessean,* January 16, 1979.

3. Larry Daughtrey and Doug Hall, "Alexander Sworn in; Blanton Pushed Out," *Tennessean,* January 18, 1979.

4. Carol Clurman and Joel Kaplan, "Blanton Guilty," *Tennessean,* June 10, 1981.

5. Fred S. Rolater, "Leonard Ray Blanton (1930-1996)," in *Tennessee Encyclopedia,* October 8, 2017, last updated May 1, 2018, https://tennesseeencyclopedia.net/entries/leonard-ray-blanton/.

6. "UT Knoxville Chancellor Emeritus Jack Reese Dies," *UT Knoxville News,* May 9, 2005, https://news.utk.edu/2005/05/09/ut-knoxville-chancellor-emeritus-jack-reese-dies/.

7. Scott Barker, "Former UT Chancellor Reese, 76, Dies," *Knoxville News Sentinel,* May 10, 2005.

8. Betsey B. Creekmore, "Jack Edward Reese: 1929–2005," in *Volopedia,* October 15, 2018, https://volopedia.lib.utk.edu/entries/jack-edward-reese/.

9. "Carbide leaving Oak Ridge," *Knoxville News Sentinel,* May 3, 1982.

10. Lisa Hood, "UT to Study Chicago Lab for ORNL Proposal," *Knoxville News Sentinel,* May 28, 1982.

11. Lisa Hood, "UT Studies Alternatives for ORNL Management," *Knoxville News Sentinel,* June 11, 1982.

12. Catherine Foster, "UT Naming Task Force to Study Operation of ORNL; 6 Months Study," *Oak Ridger,* June 23, 1982.

13. Personal memory by Jack Campbell, told to Lee Riedinger on a tennis court in April 2022.

14. The submitting universities included UT, Tennessee Tech, Vanderbilt, Murray State University, and the University of Georgia through a consortium, including itself along with Duke, Georgia Tech, North Carolina State University, Vanderbilt, and the University of North Carolina, Chapel Hill.

15. "6 Universities Aim for Oak Ridge Pact," *Tennessean*, October 20, 1982.

16. Betsey B. Creekmore, "Distinguished Scientist Program (UT/ORNL)," in *Volopedia*, September 25, 2018, last updated October 6, 2018, https://volopedia.lib.utk.edu/entries/distinguished-scientist-program-ut-ornl/.

17. Lee Riedinger, personal information as one of authors of the Distinguished Scientist program.

18. Jim Campbell, "Herman Postma, 1933–2004," https://web.ornl.gov/~cabagewh/open/postma.html.

19. Alex Zucker, "Herman Postma," *Physics Today* 58, no. 9 (2005): 76, https://doi.org/10.1063/1.2117837.

20. Lee Riedinger, as a member of the Task Force, recorded the dates and details of this chain of events.

21. Lee Riedinger, as one of the originators of this program, recorded the details of the Distinguished Scientist appointments.

22. Chas Sisk, "Howard Baker, Former Senate Majority Leader, Dies at 88," *Tennessean*, January 26, 2014, https://www.tennessean.com/story/news/2014/06/26/howard-baker-died/11406251/.

23. The Howard H. Baker Center for Public Policy, "Senator Baker," University of Tennessee, Knoxville, https://web.archive.org/web/20130419072353/http://bakercenter.utk.edu/about-us/senator-baker/.

24. Ira Shapiro, *The Last Great Senate: Courage and Statesmanship in Times of Crisis* (NY: PublicAffairs, 2012).

25. Margaret Marynowski and William Sweet, "Tennessee Distinguished Scientist Program Begins," *Physics Today* 37, no. 12 (1984): 65, https://doi.org/10.1063/1.2915995.

26. "2 Scientists Appointed to Program," *Oak Ridger*, June 12, 1984.

27. Leland Johnson and Daniel Schaffer, "Ion Implantation of Materials," *ORNL Review* 25, nos. 3–4 (1992): 218.

28. Ibid., 224–25.

29. High temperature superconductors are materials that become superconducting at or above liquid nitrogen temperature (-320°F), which is much easier to achieve than that of liquid helium (-452°F) needed for traditional superconductors.

30. June Adamson, "National Academy of Science Taps Gerald D. Mahan," *Knoxville News Sentinel*, May 7, 1995.

31. This new accelerator laboratory was named after U.S. Congressman Chet Holifield (D-California), a long-time member of the Joint Committee on Atomic Energy, as discussed in Chapter 5.

32. P. J. Twin, B. M. Nyakó, A. H. Nelson, J. Simpson, M. A. Bentley, H. W. Cranmer-Gordon, P. D. Forsyth, D. Howe, A. R. Mokhtar, J. D. Morrison, J. F.

Sharpey-Schafer, and G. Sletten, "Observation of a Discrete-Line Superdeformed Band up to $60\hbar$ in ^{152}Dy," *Physical Review Letters* 57, no. 811 (1986), https://doi.org/10.1103/PhysRevLett.57.811.

33. Mark Potts, "5-Year Government Contract," *Washington Post,* December 14, 1983, https://www.washingtonpost.com/archive/business/1983/12/14/5-year-government-contract/6b509271–4ed6–4afa-9e9c-d3d881352c76/.

34. Richard Powelson, "Reagan brings Crusade to East TN," *Knoxville News Sentinel,* September 24, 1985.

35. Homer Fisher, interview by Lee Riedinger (author), April 26, 2020.

36. "ORNL-UT Interactions Outside the Science Alliance," *ORNL Review* 24, no. 2 (1991): 62–65.

37. Gibbons started the UT Environment Center in 1973 and changed its name in 1979 to the Energy, Environment, and Resources Center.

38. Ibid.

39. Kim McDonald, "Reagan Backs Giant, $4.4-Billion Particle Accelerator; Scientists Face Major Hurdles in Promoting the Device," *The Chronicle of Higher Education,* February 11, 1987.

40. Kim McDonald, "Money for Super Conducting Supercollider Is Requested by U.S. Energy Secretary," *The Chronicle of Higher Education,* February 18, 1987.

41. Michael D. Lemonick, "A Controversial Prize for Texas," *Time*, November 18, 1988, 79.

42. Larry Lee, "ORNL in Six-Way Race to Construct Megatron Detector," *Knoxville News Sentinel,* April 15, 1990.

43. A calorimeter is a device used for measuring heat/energy resulting from a reaction (physical, chemical, or nuclear). It provides a way to measure the total energy of the hundreds of products resulting from a high-energy collision.

44. John Avery Emison, "ORNL gets $388,000 in Super Collider grants," *Oak Ridger*, April 20, 1990.

45. Michael Riordan, "A Bridge Too Far: The Demise of the Superconducting Super Collider," *Physics Today* 69, no. 10 (2016): 48, https://doi.org/10.1063/PT.3.3329.

46. Johanna L. Miller, "The Higgs Particle, or Something Much Like It, Has Been Spotted," *Physics Today* 65, no. 9 (2012): 12, https://doi.org/10.1063/PT.3.1699.

47. D. Akimov, et al., "First Measurement of Coherent Elastic Neutrino-Nucleus Scattering on Argon," *Physical Review Letters* 126, no. 1 (January 2021).

48. Richard Powelson and Larry Lee, "Welcome, President Bush," *Knoxville News Sentinel,* February 2, 1990.

49. "Bush to Learn of Science Alliance," *Knoxville News Sentinel,* February 2, 1990.

50. The new Summer School of the South was a reincarnation of a similar UT program that operated from 1902 to 1918; see: Betsey B. Creekmore, "Summer School of the South: 1902–1911," in *Volopedia,* October 15, 2018, last updated October 17, 2018, https://volopedia.lib.utk.edu/entries/summer-school-of-the-south/.

51. Larry Lee and Kaye Franklin Veal, "State's $3 Million Boosts Bush Goal," *Knoxville News Sentinel,* February 3, 1990.

7. The Decade of the 1990s

1. Robert B. Crease, Catherine Westfall, "The New Big Science," *Physics Today* 69, no. 5 (2016): 30, https://doi.org/10.1063/PT.3.3167.

2. Electa Draper, "Feds Raided Rocky Flats 25 Years Ago, Signaling the End of an Era," *Denver Post,* May 31, 2014, https://www.denverpost.com/2014/05/31/feds-raided-rocky-flats-25-years-ago-signaling-the-end-of-an-era/.

3. Eliot Marshall, "Tiger Teams Draw Researchers' Snarls," *Science* 252, no. 5004 (April 1991): 366–368.

4. Bill Cabage, "Tom Row Spearheaded Prep for a Culture-changing ES&H Audit," *ORNL Reporter* (blog), November 6, 2018, https://www.ornl.gov/blog/ornl-reporter/when-tiger-teams-prowled.

5. C. R. Richmond, Lee Riedinger, and Tom Garritano, "Science Alliance: A Vital ORNL-UT Partnership," *ORNL Review* 24, no. 2 (1991): 42–57.

6. Recollections of William Bugg, then head of the UT Department of Physics and Astronomy.

7. Ibid.

8. Recollections of Lee Riedinger, then UT vice chancellor for research.

9. Ibid.

10. Charles M. Susskind, "Alvin W. Trivelpiece: AAAS Executive Officer," *Science* 236, no. 4800 (1987): 377, DOI: 10.1126/science.236.4800.377.

11. National Research Council, *Major Facilities for Materials Research and Related Disciplines: Presentations to the Major Materials Facilities Committee* (Washington, D.C.: National Academy Press, 1984).

12. Stephen H. Stow and Marilyn Z. McLaughlin, "An Interview with Alvin Trivelpiece," The Department of Energy Oral History Presentation Program, April 3, 2003.

13. Recollections of Lee Riedinger, who played a central role in these happenings.

14. Yu, Ts. Oganessian, et al., "Synthesis of a New Element with Atomic Number Z = 117," *Physical Review Letters* 104, no. 14 (April 2010): 1–4.

15. Colin West, "The Advanced Neutron Source Reactor: An Overview," ASTM-EURATOM Symposium on Reactor Dosimetry, Strasbourg, France, August 27–31, 1990, https://www.osti.gov/biblio/6995500.

16. Jeremy Rumsey, "A History of Neutron Scattering at ORNL," *Neutron News* 29, no. 1 (April 2018): 10–16, https://doi.org/10.1080/10448632.2018.1446588.

17. John J. Rush, "US Neutron Facility Development in the Last Half-Century: A Cautionary Tale," *Physics in Perspective* 17, no. 2 (April 2015): 135–55.

18. Ibid.

19. Ibid.

20. Science News Staff, "Gore Backs New Neutron Source," *Science,* January 20, 1998, https://www.science.org/content/article/gore-backs-new-neutron-source.

21. "Vice President Gore Visits ORNL, Announces Administration Budget Request for SNS," *Ridgelines* 20, January 29, 1998, https://web.ornl.gov/info/ridgelines/gore.htm.

22. The term spall is often associated with geologic prospecting—breaking apart rock to harvest valuable ore.

23. Frank Munger, "Management Shift Likely at Oak Ridge Project—Spallation Neutron Source Gets 'Brutal' Review," *Knoxville News Sentinel*, February 13, 1999.

24. Larisa Brass, "Moncton Named New SNS Head," *Oak Ridger*, February 22, 1999.

25. Frank Munger, "Appleton Brought SNS to Brink of Success," *Knoxville News Sentinel*, February 17, 1999.

26. Frank Munger, "Moncton Named SNS Project Director—Physicist Will Report to Leader of ORNL," *Knoxville News Sentinel*, February 20, 1999.

27. "Prime and Subcontracting," U.S. Small Business Administration, https://www.sba.gov/federal-contracting/contracting-guide/prime-subcontracting.

28. Bob Honea, interview by Lee Riedinger (author), October 16, 2019.

29. U.S. General Services Administration: A Military Interdepartmental Purchase Request (MIPR) allows military agencies the opportunity to obligate funds from a finance and accounting office to federal agencies for the purpose of purchasing products and services.

30. "Joint Flow and Analysis System for Transportation (JFAST)," *GovTribe*, November 1, 2013, https://govtribe.com/opportunity/federal-contract-opportunity/joint-flow-and-analysis-system-for-transportation-jfast-htc71113rd009#.

31. Personal recollection of Lee Riedinger, who was in charge of the UT Office of Research.

32. Funding agencies use task order contracts for services that do not procure or specify a firm quantity of services (other than a minimum or maximum quantity) and that provide for the issuance of orders for the performance of tasks during the period of the contract.

33. Bob Honea, interview by Lee Riedinger (author), October 16, 2019.

34. Association of American Geographers, "Previous Anderson Medal of Honor Recipients," https://community.aag.org/appliedgeography/andersonmedal/pastrecipients.

35. Communications staff, "Duncan, Slater Break Ground for National Transportation Research Center," *ORNL News*, April 8, 1999, https://www.ornl.gov/news/duncan-slater-break-ground-national-transportation-research-center.

36. Ibid.

37. "Thomas Zacharia, Director, Oak Ridge National Laboratory," *Energy.gov*, https://www.energy.gov/contributors/thomas-zacharia.

38. The UT/ORNL Science Alliance took the lead in hiring distinguished scientists, and Lee Riedinger was the director when Dongarra was attracted to UT and ORNL.

39. Tyra Haag and David Goddard, "Dongarra Receives Prestigious Turing Award," *UTK News*, March 30, 2022, https://news.utk.edu/2022/03/30/dongarra-receives-prestigious-turing-award/.

40. "Simulating Supernovae on Supercomputers," *ORNL Review* 35, no. 1 (2002): 18–19.

41. "Astrophysics: Simulating Supernovae," *ORNL Review* 37, no. 2 (2004): 14.

42. "Supernova Discoveries," *ORNL Review* 39, no. 2 (2006): 26.

43. W. Raphael Hix, Eric J. Lentz, Eirik Endeve, Mark Baird, M. Austin Chertkow, J. Austin Harris, O. E. Bronson Messer, Anthony Mezzacappa, Stephen Bruenn, and John Blondin, "Essential Ingredients in Core-Collapse Supernovae," *AIP Advances* 4, no. 041013 (2014), https://doi.org/10.1063/1.4870009.

44. Stephen W. Bruenn, John M. Blondin, W. Raphael Hix, Eric J. Lentz3, O. E. Bronson Messer, Anthony Mezzacappa, Eirik Endeve, J. Austin Harris, Pedro Marronetti, Reuben D. Budiardja, Merek A. Chertkow, and Ching-Tsai Lee, "Chimera: A Massively Parallel Code for Core-Collapse Supernova Simulations," *The Astrophysical Journal Supplement Series* 248, no. 1 (April 2020): 1.

45. "An Astrophysicist from Knoxville, TN, Bronson Messer," *J! Archive*, May 2003, https://www.j-archive.com/showplayer.php?player_id=5325.

46. "OLCF Names New Director of Science," *OLCF in the News*, June 4, 2020, https://www.olcf.ornl.gov/2020/06/04/olcf-names-new-director-of-science/.

47. "Vanished into Thin Air: The Search for Children's Fingerprints," *Analytical Chemistry* 67, no. 13 (July 1995): 435A–438A.

48. Deborah Noble, "Mystery Matters: The Disappearing Fingerprints," *Chem Matters*, (February 1997): 9-12, https://teachchemistry.org/chemmatters/february-1997/mysterymatters-the-disappearing-fingerprints.

49. Frank Munger, "ORNL Scientist Enjoys High Profile Forensics but Wants to Do More," *Knoxville News Sentinel*, December 14, 2008, https://archive.knoxnews.com/news/local/ornl-scientist-enjoys-high-profile-forensics-work-but—wants-to-do-more-ep-410619267-359669271.html/.

50. Arpad A. Vass, "Dust to Dust: The Brief, Uneventful Afterlife of a Human Corpse," *Scientific American* 303 (September 2010): 56–59.

51. Arpad A. Vass, "Beyond the Grave: Understanding Human Decomposition," *Microbiology Today* 28 (November 2001): 190–192, http://www.academia.dk/BiologiskAntropologi/Tafonomi/PDF/ArpadVass_2001.pdf.

52. L. V. Hamilton, L. W. McMahon, and L. G. Shipe, "Site and Operations Overview," Chapter 1 in *Oak Ridge Reservation Annual Site Environmental Report for 1995*, DOE report ES/ESH-69 https://doeic.science.energy.gov/ASER/aser95/ch-1.htm 53.

53. Frank Munger, "UT Prepares for Possible ORNL Management Bid," *Knoxville News Sentinel*, June 13, 1995.

54. Frank Munger, "DOE to Split Ridge Pacts into New Bids," *Knoxville News Sentinel*, August 7, 1996.

55. "Alex Zucker," *Oak Ridger*, November 9, 2017, https://www.oakridger.com/story/news/2017/11/13/retired-ornl-director-alex-zucker/17066193007/.

56. D. Ray Smith and Carolyn Krause, "Alex Zucker—Pillar at Lab and in

Community," Historically Speaking column, *Oak Ridger*, March 3, 2014, http://smithdray1.net/historicallyspeaking/archive.html.

57. D. Ray Smith, "Just 'Another Day' at Y-12 Plant, the National Lab," *Oak Ridger, Y-12: Oak Ridge Treasure, National Resource* column, March 27, 2009, https://www.oakridger.com/story/opinion/columns/2009/03/27/just-another-day-at-y/63372516007/.

58. Homer Fisher, interview by Lee Riedinger (author), April 26, 2020.

59. Ibid.

60. Munger, "DOE to Split Ridge Pacts into New Bids."

61. Frank Cagle, "More Embarrassing than Florida Loss," *Knoxville News Sentinel*, June 20, 1998.

62. Frank Munger, "Who's on First?—Spallation Neutron Source," *Knoxville News Sentinel*, August 31, 1998.

63. Frank Munger, "ORNL Contract Stirs the Rumor Mill," *Knoxville News Sentinel*, October 21, 1998.

64. Frank Munger, "DOE Reveals Timing to Rebid Management Contract at ORNL," *Knoxville News Sentinel*, September 29, 1998.

65. Betsey B. Creekmore, "Joseph Edwin Johnson," in *Volopedia*, September 25, 2018, last updated March 1, 2019, https://volopedia.lib.utk.edu/entries/joseph-edwin-johnson/.

66. Jennifer Sicking, "Joe Johnson: A University Icon," *Torchbearer*, University of Tennessee, September 25, 2019, https://torchbearer.utk.edu/2019/09/joe-johnson-a-university-icon/.

67. Munger, "ORNL Contract Stirs the Rumor Mill."

68. Frank Cagle, "Will We Go for It—or Will We Punt?" *Knoxville News Sentinel*, October 17, 1998.

69. Nicole Allen and Jay Mayfield, "UT's Homer Fisher Receives 'Muddy Boot'" Economic Award," *UT News*, December 2, 2005, https://news.utk.edu/2005/12/02/uts-homer-fisher-receives-muddy-boot-economic-award/.

70. Homer Fisher, interview by Lee Riedinger (author), April 26, 2020.

71. "Battelle— Pushing Forward Today for a Better Tomorrow," Battelle, https://www.battelle.org/.

72. "Pro2Serve Founder Retiring, New CEO Named to Oak Ridge Company," *Oak Ridger*, October 3, 2021, https://www.oakridger.com/story/news/2021/10/04/pro-2-serve-founder-retiring-new-ceo-named-oak-ridge-company/5946754001/.

73. Frank Munger, "UT Launches Bid to Manage ORNL," *Knoxville News Sentinel*, January 20, 1999.

74. "William Madia Biography," https://www.howold.co/person/william-madia/biography.

75. Frank Munger, "State Supports UT's Bid to Manage ORNL—Offers to Spend Millions on Facilities," *Knoxville News Sentinel*, April 23, 1999.

76. Frank Munger, "Update from ORNL: Contract Bidding, Security Crackdowns," *Knoxville News Sentinel*, April 5, 1999.

77. Frank Munger, "DOE Extends Timeline on ORNL Contract Proposals," *Knoxville News Sentinel*, April 13, 1999.

78. Frank Munger, "University Group is Poised to Submit Bid on ORNL Contract," *Knoxville News Sentinel*, April 22, 1999.

79. Frank Munger, "Lockheed Martin Joins Consortium URA in Bidding to Manage ORNL," *Knoxville News Sentinel*, May 5, 1999.

80. Ron Bridgeman, "Postma: Lockheed Martin on the 'Bad' List," *Oak Ridger*, July 15, 1999.

81. Frank Munger, "ORNL Management Hopefuls Set Proposals for DOE—It's UT-Battelle vs. Lockheed Martin-Universities Research," *Knoxville News Sentinel*, August 3, 1999.

82. Frank Munger, "State Puts up $18 Million on UT-Battelle Bid for ORNL—Move Represents $6 Million Jump over Previous Offer," *Knoxville News Sentinel*, August 25, 1999.

83. Frank Munger, "Impressive Addition Suddenly Joins ORNL Contract-Bid Mix," *Knoxville News Sentinel*, August 9, 1999.

84. Frank Munger, "ORNL Pact Called Top Coup in UT's History—Lockheed Due to Hand over Reins in April," *Knoxville News Sentinel*, October 21, 1999.

85. Frank Munger, "UT Team's Approach One of Keys in Winning Lab Pact," *Knoxville News Sentinel*, October 25, 1999.

86. "Department of Energy Contract Management," *United States Government Accountability Office*, GAO/HR-93-9, December, 1992, 3.

87. Negotiations of contract extensions determine fee increases. The total available annual performance fee is now $14.9 million (versus $7 million in the initial contract), but corporate expenses have increased as well, so the financial rewards are still at the level of a few million dollars to each partner.

88. Frank Munger, "Ground Broken for SNS—Gore Calls Project Bipartisan Outlay in Nation's Future," *Knoxville News Sentinel*, December 16, 1999.

8. *The Decade of the 2000s*

1. Frank Munger, "Wamp All for Renovations at ORNL," *Knoxville News Sentinel*, April 20, 2000.

2. David Millhorn, Stacey Patterson, and Billy Stair, *Breaking the Mold: The University of Tennessee, Battelle, and the Resurgence of Oak Ridge National Laboratory*, (Knoxville: University of Tennessee, 2013), 19.

3. U. S. General Accounting Office, "Uncertain Progression Implementing National Laboratory Reforms," Report GAO/RCED-98–197, September 10, 1998.

4. Charles B. Curtis, John P. McTague, and David W. Cheney, "Fixing the National Laboratory System," *Issues in Science and Technology* 13, no. 3 (Spring 1997): 49-52.

5. U.S. House of Representatives, Committee on Science, Subcommittee on Basic Research and Subcommittee on Energy and Environment, "Restructuring

on the Federal Scientific Establishment: Future Missions and Governance for the Department of Energy (DOE) National Labs," Joint Hearing on H.R. 87, H.R. 1510, H.R. 1993 (Title II), and H.R. 2142, Sept. 7, 1995 (Washington, D.C.: U.S. Government Printing Office, Committee Print No. 30, 1996).

6. A beam of electrons confined in a circular accelerator emits electromagnetic radiation at various wavelengths which can be selected (e.g., x rays or visible light) for particular experiments on the property of materials. Such a device is called a "light source" from a synchrotron, a type of circular accelerator.

7. ANS was canceled in part due to the proposed use of highly enriched uranium (HEU), which has a 20-percent or higher concentration of fissionable ^{235}U. This is common for research reactors in order to have a high enough flux of neutrons in the core to carry out experiments. However, new research reactor construction with HEU is discouraged in the era of proliferation concerns.

8. Frank Munger, "Oak Ridge Reactor Violations Worry DOE—Official Calls HFIR 'failures' Serious," *Knoxville News Sentinel*, November 18, 1998.

9. Frank Munger, "UT-Battelle Finds Home in 'Winter Palace,'" *Knoxville News Sentinel*, January 19, 2000.

10. Frank Munger, "Tearing Down the Lab to Help Build It up," *Knoxville News Sentinel*, May 3, 2000.

11. Frank Munger, "Bigger, Better Plans for ORNL—New Leadership Sees Plenty of Promise," *Knoxville News Sentinel*, March 26, 2000.

12. John Huotari, "UT-Battelle Gets Five-year Extension to Manage ORNL," *Oak Ridge Today*, April 5, 2020, https://oakridgetoday.com/2020/04/05/UT-Battelle-gets-five-year-extension-to-manage-ornl/.

13. "DOE Prime Contract with UT-Battelle, LLC, for the Management and Operation of Oak Rdge National Laboratory DE-AC05-00)R22725," ORNL Prime Contract, Oak Ridge National Laboratory, https://primecontract.ornl.gov/.

14. Kai-Henrik Barth, "The Department of Energy's Spallation Neutron Source Project: Description and Issues," CRS Report for Congress, Order Code RL30385, December 10, 1999, 5.

15. John J. Rush, "US Neutron Facility Development in the Last Half-Century: A Cautionary Tale," *Physics in Perspective* 17 (April 2015): 135–55.

16. "How SNS Works," *Neutron Sciences Directorate,* Oak Ridge National Laboratory, n.d., https://neutrons.ornl.gov/content/how-sns-works.

17. Barth, "The Department of Energy's Spallation Neutron Source Project: Description and Issues," 13.

18. Ibid., 10.

19. Millhorn, Patterson, and Stair, *Breaking the Mold*, 20.

20. Ibid., 19.

21. Ibid.

22. Frank Munger, "Spallation Neutron Source Project Wins Opponent's OK," *Knoxville News Sentinel*, March 9, 2000.

23. Frank Munger, "Spallation Director Opts to Leave Oak Ridge," *Knoxville News Sentinel,* Jan. 18, 2001, C1.

24. Frank Munger, "Moncton Put SNS on Path to Succeed," *Knoxville News Sentinel,* January 22, 2001.

25. Frank Munger, "4 in Running to Head Spallation Neutron Source," *Knoxville News Sentinel,* February 10, 2001.

26. Frank Munger, "Mason to Direct Spallation Neutron Source—ORNL Chief Lauds 36-year-old Scientist," *Knoxville News Sentinel,* February 23, 2001.

27. Frank Munger, "Bredesen Visits ORNL to Dedicate Institute," *Knoxville News Sentinel,* December 4, 2010.

28. Frank Munger, "State's Tab for Joint Institutes at ORNL: $32.1M," *Knoxville News Sentinel,* December 6, 2010.

29. Linda Magid, "SNS Neutron Instrumentation Workshop and Oak Ridge Neutron Users Meeting, Knoxville, Tennessee, November 9–11, 1998," National Science Foundation award #9819471, $150,000, November 1, 1998.

30. Linda Magid, "NSFCHEMBIO Workshop: Neutron Scattering for Chemistry and the Chemistry/Biology Interface," National Science Foundation award #0335614, $72,880, July 8, 2003.

31. Linda Magid, "SENSE Workshop: Sample Environments for Neutron Scattering Experiments in Tallahassee, FL, September 24–26, 2003," National Science Foundation award #0335615, $38,850, July 15, 2003.

32. John Larese and Bruce Hudson, "IMR-MIP: VISION-CED for a Neutron Vibrational Spectrometer for SNS," National Science Foundation award #0412231, $1,500,000, August 23, 2004.

33. Peter Liaw, "Advanced Neutron Scattering Network for Education and Research with a Focus on Mechanical Behavior," National Science Foundation award #0231320, $4,250,000, February 1, 2003.

34. Peter Liaw, "MRI: Development of an In-Situ Neutron-Scattering Facility for Research and Education in the Mechanical Behavior of Materials," National Science Foundation award #0421219, $1,999,996, September 1, 2004.

35. Leland Johnson and Daniel Schaffer, "Information Acceleration," *ORNL Review* 25, nos. 3–4 (1992): 82.

36. Michael T. Heath, "Parallel Computing at ORNL," *ORNL Review* 18, no. 4 (1995): 3.

37. Kenneth Kliewer, "The Center for Computational Sciences: High-Performance Computing Comes to ORNL," *ORNL Review* 30, nos. 3–4 (1997): 9.

38. Alvin Trivelpiece, "State of Laboratory," *ORNL Review* 28, no. 1 (1995): 105.

39. Jack J. Dongarra, G. A. Geist, Robbert Manchek, V. S. Sunderam, "Integrated PVM Framework Supports Heterogeneous Network Computing," *Computers in Physics* 7, no. 2 (April 1993):166–75.

40. Thomas Zacharia, "Center for Computational Sciences: A Leadership-Class Facility," *ORNL Review* 37, no. 2 (2004): 16.

41. "Thomas Zacharia Interview: Amazing Journey," *ORNL Reporter,* August 17, 2017, https://www.ornl.gov/blog/ornl-reporter/thomas-zacharia-interview-amazing-journey.

42. Bechtel Jacobs was succeeded in 2011 by UCOR, the URS-CH2M Oak Ridge team.

43. Brian Katulis and Peter Juul, "The Lessons Learned for U.S. National Security Policy in the 20 Years Since 9/11," Center for American Progress, September 10, 2021, https://www.americanprogress.org/article/lessons-learned-u-s-national-security-policy-20-years-since-911/.

44. David Koenig, "How 9/11 Changed Air Travel: More Security, Less Privacy" *AP News*, September 6, 2021, https://apnews.com/article/how-sept-11-changed-flying-1ce4dc4282fb47a34c0b61ae09a024f4.

45. Frank Munger, "ORNL Restricts Truck Traffic from Its Main Roadway," *Knoxville News Sentinel,* October 31, 2001.

46. Frank Munger, "Part of Oak Ridge road closing - ORNL security is cited as the reason," *Knoxville News Sentinel,* December 5, 2001.

47. "ORNL Employees Honored," *Oak Ridger*, November 16, 2000.

48. "UT-Battelle honors 300+ at Its first Awards Night," *ORNL Reporter,* November 21, 2000, https://web.ornl.gov/info/reporter/no21/nov00.htm.

49. Frank Munger, "Private Funds to Build 3 ORNL Facilities—UT-Battelle Lays out Bidding Rules," *Knoxville News Sentinel,* April 25, 2001.

50. Frank Munger, "Bringing Private Sector into ORNL Development Gets Positive Reaction," *Knoxville News Sentinel,* April 30, 2001.

51. Frank Munger, "S.C. Firm to Design, Build Three Research Facilities at ORNL," *Knoxville News Sentinel,* August 30, 2001.

52. Thomas Zacharia, "A National Institute for Computational Sciences to Provide Leading-Edge Computational Support for Breakthrough Science and Engineering Research," National Science Foundation award #0711134, $64,442,171, October 1, 2007.

53. Frank Munger, "UT Names VP for Science/Technology," *Knoxville News Sentinel,* Nov. 6, 2007, 11.

54. Frank Munger, "UT in Line to Be among Elite in Supercomputing," *Knoxville News Sentinel,* August 9, 2007.

55. Frank Munger, "UT-ORNL Partnership Made Computer Possible," *Knoxville News Sentinel,* January 28, 2009.

56. Frank Munger, "Off to Super Fast Start," *Knoxville News Sentinel,* January 24, 2009.

57. Frank Munger, "ORNL Computer is World's Fastest," *Knoxville News Sentinel,* November 16, 2009.

58. Millhorn, Patterson, and Stair, *Breaking the Mold*, 73.

59. "Research Teams Use Summit Supercomputer to Win Gordon Bell Prize," *TechMonitor,* November 20, 2018, https://techmonitor.ai/technology/emerging-technology/summit-supercomputer.

60. One nanometer (nm) is defined as a one billionth of a meter; the diameter of a helium atom is about 0.06 nm. Human hair widths are between 17 and 180 thousand nm.

61. Shiwang Cheng, Shi-Jie Xie, Jan-Michael Y. Carrillo, Bobby Carroll, Halie Martin, Peng-Fei Cao, Mark D. Dadmun, Bobby G. Sumpter, Vladimir N. Novikov, Kenneth S. Schweizer, and Alexei P. Sokolov, "Big Effect of Small Nanoparticles: A Shift in Paradigm for Polymer Nanocomposites," *ACS Nano* 11, no. 1 (January 2017): 752–759, https://doi.org/10.1021/acsnano.6b07172.

62. David Keim, "UT Official under Scrutiny," *Knoxville News Sentinel,* May 25, 2001.

63. Ron Leadbetter, *Big Orange, Black Storm Clouds and More* (Scotts Valley, CA: CreateSpace, 2015).

64. Jennifer Lawson and Bill Brewer, "UT President Resigns—Gilley Cites Personal Reasons," *Knoxville News Sentinel,* June 2, 2001.

65. Gilley's successors, John W. Shumaker and John D. Petersen, both engendered considerable controversy; Shumaker resigned from the office in 2003 and Petersen in 2009.

66. David Keim, "UT Administrator Reed Resigns—She Cites Concern," *Knoxville News Sentinel,* June 14, 2001.

67. David Keim, "Reed Claims Love for Gilley," *Knoxville News Sentinel,* June 14, 2001.

68. David Keim, "Gilley E-mail Reveals Sexual Liaison with Reed," *Knoxville News Sentinel,* July 20, 2001.

69. Michael Cass, "Vanderbilt May Join Oak Ridge Consortium," *Tennessean,* September 29, 2002.

70. ORNL News Release, September 30, 2004, "Vanderbilt Joins ORNL Core Universities," https://www.ornl.gov/news/vanderbilt-joins-ornl-core-universities#:~:text=Vanderbilt%20joins%20Duke%2C%20Florida%20State,in%20a%20variety%20of%20ways.

71. Scott Barker, "Shumaker to Repay UT," *Knoxville News Sentinel,* July 17, 2003.

72. J. J. Stambaugh, "Stakes High in Shumaker Split," *Knoxville News Sentinel,* August 2, 2003.

73. J. J. Stambaugh, "Pressure Builds at UT," *Knoxville News Sentinel,* August 6, 2003.

74. Ina Hughes, "Shumaker's Touch," *Knoxville News Sentinel,* October 7, 2003.

75. Tom Humphrey, "Shumaker Lied, Says Governor," *Knoxville News Sentinel,* October 10, 2003.

76. Shayla Byrd, "Shumaker Planned to Ditch Orange Carpet for Red," *Nashville Post,* September 10, 2003, https://www.nashvillepost.com/home/article/20446062/shumaker-planned-to-ditch-orange-carpet-for-red.

77. Randy Kenner, "Members of Council Take Closer Look at Finalists," *Knoxville News Sentinel,* April 12, 2004.

78. Randy Kenner, "Letter Incident Handled Wrongly, UT Candidate Says," *Knoxville News Sentinel,* May 19, 2004.

79. Randy Kenner, "Positives for Petersen—Trustees Recommend," *Knoxville News Sentinel,* June 21, 2006.

80. It may be useful to explain the university's administrative structure. While Petersen, as UT president, was the top executive for the UT system, including all UT campuses statewide, Crabtree, as chancellor, was the chief executive of the Knoxville campus. Though the UT, Knoxville chancellor enjoys a considerable degree of independence, ultimately, they report to the president of the UT system.

81. Hayes Hickman, "UT Chancellor, Chief at Odds," *Knoxville News Sentinel,* September 8, 2007.

82. Chloe White, "UT Staff Take Pay Cut," *Knoxville News Sentinel,* December 19, 2008.

83. Bryan Mitchell, "Colorado State Provost Headed to UT for Same Post," *Knoxville News Sentinel,* March 9, 2001.

84. Darren Dunlap, "A 'Somber' Reception," *Knoxville News Sentinel,* January 18, 2008.

85. Darren Dunlap, "UT Faculty Voice Discontent with Petersen," *Knoxville News Sentinel,* January 23, 2008.

86. Darren Dunlap, "Petersen Focus of Faculty Senate," *Knoxville News Sentinel,* January 29, 2008.

87. Frank Munger, "Computing Collaboration," *Knoxville News Sentinel,* April 4, 2008.

88. Tom Humphrey, "State to Cut Jobs, Spending," *Knoxville News Sentinel,* May 8, 2008.

89. Chloe White, "Incident Mars Social Event," *Knoxville News Sentinel,* November 25, 2008.

90. Chloe White, "Key UT Donor Returns to Resigned Position," *Knoxville News Sentinel,* December 11, 2008.

91. Chloe White and David Keim, "Petersen First to Be Given 5-year Review," *Knoxville News Sentinel,* February 8, 2009.

92. Chloe White, "'My Choice' to Leave—John Petersen Resigns as UT President," *Knoxville News Sentinel,* February 19, 2009.

93. Karen Doss Bowman, "The Erosion of Presidential Tenure," Public Purpose, American Associaiton of State Colleges and Universities, Summer 2017, https://www.aascu.org/MAP/PublicPurpose/2017/Summer/TheErosion.pdf.

94. "The Founding Father of JIAM: Ward Plummer (1940-2020)," Department of Physics and Astronomy, University of Tennessee, July 20, 2020, http://www.phys.utk.edu/news/2020/ward-plummer.html.

95. University of Tennessee, Knoxville, "Big Orange Big Ideas: Where Minds Meet," video, 0:30, September 3, 2015, https://www.youtube.com/watch?v=u8IsODwjg9c.

96. Millhorn, Patterson, and Stair, *Breaking the Mold,* 75.

97. Jennifer Hutchins and Jennifer Burke, "DDCE, University of Tennessee/Genera Energy Demonstration Plant Grand Opening," *US Fed News*, January 8, 2010.

98. "West Tennessee Solar Farm," The University of Tennessee System, https://solarfarm.tennessee.edu/about/.

99. "Biophysicist Named first UT-ORNL Governor's Chair," *Oak Ridger*, June 22, 2006.

100. Rachel McDowell, "ORNL Team Enlists World's Fastest Supercomputer to Combat the Coronavirus," Leadership Computing Facility, Oak Ridge National, Laboratory, https:// www.olcf.ornl.gov/2020/03/05/ornl-team-enlists-worlds-fastest-supercomputer-to-combat-the-coronavirus/.

101. "Governor's Chairs," The University of Tennessee System, https://research.tennessee.edu/governors-chairs/.

102. J.R. Minkel, "The 2003 Northeast Blackout—Five Years Later," *Scientific American,* August 13, 2008, https://www.scientificamerican.com/article/2003-blackout-five-years-later/.

103. Zhian Zhong, Chunchun Xu, B. J. Billian, Li Zhang, S.-J. S. Tsai, R. W. Conners, V. A. Centeno, A. G. Phadke, Yilu Liu, "Power System Frequency Monitoring Network (FNET) Implementation," *IEEE Transactions on Power Systems* 20, no. 4 (November 2005): 1914–1921.

104. Yilu Liu and Chris J. O'Reilley, "FNET/GridEye: A Tool for Situational Awareness of Large Power Interconnetion Grids," *2020 IEEE PES Innovative Smart Grid Technologies Europe,* Institute of Electrical and Electronics Engineers, Washington, D.C., October 2020, 379–383.

105. Rachel Ohm, "Millhorn, UT Vice President Emeritus, Adviser to ORNL, Dies," *Knoxville News Sentinel,* December 20, 2017.

106. The Joint Institute for Advanced Materials is now called the Institute for Advanced Materials and Manufacturing.

107. Monica Kast, "UT Partners with VW, ORNL—Research Aims for Development of Electric Vehicles," *Knoxville News Sentinel*, January 17, 2020.

9. The 2010s

1. Located at the Manufacturing Demonstration Facility, halfway between the UT and ORNL campuses.

2. "Biophysicist Named First UT-ORNL Governor's Chair," *Oak Ridger*, June 22, 2006.

3. "Bioenergy Facility Opens in Knoxville-Oak Ridge Innovation Valley," *PR Newswire*, February 1, 2010.

4. Roger Harris, "Tennessee Solar Institute Offers $14.5M in Grants," *Knoxville News Sentinel*, August 25, 2010.

5. "The 50 greatest Yogi Berra quotes" *USA Today Sports*, https://ftw.usatoday.com/2019/03/the-50-greatest-yogi-berra-quotes.

6. Ibid.

7. Roger Harris, "Program Expands UT's Ties to ORNL," *Knoxville News Sentinel*, January 9, 2010.

8. John Huotari, "Bredesen Reiterates Oak Ridge-UT Plan During AC Visit," *Oak Ridger*, January 11, 2010.

9. Tom Humphrey, "Bredesen to Budget $6 Million for New UT Program," *Knoxville News Sentinel*, January 8, 2010.

10. "Complete College Tennessee Act of 2010," Senate Bill No. 7006, https://www.tbr.edu/sites/default/files/media/2015/01/Complete%20College%20TN%20Act%202010%20-%20signed.pdf.

11. "Bredesen Signs Landmark Education Bills into Law," *Murfreesboro Vision*, February 4, 2010,

12. "A Look at Governor's Special Education Proposal," *Associated Press: Clarksville Metro Area*, January 12, 2010.

13. Frank Munger, "UT-ORNL Deal Could Bring Youthful Enthusiasm, (Relatively) Cheap Labor and More Science," *Knoxville News Sentinel*, January 14, 2010.

14. Whitney Holmes, "UT Knoxville, ORNL Name Director for Interdisciplinary Research Center," *US Fed News*, August 26, 2010.

15. Both consultants had worked closely with Riedinger when he was deputy director at ORNL, and both knew well the importance of fostering academic programs at the laboratory and its partner universities.

16. Megan Boehnke, "UT to Name Joint Center at ORNL after Bredesen," *Knoxville News Sentinel*, October 28, 2011.

17. "UT Names New Interdisciplinary Center after Former Governor Bredesen," *PR Newswire*, October 28, 2011.

18. Georgiana Vines, "Bill Richardson to Lecture in Knoxville," *Knox News*, April 20, 2013.

19. Georgiana Vines, "Former Governors Give Practical Advice During UT Visit," *Knox News*, April 29, 2013.

20. Tom Humphrey, "Bredesen to UT Students: 50-Cent Gas Tax Increase Not a Great Idea," *Humphrey on the Hill*, April 29, 2013.

21. Chloe White, "UT Chancellor Will Try to Protect Classrooms," *Knoxville News Sentinel*, February 3, 2009.

22. This second Bredesen Center PhD program, in data science and engineering, would be created in 2017.

23. Carly Harrington, "UT's Campus Getting $1-billion Facelift," *Chattanooga Times Free Press*, June 7, 2015.

24. Frank Munger, "ORNL's Big Turnaround: UT Will Fight to Keep Lab Management Contract," *Knoxville News Sentinel*, June 1, 2013.

25. M. J. Slaby, "UT Chancellor Will Return to Teaching—Cheek Stepping Down for More Family Time," *Knoxville News Sentinel*, June 22, 2016.

26. Jimmy Cheek, "UT Chancellor Appreciates Hard Work," *Knoxville News Sentinel*, February 12, 2017.

27. “Student Serves on Panel with Nobel Winners at Prestigious Conference,” *US Fed News*, August 5, 2015.

28. “Grow Bioplastics Wins Vol Court Pitch Competition,” *Targeted News Service*, November 1, 2014.

29. “Student at Bredesen Center Wins $50,000 ORNL Prize,” *US Fed News*, May 5, 2014.

30. Frank Munger, “ORNL Director Likes Grad Program,” *Knoxville News Sentinel*, January 18, 2010.

31. Frank Munger, “Grad Students Galore at ORNL,” *Knoxville News Sentinel*, January 11, 2011.

32. “UT Board Approves Big Data Doctoral Program,” *US Fed News*, April 2, 2017.

33. “ORNL, UT Launch New Doctoral Program in Data Science,” *Oak Ridger*, May 17, 2017.

34. “Impact of New UT-ORNL Big Data Doctorate to Be Felt across State,” *US Fed News*, May 16, 2017.

35. Frank Munger, “Jaguar Supercomputer to Be Upgraded to Titan,” *Knoxville News Sentinel*, October 12, 2011.

36. Jacobson grew up in Oak Ridge, worked for fourteen years in South Africa, and brought with him four graduate students to enroll in the Bredesen Center PhD programs.

37. Deborah Weighill, et al., “Pleiotropic and Epistatic Network-based Discovery: Integrated Networks for Target Gene Discovery,” *Frontiers in Energy Research* 11 (May 2018): https://doi.org/10.3389/fenrg.2018.00030.

38. Jonathan Hines, “Genomics Code Exceeds Exaflops on Summit Supercomputer,” *OLCF in the News*, June 8, 2018, https://www.olcf.ornl.gov/2018/06/08/genomics-code-exceeds-exaops-on-summit-supercomputer/.

39. Megan Boehnke, “Obama’s Promise,” *Knoxville News Sentinel*, January 10, 2015.

40. Josh Flory, “Carbon Collaboration,” *Knoxville News Sentinel*, January 13, 2015.

41. Megan Boehnke, “UT-ORNL Hire Advanced Manufacturing Expert as 11th Governor’s Chair,” *Knoxville News Sentinel*, May 8, 2013.

42. “Advanced Composites Expert Named Newest UT-ORNL Governor’s Chair,” *Oak Ridger*, June 17, 2015.

43. “New Vice Chancellor for Research and Engagement hired,” *US Fed News*, August 27, 2012.

44. Bill Cabage, “Charmed Half-Life: Target Used to Discover Element 117 Took a Circuitous Route to Russia,” *ORNL Review* 49, no. 3 (2016): 26–27, https://www.ornl.gov/blog/ornl-review/charmed-half-life-target-used-discover-element-117-took-circuitous-route-russia.

45. Yu. Ts. Oganessian, et al., “Synthesis of a New Element with Atomic Number Z = 117,” *Physical Review Letters* 104 (2010): 142502.

46. "IUPAC Is Naming the Four New Elements Nihonium, Moscovium, Tennessine, and Oganesson," International Union of Pure and Applied Chemistry, June 8, 2016, https://iupac.org/iupac-is-naming-the-four-new-elements-nihonium-moscovium-tennessine-and-oganesson/.

47. The Strategic Defense Initiative, announced by President Ronald Reagan on March 23, 1983, sought to develop land- and space-based defense systems to destroy incoming Soviet intercontinental ballistic missiles at various stages of their flights, before they could reach U.S. soil.

48. These relatively short-range missiles carry either nuclear or conventional explosives and are deployed to protect a strategic region or battle zone. Many of these weapons are designed to intercept and destroy incoming ballistic missiles.

49. Bulent Yusuf, "US Navy 3D Prints First Submersible Hull in Just Four Weeks," *All3DP*, July 21, 2017, https://all3dp.com/us-navy-3d-print-submersible-hull/.

50. Timothy A. Delk, "ORNL Military Fellows Program," *ORNL News*, June 16, 2022, https://www.ornl.gov/content/ornl-military-fellows-program.

51. "UT Researcher's Nonwoven Fabrics Protect the Health of More Than a Billion People," *UT Research Foundation News*, August 27, 2019, https://utrf.tennessee.edu/ut-researchers-nonwoven-fabrics-protect-the-health-of-more-than-a-billion-people/.

52. Vincent Gabrielle, "The Humble Knoxville Scientist Who Became a Worldwide Hero During the Pandemic," *Knoxville News Sentinel*, June 19, 2020.

53. Kathryn King, "UT, CNS Sign $9.5-million Agreement," Y-12 press release, June 20, 2022, http://www.y12.doe.gov/news/press-releases/ut-cns-sign-95-million-agreement.

54. "UT-Battelle Donates $100,000 to Children's Hospital Expansion," *Oak Ridge Today*, December 22, 2014, https://oakridgetoday.com/2014/12/22/ut-battelle-donates-100000-childrens-hospital-expansion/.

55. Benjamin Pounds, "UT-Battelle Gives City $500K for Airport," *Oak Ridger*, March 23, 2021, https://www.msn.com/en-us/news/us/UT-Battelle-gives-city-500k-for-airport/ar-BB1eTkjU.

56. Brittany Crocker, "New UT Chancellor Donde Plowman Announces Office Hours, ORNL Partnership Plans," *Knoxville News Sentinel*, July 10, 2019.

57. Monica Kast, "University of Tennessee and ORNL Name Director for 'Truly Visionary' Oak Ridge Institute," *Knoxville News Sentinel*, February 16, 2021.

58. Tom Ballard, "David Sholl named Interim Executive Director of UT-Oak Ridge Innovation Institute," https://www.teknovation.biz/david-sholl-named-interim-executive-director-of-ut-oak-ridge-innovation-institute/.

Epilogue

1. Chiles Coleman, "Atomic Super-Bomb Made in Oak Ridge, Strikes Japan—Oak Ridge Has More than 425 Buildings," *Knoxville News Sentinel*, August 6, 1945.

2. See Appendix 2 for a listing of the twenty-three identified critical connections.

3. Thomas Wellock, ""Too Cheap to Meter": A History of the Phrase," U.S. Nuclear Regulatory Commission, September 24, 2021, https://www.nrc.gov/reading-rm/basic-ref/students/history-101/too-cheap-to-meter.html.

4. "Gates Myth No 6:- "Gates said 640k of Memory was enough," http://www.billgatesmyths.org.uk/memory.html.

5. Robert Strohmeyer, "The 7 Worst Tech Predictions of All Time," *PCWorld*, December 31, 2008, https://www.pcworld.com/article/532605/worst_tech_predictions.html.6. Wilcox, *An Overview of the History of Y-12, 1942–1992*, 19–22.

7. Louis T. Iglehart, "Impact of ORNL's Atomic Energy Research Program Felt Vividly at UT Where 'Atom Shines in Academic Sun,'" *ORNL—News*, June 29, 1951.

8. Alvin M. Weinberg, "Federal Laboratories and Science Education," *Science* 136, no. 3510 (April 1962): 27–30.

9. Becca Wright, "University of Tennessee Relaunches Advanced Materials Science Center to Turn Research into Reality," *Knoxville News Sentinel*, April 23, 2022; https://www.knoxnews.com/story/news/education/2022/04/23/university-tennessee-relaunches-advanced-materials-science-center/7384472001/.

10. Danielle Farrie and David G. Sciarra, "Making the Grade 2021," Education Law Center, January 3, 2002, https://edlawcenter.org/assets/MTG%202021/2021_ELC_MakingTheGrade_Report_Dec2021.pdf.

11. "State Support for Higher Education per Full-Time Equivalent Student," Science & Engineering State Indicators, U.S. National Science Foundation, June 30, 2022; https://ncses.nsf.gov/indicators/states.

12. Erin Chapin and Heather Peters, "Faculty Named in 2020 Most Cited Researchers Report," *UTK News,* December 14, 2020, https://news.utk.edu/2020/12/14/faculty-named-in-2020-most-cited-researchers-report/.

13. Clarivate, "Highly Cited Researchers," https://recognition.webofscience.com/awards/highly-cited/2020/.

14. Erin Chapin, "World's Top 2 Percent of Scientists List Includes More than 150 UT Faculty," *UTK News,* January 6, 2021, https://news.utk.edu/2021/01/06/worlds-top-2-percent-of-scientists-list-includes-more-than-150-ut-faculty/.

15. Monica Kast, "UT Tuition Hikes, Oak Ridge Institute OK'd," *Knoxville News Sentinel*, June 22, 2019.

16. Monica Kast, "$20M Grant Will Help Make Oak Ridge Institute One of the Top Research Facilities Worldwide," *Knoxville News Sentinel*, June 17, 2020.

17. Victor Ashe, "Sen. Alexander Cements Legacy with Support of Smokies, Oak Ridge Institute," *Knoxville News Sentinel*, July 8, 2020.

18. "U.S. Department of Energy Awards $20 Million to New Oak Ridge Institute at the University of Tennessee," *UT System News,* June 17, 2020, https://news.tennessee.edu/2020/06/17/doe-awards-20-million-to-new-oak-ridge-institute/.

19. Monica Kast, "Oak Ridge Institute Gets $20M Boost—Grant Puts It on Path to Be among Top Centers in the World," *Knoxville News Sentinel*, June 18, 2020.

20. Randy McNally and Ken Yager, "Institute Will Help Create a Leader in STEM—UT-ORII Will Train More Tennesseans for High-paying Jobs in Manufacturing, Energy and Data Science," *Knoxville News Sentinel,* May 1, 2022.

21. Monica Kast, "Institute Designed for Big Thinking—Oak Ridge to Promote Students' Talents," *Knoxville News Sentinel,* November 12, 2019.

22. Monica Kast, "University of Tennessee and ORNL Name Director for 'Truly Visionary' Oak Ridge Institute," *Knoxville News Sentinel,* February 16, 2021; https://www.knoxnews.com/story/news/education/2021/02/16/ut-ornl-name-director-oak-ridge-institute-research-partnership/6751645002/.

23. "Oak Ridge National Laboratory Hires Search Firm to Find its Next Director," *Knoxville News Sentinel,* January 22, 2023.

24. Vincent Gabrielle, "Frontier, an Oak Ridge Supercomputer, is now the Fastest on Earth. Here's Why It is Astounding—," *Knoxville News Sentinel,* June 2, 2022, https://www.knoxnews.com/story/news/2022/06/01/frontier-supercomputer-tennessee-ranked-fastest-earth-breaks-exaflop-barrier-oak-ridge/7452600001/.

25. "U.S. Department of Energy Announces $61 Million to Advance Breakthroughs in Quantum Information Science," *Department of Energy News*, August 19, 2021, https://www.energy.gov/articles/us-department-energy-announces-61-million-advance-breakthroughs-quantum-information.

26. "Quantum Leap for Appalachia," *Cross Sections* (newsletter), Department of Physics and Astronomy, University of Tennessee, Fall 2019, http://www.phys.utk.edu/news/xsections/fall-2019/quantum-leap.html.

27. "Volkswagen, UT, ORNL Announce Collaboration, Innovation Hub," *UT System News*, January 17, 2020, https://news.tennessee.edu/2020/01/17/volkswagen-ut-ornl-innovation-hub/.

28. "Electrifying Vehicles, Making Light Components 1st Hub Aim," *Oak Ridger,* January 22, 2020.

29. Monica Kast, "UT Partners with VW, ORNL—Research Aims for Development of Electric Vehicles," *Knoxville News Sentinel,* January 17, 2020.

30. Karen K. Dunlap, "Liquid-to-solid Battery Electrolyte Technology Licensed Exclusively to Safire," *ORNL News*, November 18, 2022, https://www.ornl.gov/news/liquid-solid-battery-electrolyte-technology-licensed-exclusively-safire.

31. Elyssa Haynes, "Safire Technology Group Expands with a New R&D Laboratory in Knoxville at UT's Spark Innovation Center," *Business Wire,* January 24, 2023, https://www.businesswire.com/news/home/20230123005801/en/.

32. Silas Sloan, "Knoxville Institute Gets Millions to Make Key Piece in EV Tech," *Knoxville News Sentinel,* April 12, 2023.

33. "Second Target Station: Additional Neutron Source Will Meet Emerging Science Challenges," *Neutron Sciences Directorate,* Oak Ridge National Laboratory, n.d., https://neutrons.ornl.gov/sts.

34. Paul Boisvert, "SNS Achieves Record 1.55 MW Power to Enable More Scientific Discoveries," *Neutron Sciences Directorate,* Oak Ridge National Laboratory,

February 23, 2023, https://neutrons.ornl.gov/content/sns-achieves-record-155-mw-power-enable-more-scientific-discoveries.

35. Leo Williams, "SNS Upgrades Will Benefit researchers," *ORNL Review* 53, no. 1 (2020): 12.

36. After the Manhattan Project closed, five additional buildings were constructed at the K-25 site for more output of enriched uranium by gaseous diffusion. One of these was K-33, built in 1954, decommissioned in the 1980s, and deconstructed and decontaminated in the last decade.

37. "Kairos Power Announces Investment in Oak Ridge—Company Will Infuse $100 Million, Create 55 Jobs," *Knoxville News Sentinel,* July 24, 2021.

38. https://www.usnc.com/ultra-safe-nuclear-corporation-sites-pilot-fuel-manufacturing-facility-in-oak-ridge-tenn/.

39. "3D-Printed Nuclear Reactor Promises Faster, More Economical Path to Nuclear Energy," *ORNL News*, May 11, 2020, https://www.ornl.gov/news/3d-printed-nuclear-reactor-promises-faster-more-economical-path-nuclear-energy.

40. Brian J. Ade, Benjamin R. Betzler, Aaron J. Wysocki, Michael S. Greenwood, Phillip C. Chesser, Kurt A. Terrani, Prashant K. Jain, Joseph R. Burns, Briana D. Hiscox, Jordan D. Rader, Jesse J. W. Heineman, Florent Heidet, Aurelien Bergeron, James W. Sterbentz, Tommy V. Holschuh, Nicholas R. Brown, and Robert F. Kile, "Candidate Core Designs for the Transformational Challenge Reactor," *Journal of Nuclear Engineering* 2, no. 1 (2021): 74–85, https://doi.org/10.3390/jne2010008.

41. Jennifer McDermott, "Small Nuclear Reactor Certified for Use," *Knoxville News Sentinel*, January 22, 2023.

42. East Tennessee Economic Council, "NextGen Nuclear in the Oak Ridge Corridor," 2023; https://www.eteconline.org/nuclear-industry-hub/.

43. "Gov. Lee Names Tennessee Nuclear Energy Advisory Council Appointees," July 13, 2023, https://www.tn.gov/governor/news/2023/7/13/gov-lee-names-tennessee-nuclear-energy-advisory-council-appointees.html.

44. R. H. Goulding, J. B. O. Caughman, J. Rapp, T. M. Biewer, T. S. Bigelow, I. H. Campbell, J. F. Caneses, D. Donovan, N. Kafle, E. H. Martin, H. B. Ray, G. C. Shaw, and M. A. Showers, "Progress in the Development of a High Power Helicon Plasma Source for the Materials Plasma Exposure Experiment," *Fusion Science and Technology* 72, no. 4 (2017): 588–594.

45. In this work and publication on MPEX, Holly Ray, Guin Shaw, and Missy Showers were all UT Bredesen Center graduate students.

46. Adrian Cho, "U.S. Physicists Rally around Ambitious Plan to Build Fusion Power Plant," *Science,* 370 (December 2020): 1258, https://dx.doi.org/10.1126/science.abg0706.

47. "Bringing Fusion to the U.S. Grid," National Academies of Sciences, Engineering, and Medicine 2021; The National Academies Press. https://doi.org/10.17226/25991.

48. Maurizio Di Paolo Emilio, "Type One Energy announces the opening of new offices in Oak Ridge, Tennessee," Power Electronics News, August 3, 2023;

https://www.powerelectronicsnews.com/type-one-energy-announces-the-opening-of-new-offices-in-oak-ridge-tennessee/. A stellarator is an alternative confinement device compared to the tokamak used in the large ITER development in France.

49. Neil J. Williams, Charles A. Seipp, Flavien M. Brethomé, Ying-Zhong Ma, Alexander S. Ivanov, Vyacheslav S. Bryantsev, Michelle K. Kidder, Halie J. Martin, Erick Holguin, Kathleen A. Garrabrant, Radu Custelcean, "CO_2 Capture via Crystalline Hydrogen-Bonded Bicarbonate Dimers," *Chem* 5, no. 3 (March 2019): 719–30.

50. Dawn Levy, "Carbon Dioxide Capture," *ORNL Review* 52, no. 2 (2019): 30.

51. "Secretary Granholm Breaks Ground on Isotope Research Center to Advance Life-saving Medical Applications and Strengthen America as a Global Scientific Leader," *Department of Energy News,* October 24, 2022, https://www.energy.gov/articles/secretary-granholm-breaks-ground-isotope-research-center-advance-life-saving-medical.

52. Sara Shoemaker, "Energy Secretary Breaks Ground on New Isotope Facility at ORNL," *ORNL News,* October 24, 2022, https://www.ornl.gov/news/energy-secretary-breaks-ground-new-isotope-facility-ornl.

53. "Texas A&M System and the University of Tennessee Join Forces in Bid for Contract at Pantex, Y-12," *UT System News,* September 17, 2020, https://news.tennessee.edu/2020/09/17/tamus-tennessee-join-forces-in-dept-energy-bid/.

54. Monica Kast, "UT, Texas A&M Systems Hope to Manage Y-12 National Security Complex and the Pantex Plant," *Knoxville News Sentinel,* December 7, 2020.

55. Brenna McDermott, "New Y-12 Management Firm Starts in December under a $2.8 Billion Contract," *Knoxville News Sentinel,* November 29, 2021.

56. "NNSA Cancels Y-12, Pantex Contract Award, to Hold New Competitions for Separate Site Management Contracts," *NNSA News,* May 16, 2022, https://www.energy.gov/nnsa/articles/nnsa-cancels-y-12-pantex-contract-award-hold-new-competitions-separate-site.

57. Vincent Gabrielle, "National Nuclear Security Administration Ends Y-12 contract," *Knoxville News Sentinel,* May 17, 2022.

58. National Nuclear Security Administration, "NNSA announces contract extension to Consolidated Nuclear Security for the Management and Operation of the Y-12 National Security Complex and the Pantex Plant," October 3, 2022, https://www.energy.gov/nnsa/articles/nnsa-announces-contract-extension-consolidated-nuclear-security-management-and.

59. "UT, CNS Sign $9.5 Million Agreement," *Y-12 News,* June 20, 2022, http://www.y12.doe.gov/news/press-releases/ut-cns-sign-95-million-agreement.

60. Robert D. Atkinson, Mark Muro, and Jacob Whiton, "The Case for Growth Centers: How to Spread Tech Innovation across America," *Brookings Institution* (report), December 9, 2019.

61. Raphael Santos, "California, New York, Massachusetts, Texas and Washington Lead the List but There Are Some Surprise Tech Hubs Emerging," *Airswift,* January 19, 2022, https://www.airswift.com/blog/top-us-states-vc-backed-2021.

62. Stephanie G. Seay, "ORNL Welcomes Innovation Crossroads Entrepreneurial Research Fellows," *EurekAlert!, AAAS,* May 18, 2017, https://www.eurekalert.org/news-releases/789998.

63. "ORNL Welcomes Sixth Cohort of Innovation Crossroads Clean Energy Entrepreneurs," *EurekAlert!, AAAS,* July 20, 2022, https://www.eurekalert.org/news-releases/959452#.Yvqa57veac8.mailto.

64. Silas Sloan, "Keeping Innovation Alive in Knoxville—Growing Tech Companies Take Root in Tennessee," *Knoxville News Sentinel,* July 5, 2022.

65. Vincent Gabrielle, "Local Startup Finalist for Musk Carbon Removal Prize—Moment of pride to be able to Represent Tenn.," *Knoxville News Sentinel,* April 21, 2022.

66. Rebecca Wright, "Going Clean—University of Tennessee Startup Accelerator Will Develop Climate Tech," *Knoxville News Sentinel,* February 19, 2022.

67. "Assessment of the Entrepreneurship Ecosystem of the Greater Knoxville Metropolitan Area," Jan. 2021; https://www.knoxmetroassessment.com/.

68. Brenna McDermott, "The New Silicon Valley? UT, ORNL and TVA Join Together to Launch Tech Startup Accelerator," *Knoxville News Sentinel,* June 3, 2021, https://www.knoxnews.com/story/money/2021/06/03/university-tennessee-oak-ridge-labs-tva-launch-techstars-startup-accelerator/5289257001/.

69. Silas Sloan, "Changing the World, One Solution at a Time—Techstars Startups are Creating Seriously Futuristic Tech," *Knoxville News Sentinel,* June 3, 2022.

70. Silas Sloan, "Angel Investors in Hiding—Some Say More Must Be Done to Encourage Practice in Knoxville," *Knoxville News Sentinel,* April 10, 2022.

71. Silas Sloan, "Knoxville Sees Benefits in Early Days of Techstars—2 Startups in Accelerator Plan to Make Area Home," *Knoxville News Sentinel,* May 26, 2022.

72. Jon Gertner, *The Idea Factory: Bell Labs and the Great Age of American Innovation* (NY: Penguin Books, 2012).

Index